Alcohol Blending and Accounting
Volume 1

U.S. 60° F. Gauging, Blending and the International Alcohol Tables

Payton D. Fireman, JD

Morgantown, West Virginia

Contents

Acknowledgements

When I started the first legal distillery to make moonshine in West Virginia, I knew very little about how to blend alcohol. I was so preoccupied with just getting the permits to operate that the mechanics of blending alcohol were a low priority. My good friend, Dan Nagowski, wrote a few simple algebraic formulas that we began with. Later, when I learned more about the TTB tables and algorithms prescribed by them, I transcribed the tables to Microsoft Excel worksheets and created an Excel-based program to execute them; Brian Corcoran gave generously of his time and energy to help design the lookups and data retrieval methods for my ABS Software program. In looking for comparable tables or some means of validating my results, I introduced myself to Harvey Wilson, who runs Katmar Software and who has pioneered the alcohol blending software genre with his AlcoDens program. He has mentored me over the years and his observations and knowledge are frequently mentioned in this work. As my work on the TTB tables progressed, Harvey and I were able to compare each of our separate means of accomplishing the same general object and, in this comparison, we have both learned more about the subject. I would also like to acknowledge the advice and assistance of Dr. Brooke Reaser for her help in solving numerous problems and Joe Ervin for his programming help, proofreading, and overall support of this project.

Preface

This is a reference book or textbook of alcohol blending procedures, formulas and the tables needed to operate them. As such, it is not an easy book to read. From a commercial standpoint, no distillery needs to use or apply all of the algorithms described in this book, but they're all here. It is good to know that these things exist, but there's no need to worry about all of them. There is text, commentary and advice but mostly the book is a series of mathematical procedures set out to demonstrate various blending formulas. It is in effect the users' manual to the Excel workbook and software (ABS Software) that I developed to simplify the execution of these procedures and to document them so that others could reproduce them independently. I've tried to make the material readable, but the subject is quite technical and one has to follow numbers and text at the same time by referring to the numerous spreadsheet diagrams that are explained.

My self-education in this area allowed me to identify many of the procedures that have been used over time to gauge and manipulate alcohol; each has its strengths and weaknesses. I ended up reproducing all of these procedures and finding which ones would do what, and to what degree of accuracy. Eventually I learned to master the methods devised by the International Organization of Legal Metrology and published as the International Alcohol Tables and the General Formula which was used to create those tables. By modifying the application of the formula, I was able to build accurate tables at any calibration temperature and also to solve, without using tables at all, using a nonlinear reduction process. The question then became whether to publish the older methods, or to just publish the final methods derived. My conclusion is that all of the methods described do work, within their parameters of accuracy and refinement, and this will provide a comprehensive compilation of procedures that have been developed to address alcohol blending in the context of operating a distillery—something the International Alcohol Tables and procedures by themselves cannot do.

The book begins by explaining all of the procedures available within the context of the regulations and tables prescribed by the Bureau of Alcohol Tobacco and Firearms, Tax and Trade Bureau of the United States Government (hereinafter TTB). The TTB tables and algorithms have strengths and weaknesses. Their strengths are that they are comprehensive, accurate, and only require basic math in order to use. Their weaknesses are not internal: the tables are an excellent set of tools that provide commercially precise control over beverage alcohol blending and accounting and have proved their worth for more than one hundred years. The problem arises when one wants to compare the TTB tables to other published literature on the subject. The TTB tables rely on an awkward calibration temperature of 60°F (15.556°C, but it can easily be adapted to 15°C), the published densities are also presented in atmospheric density and mass rather than vacuum or absolute densities, and the mass and volume units are pounds avoirdupois and US fluid gallons of 231 cubic inches rather than metric units.

For those readers not under the TTB regimen, the TTB procedures can show the logic behind the procedures and they can be applied in a metric environment as well. Parts of the book deal with metric-based methods centered around the International Alcohol Table methods, but most of that material has been allocated to a second volume on that topic and other more advanced topics.

My efforts to try and keep this book to a reasonable length (and to keep the size of the diagrams large enough to be legible) has resulted in some variations from normal book formatting arrangements. These deviations from standard publishing practices are entirely my doing. I have also used a rather idiosyncratic system of notation and abbreviation in the diagrams for degrees Celsius and various density regimens such as kilograms per meter cubed which is abbreviated to "kg m3," rather than superscript. I have also abbreviated the word vacuum to "Vac or vac" repeatedly and, because the TTB regulations specify the American spelling of liter, I have used that spelling rather than the SI version of litre.

I will often use more decimal places than are strictly necessary given the general nature of the formulas. Part of this is just so I can see the small variations that are being dealt with, and also so that it makes it easier to pick out the values where they are identified in the worksheets. I have not generally highlighted numerical values in the tables that are referred to in the text. I encourage users to circle or highlight those values to make them easier to identify the next time the book is needed.

To reduce the overall length of the book I have not labeled all of the diagrams with the source worksheet from which they originate. However, each section relates to one worksheet and it is identified somewhere in that section above or below the unreferenced one.

This turned out to be a two-volume work and the bibliography for both volumes is included in each. Some references are not used in each individual volume but almost all references are footnoted when they occur in.

Volume 2 is referenced as: *Alcohol Blending and Accounting, Volume 2, Advanced Topics and International Alcohol Tables*, ISBN: 978-1-7320124-1-7

For those wishing to run their own calculations using the procedures described in this book, I've published an associated software program, containing working versions of all of the worksheets, formulas and procedures outlined in the work. The program is presently named Alcohol Manual 4.8 but later versions may carry a different designation. Also available is ABS (Alcohol Blending Software) which is currently in development as version 3.5. Both are available at alcoholblending.com; and mountainmoonshine.com.

Gauging and Proofing

Units of Measurement

In order to carry out alcohol gauging and blending operations using the TTB tables as our guide, we must use the units of measurement for volume, alcohol content and temperature that those tables prescribe.

The standard volume unit we will use is the *wine gallon*. A wine gallon is simply a US fluid gallon comprised of 231 cubic inches (3,785.4118 ml) of pure water at a temperature of 60° F (15.556°C). This volume is also known as a wine gallon (WG) and has an assigned specific gravity of 1.0 under atmospheric conditions (not by using absolute densities in a vacuum). Several values for the mass, under standard atmospheric conditions, have been published and I have developed my own value using the *Handbook of Chemistry and Physics*, 89th edition (HCP).

Compare All Water Values in Air	Lbs. per WG	WG per Lb.	
HCP 72nd Lbs. per Gal. Air 15.556	8.328264	0.12007304	
Lbs. per Gal 60° F. § 30.66 Old §186.66 & T5	8.328230	0.12007353	
HCP 89th & Kaye & Laby Equation Vac to Air	8.328201	0.12007395	Book & ABS Std. Value
Lbs. per WG §30.41 & Tycos Tables	8.328200	0.12007400	

Constants

The standard alcohol content unit we will be using to measure our alcohol with is the *proof gallon* (PG). A proof gallon has two characteristics: the first relates to its alcohol content and the second defines its volume with respect to temperature. The alcohol content of a proof gallon is defined as 50% alcohol by volume, or 100 proof; the gallon itself conforms to the 231 cubic inches at 60°F standard.[1]

Comparing the weight of water per unit of volume to that of absolute alcohol (200 proof) shows that there is a 1.718 lb.-per-gallon difference and 205.99 kilograms-per-meter-cubed (kg/m^3) difference in the weight per unit volume of these two substances: that amounts to a total percentage change of 20.6% in each unit of volume.

Water to Absolute Alc.

Lbs. / Gal Air	Kg m3 Vac
8.3282	999.016
6.6098	793.03
1.718	205.99
0.794	0.794
20.6%	20.6%

Constants

Using the proof scale allows for 2,000 places between the density of water and absolute alcohol. With essentially 206 kg/m^3 difference between them, this amounts to an average of 103 grams-per-meter-cubed (g/m^3) difference between each individual slot in the table. This is roughly 4 avoirdupois ounces at 28 grams per ounce. Now this is an average. With low-proof alcohol, density will change very slowly with increasing proof. Forty-proof alcohol changes by only 63 g/m^3 for each 0.1 proof increment, while the incremental change for 190-proof alcohol is much greater at 192 g/m^3 per 0.1 proof increment.

Managing the respective difference in density between alcohol and water is the primary concern in alcohol blending. The entire alcohol gauging regimen prescribed by the regulations is predicated upon all components being compared to the volume and density they would have if they were at the temperature of 60° F, (15.556°C). This is known in the regulations as the True Proof at 60°F.

---

[1] 27 CFR Subchapter A - Liquors Sections 1 to end. Subpart B Section 5.11 Meaning of Terms A Proof Gallon = 1 U.S. Gal containing 50% by Volume of Ethyl Alcohol which at 60° F has a specific gravity of 0.7939 referenced to water at 60°F. This value is in reference to the specific gravity of alcohol in a vacuum rather than in air. Table 6 lists the specific gravity of alcohol at 60°F as 0.79389 in a vacuum and 0.79365 in air with reference to water at the same temperature. Distilled Spirits are Ethyl Alcohol , Hydrated Oxide of Ethyl Section 5.11 A proof gallon is one liquid gallon of alcohol that is 50% alcohol at 60 degrees F. See Water Analysis section for harmonization of TTB and modern SG values.

This construct exists only in an idealized world where everything is 60°F. All blending actions occur in this world, and in order to enter it you have to normalize or adjust the apparent proof and actual temperature of the alcohol into this construct. To do so is to determine the true proof at 60°F of the alcohol. This 60°F requirement also applies to the water component when blending a batch of alcohol to a target proof, so adding water at temperatures other than 60°F (without conducting a volume correction calculation) can lead to blends that are out of tolerance as to alcohol content.

This regimen requires that one obtain accurate information about three qualities of the alcohol to be investigated: the temperature, apparent proof (specific gravity or relative density), and either a precise weight or volume. If you blend by volume you must run a Table 7 volume correction before adding the component if the temperature has changed between the time you establish its alcohol content and the time it is subsequently used. Then your volumetric metering must be precise enough to measure that corrected volume. The easiest and most accurate way to measure alcohol and water is by weight.

If your measurements of temperature and proof are not precise with respect to the spirits being gauged, then the conversion to the realm of true proof at 60°F will be inaccurate.

Measuring Density

With our temperature, volume and alcohol content standards defined, we can discuss how we obtain accurate measurements of density, temperature, and either weight or volume of the alcohol we will use in our operations. Density is measured with a hydrometer. The hydrometers used by distillers are designed to read accurately on their face the proof of liquid alcoholic solutions at 60°F; they read on a scale that runs from 0, for water, to 200, for absolute pure alcohol when the liquid being measured is 60°F. Because of overlap in proof on each hydrometer there are 11 hydrometers needed to cover this entire range.

correct reading of
hydrometer at
dotted line,
avoiding false
reading at
meniscus curve

Picture of hydrometer set   *A helpful image for hydrometer use*[2]

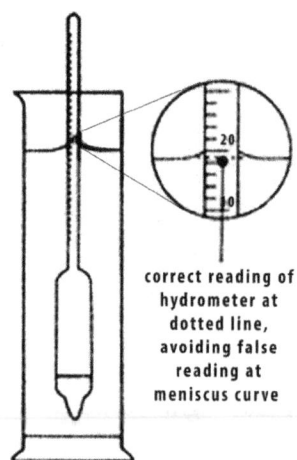

When alcohol is colder than 60°F, it is denser than it would be if it were 60°F. Therefore, the alcohol occupies a smaller volume than it would at 60°F. Because of this increased density, the hydrometer will not sink as far into the alcohol. The colder (denser) the alcohol becomes, the more the hydrometer is "squeezed" by the fluid itself and consequently the more buoyant it becomes and the less it will sink into the alcohol. In this condition, more of the hydrometer stem remains above the fluid and less is below. Because the hydrometer is graduated from low proof at the bottom of the stem to high proof at the top of the stem, the more the hydrometer stem sticks out of the solution, the lower the proof reading it provides, since lower proof is more dense than higher proof.

This increased density (lower proof) has the effect of increasing the specific gravity or relative density of the alcohol with respect to water. Because alcohol is lighter than water (0.79365 for absolute alcohol, in air, and 1.0

---

[2] C. Joseph (gen. ed.), *A Measure of Everything*, (Firefly Books, 2005), page 189.

for water),[3] any apparent (temperature-based) decrease in proof will show up as an increase in density if a specific gravity hydrometer is used. The specific gravity will always move closer to 1.0 when the spirits are less than 60°F.

If the temperature of the alcohol is raised, the addition of the heat energy necessary to do so will cause the mass of the alcohol to expand. This expansion is expressed as an increase in volume. The alcohol will become less dense and take up more volume. As a result, the hydrometer stem will sink farther into the alcohol and the intersection of stem with the alcohol surface will indicate a higher apparent proof. As density goes down, so does specific gravity (farther away from 1.0 and towards 0.79365, which is the relative density of 200-proof absolute alcohol with reference to water density). Only at 60°F will the hydrometer indicate the true proof or specific gravity of the alcohol.

The selection of 60°F as the standard is not simply arbitrary. It is a common enough ambient temperature so as to require the least amount of correction above and below that point. Within 20°F of the standard is 40°F and 80°F. This temperature range covers most of the temperatures that distillers will find in their plants and it requires the least amount of hydrometer thermal correction to be used for the majority of applications.

You might ask why not just measure all the alcohol in a 60°F box where the alcohol source, the sample and instruments are all the same temperature? This will always result in obtaining the true proof of the alcohol at 60°F directly from the hydrometer reading. While this is true and will result in true proof being obtained, it is impractical. The purpose of the TTB tables and procedures is to translate the present temperature and apparent proof into true proof at 60° F, whatever that temperature happens to be at the time the operation is carried out.

Once alcohol is gauged or "proofed" you can use it at any temperature and its true proof will not change even though its apparent proof will change with temperature. If it is proofed again at a different temperature its true proof will remain the same.

However, the temperature of the alcohol source that is being gauged is critical for determining the volume of the batch being sampled. If you are blending by volume, it is vitally important that the sample temperature match that of the reservoir, and also that this temperature be taken as close as possible to the time the spirits will actually be metered, so as to minimize the variable of temperature change between the time the reading is taken and the time the volume is metered out.

When blending by weight, it is less vital that the sample temperature match that of the source. So long as alcohol is not allowed to evaporate from the sample (which it will do, and quickly, particularly if it is warm), its temperature can change and the tables will accurately correct for the true percent of proof at 60°F. It is still important that the sample temperature and density be measured simultaneously. One may not take a temperature reading and then one minute later take a density reading and expect the results to be accurate because changes in temperature in the interim will throw off the calculation of true proof.

Procedures for Taking Readings.

The following procedures for taking accurate readings of apparent alcohol proof and temperature are somewhat extended and overinvolved but, in order to highlight all the factors that can affect the measurements, it is worth the two pages to set forth a full description. In general, good shop practices are all that are required and with practice some of the finer points discussed can be left to fend for themselves. When proofing alcohol, always start by using a higher-proof hydrometer than you think the alcohol might be. That way the hydrometer will not sink rapidly to the bottom of the beaker and possibly break.

1. When trying to determine the volume of a tank one should try to collect a sample from the middle of the tank. Most tanks and drums will develop a temperature gradient of colder at the bottom and warmer at the top. The top will be warmer than the bottom, particularly if they are sitting on the ground. With respect to blending by weight it is not nearly as important where the sample originates. Alcohol and water mixtures once blended will not disassociate (at least on human time scales of days, years or decades) and while the temperature may vary from top to bottom of a tank the true proof will not.

---

[3] Section 5.11, places the value at 0.7939 at 60 degrees F. Table 6 lists a value of 0.79365 in atmosphere and 0.79389 in vacuum.

2. The hydrometers, thermometers and testing beakers should be kept clean and free of any oily substance.

3. The hydrometer, thermometer and beaker should be as close as possible to the same temperature as the alcohol to be measured. This is particularly necessary when trying to fix the temperature of the larger tank being sampled for purposes of determining its volume. One does not want the temperature of the sample being increased or decreased by the temperature of the hydrometer, beaker, or thermometer. To get an accurate temperature reading, don't bring instruments from the heated side of the shop into the unheated alcohol storage area and expect to get accurate temperature readings of the alcohol right away. Bring them in a half hour earlier and set them on the drums.

After a while, you will get to know the temperature of your shop in different seasons. There is a pretty steady base ground temperature that rises in the summer and falls in the winter. One can measure this temperature by measuring the temperature of the water coming out of the cold-water tap in your shop. My shop water temperature fluctuates from a low of 45°F in the winter to a high of 72°F in the summer. Things sitting on the floor acquire this temperature by virtue of being in contact with the ground. Try to match your instrument temperature to the ambient temperature of your shop.

4. Take your sample of spirits. When taking the temperature of the sample, keep your hands off the beaker as much as possible. Holding the beaker with your hand will conduct heat out of your hand and into the sample, assuming you hand is warmer than the sample Also, the table the beaker is sitting on will add or remove heat from the sample depending on the temperature of the table.

5. Pour your sample into the beaker while introducing as little air as possible into the sample. You don't want to spill alcohol on the outside of the beaker. Any alcohol evaporating on the outside of the beaker will carry away heat and reduce the temperature of your sample below that of the alcohol being measured. Dry the outside of the beaker if any alcohol or water is on it.

6. Insert the thermometer as soon as the alcohol is placed in the beaker, while any air bubbles are rising to the surface and dissipating into the atmosphere. You need to get a fix on temperature as soon as possible so that it is as close to that of the alcohol in the larger reservoir from which you drew the sample. You should take two readings about 10 seconds apart to determine if the thermometer is still adjusting to the alcohol temperature. This is where having your instruments as close as possible to the same temperature as the source alcohol comes in handy because you will have less temperature adjustment to overcome and you should be able to get a true reading quickly.

Because of the large surface area of the beaker with respect to the sample size, the sample will tend to absorb heat or lose heat with respect to the ambient temperature. Also, the large surface area of the beaker makes it susceptible to cooling by drafts of air. Combined with any alcohol spilled on the outside of the beaker, this will accelerate any evaporative cooling of the sample. So, keep the outside of the beaker dry and away from the fan, air conditioner, or heater vent. In the summer, if you can arrange to do this in the morning when the ambient temperature is closest to the ground temperature, so much the better.

When dealing with volume, remember, with this temperature measurement you are not measuring the sample as an end in itself; you are measuring the temperature of the reservoir of spirits being sampled. It is the temperature of the reservoir or source of the spirits one is trying to determine. One could do this by taking the temperature of the reservoir itself, but what if you break a mercury thermometer in the reservoir? Not only will you be out a thermometer but you will have to pitch out the entire reservoir of alcohol itself. Good luck cleaning up from that mess—and there is very little forgiveness of tax due as a result of your mistake.

Once you get an accurate reading of the sample as indicative of the alcohol reservoir temperature, you can use that temperature for volume correction, but subsequent measurements of the sample can be used when you read the hydrometer and thermometer to determine apparent proof for input into the tables to convert to true proof. Since you have fixed the source temperature with your original reading, the tables will still take care of finding true proof based on the subsequent sample temperature so long as no alcohol has evaporated from the sample in the intervening time period. The source temperature will give you your value for volume correction. The proofing temperature and apparent proof will be adequate for translation into true proof. In the best case, the temperature will not change before you can take your hydrometer reading.

7. Put the hydrometer into the sample and see if the temperature of the sample changes by looking at the thermometer. The thermal mass of the hydrometer, if it is a different temperature than the alcohol, will affect the temperature of the sample. Dry the stem so it does not cool by evaporation or "weigh down" the hydrometer. Try not to let the hydrometer rest against the side of the beaker. Take a second temperature reading of the sample from the thermometer.

8. To determine where the hydrometer stem is intersecting the alcohol, look from below the surface of the alcohol at the point where the surface of the alcohol intersects the hydrometer stem. You should see an ellipse where the glass rises through the surface. This is the *meniscus* and it is caused by surface tension (or capillary action) of the alcohol as it clings to the hydrometer stem. We don't want to read the meniscus value. If we do read at the meniscus, the reading will always result in our reading a higher proof than the actual alcohol sample.

If you read the meniscus (and overestimate the proof) when gauging alcohol to be deposited into the storage account, you will measure more alcohol content than is actually present in the mixture. The distiller will report more proof gallons on premises than are actually present and this will lead to eventually having to report losses in the storage account that are entirely avoidable.

When blending alcohol to achieve a target proof, the distiller who has overestimated the proof will end up adding fewer proof gallons than are necessary to achieve the target proof. This will cause the blend to be contain less alcohol and be of a lower proof than intended.

Returning to the gauging procedures and hydrometer reading, after identifying the meniscus now raise your eyes and the shape of that ellipse will begin to change. It will get smaller and eventually become a straight line. If you raise your eyes too far it will begin to become an ellipse again and become bigger. The point of smallest diffraction is the correct hydrometer reading.

As usual a picture is better than 1,000 words:

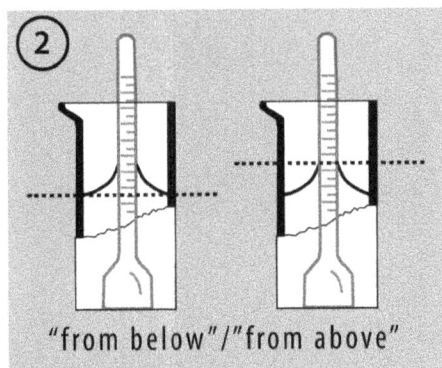

"from below"/"from above"

Because our fluid is clear, we read from below. With opaque fluids one must read from above.[4]

The regulations prescribe that hydrometer readings should be made to the nearest 0.05 degree of proof. Hydrometers are graduated in 0.2° increments. Therefore, according to the regulations, one should attempt to read the hydrometer to an accuracy of one-fourth of the instrument's demarcation. (0.05 * 4 = 0.2). I find this to be an unreasonable standard. I am only able to read hydrometers accurately to 0.1 proof or one-half of the marked values on the hydrometer stem.

9. Thermometer readings should be made to the nearest 0.1°F. Take a third temperature reading at as close to the same time as the hydrometer reading as possible. If the temperature reading has not changed since the first time it was taken in order to determine the source alcohol temperature or the second time that you read the sample temperature, you will have an additional assurance that the spirits being measured, the proofing sample, and the instruments are all at the same temperature. If the temperature has changed, then the sample temperature has changed from that of the spirits being sampled. The true proof can be determined only at the current temperature, whatever temperature happens to be.

---

[4] Figure from Brand.de, http://www.brand.de/en/laboratory-instruments-liquid-handling-life-science/.

I would be remiss if I didn't point out that the regulations indicate that a second sample "should" be taken and that you should repeat the entire procedure for a second sample and to obtain identical results before you proceed. This seems like overkill and frankly it is hard to get a second sample to read within 0.1 proof of the first. The variables are many and the instruments are only so precise.

I have found that thermometers and hydrometers are most accurate near their calibration temperatures. If time allows, proof the alcohol at its present temperature. Then seal the sample in a container and allow it to rise or fall to a temperature near the calibration temperature. See if the same true proof can be read at the changed temperature. This will give an idea of how accurate your instruments are over a range of temperatures. Also, use two thermometers and see if you can get any two to read the same temperature. Mine always read slightly different temperatures, but I label one of them as the standard and don't try to interpolate the difference between two different thermometers. Once you choose a standard thermometer, use it and no other so your measurements will be consistent.

Good shop procedures are always a plus and applying them assists in getting accurate readings without allowing alcohol to evaporate and throw off your calculations or allowing temperature change of the sample to do the same.

Having acquired the present temperature and apparent proof (relative density) of our sample, we can turn our attention to the calculations to determine the true proof of the alcohol at 60°F (15.56°C).

Determining True Proof

The following is an example contained in the gauging manual in section §186.23 (old style) and §30.23 (new style) for determining true proof, using Table 1 with temperature and proof interpolation between whole proof values. There are nineteen calculations and four lookups in this procedure.

Temperature of Alcohol ° F.	72.10	0.050	Thermometer Correction Factor
Apparent Proof of Alcohol	192.85	-0.030	Hydrometer Correction Factor

True Proof Rounded to Nearest 1/100 proof	189.97		Alcohol Temperature Results
True Proof Rounded to 1/10 Proof for Table 4	190.00	72.15	Corrected Temperature
True at Nearest Whole Proof for For Table 3,5,6 & 7	190.00	72.20	Corrected Temperature Rounded to 1/10th °F

Find True Proof @ 60° F. Using §186.23 & §30.23 Method of Apparent Proof and Temperature Correction

Hydrometer Reading of Apparent Proof	192.85	
Correction Factor for This Hydrometer	-0.030	
Add Correction (+ or - value) to Aparent Reading =	192.82	Corrected Hydrometer Reading
Find Fractional Proof Value of Corrected Reading	0.82	

	Proof of Lookup	Temp. of Lookup		
Round Reading Down to Whole Proof Lower.	192.00	72.00	189.10	Lookup Table 1 Value for Whole Proof Lower
Add 1 to obtain 1 proof higher	193.00	72.00	190.20	Lookup Table 1 Value for Whole Proof Higher
Subtract Lower Proof Value from Higher proof Value			1.10	Hydrometer Based Table 1 Proof Difference

Apparent Temperature Reading	72.10	
Correction Factor for This Thermometer	0.05	
Add Correction (+ or - value) to Aparent Reading =	72.15	Corrected Temperature
Corrected Temperature Rounded to 1/10 ° F.	72.20	
Find Fractional Degree F. of Corrected Reading	0.15	

	Temp of Lookup	Proof of Lookup		
Round Corrected Reading Down to Whole ° F. Lower	72.00	192.00	189.10	Lookup Table 1 Value for Whole Proof Lower
Add 1 to obtain 1 ° F. Higher	73.00	192.00	188.90	Lookup Table 1 Value for Whole Proof Higher
Subtract Higher Temp Table 1 Proof from Lower Temp Table 1 Value.			0.20	Thermometer Based Proof Difference

Hydrometer Based Table 1 Proof Difference	1.10	
Fractional Degree of Hydrometer Reading	0.82	
Proof Dif. * Fractional Degree of Hydrom. Reading =	0.902	Result (T)

Thermometer Based Proof Difference	0.20	
Times Fractional Degree of Thermometer Reading	0.15	
Temp Dif. * Fractional Degree to Temp. Reading	0.03	Result (B)

Finding Hydrometer Corrected Proof from Sample Temperature to 60° F

Lower Whole Proof Table 1 Value Restated	189.10	
Result (T) Restated	0.902	
Lower Whole Proof + Result (T) =	190.002	Result
Result (B) Restated	0.030	
Lower Whole Proof - Result (B) + Result (T) =	189.972	True Proof @ 60° F.
True Proof Rounded to Nearest 1/100 proof	189.97	
True Proof @ 60° F. Rounded to 1/10th proof	190.0	For Table 4
True Proof Rounded to Nearest Whole Proof	190.0	For Tables 3,5,6 & 7

Alcohol Properties Expanded to 1/10 Proof

This methodology is required any time that either temperature or apparent proof are not whole numbers. See Volume 2 for additional methods of gauging alcohol at 60°F and other calibration temperatures.

Most hydrometers and thermometers are not "perfect" and will have some correction factor associated with them. You may have to ask your thermometer or hydrometer provider to tell you what these particular values are for the instruments you order from them. Periodically it is advisable to get the instruments checked to see that they still read accurately.

Retrieving table values associated with this proof shows that the following information about the sample can be obtained.

Properties of Alcohol Corrected to 60° F. True Proof Rounded to 1/10 Proof	Proof Gal. per Lb.	Wine Gal. per Lb.	Lbs. per Wine Gal.	Lbs. per Proof Gal.	Parts Alc.	Parts Water	Total Parts	SG Air 60/60	SG Vac 60/60	Kg. m3 Vac	Kg. m3 Air	Mass % (Air)	Vac Mass %
190.0	0.27966	0.14719	6.79389	3.57577	95.00	6.19	101.19	0.81577	0.81599	815.191	814.09	92.414	92.418

Alcohol Properties Expanded to 1/10 Proof

Each of these various values will provide a basis for blending and accounting for alcohol.[5]

Alcohol Data Available

We need to examine more closely the data that is available in the TTB tables regarding various qualities of alcohol. For our example we will use a true proof of 158.7.

Find True Proof @ 60° F. Using §186.23 & §30.23 Method of Apparent Proof and Temperature Correction

Temperature of Alcohol ° F.	72.90	0.000	Thermometer Correction Factor
Apparent Proof of Alcohol	163.00	0.000	Hydrometer Correction Factor

True Proof Rounded to Nearest 1/100 proof	158.73		Alcohol Temperature Results
True Proof Rounded to 1/10 Proof for Table 4	158.70	72.90	Corrected Temperature
True at Nearest Whole Proof for For Table 3,5,6 & 7	159.00	72.90	Corrected Temperature Rounded to 1/10th °F

Properties of Alcohol Corrected to 60° F. True Proof Rounded to 1/10 Proof	Proof Gallons per Lb. T4	Wine Gal. per Lb. T4	Lbs. Per Wine Gal Table 5 PF	Lbs. per Proof Gal. Table 5 PF	Parts Alcohol	Parts Water	Total Parts	Specific Gravity in Air	Specific Gravity in
158.7	0.22018	0.138740	7.20762	4.54171	79.35	23.57	102.92	0.86544	0.86559
Quotient Values Using 1/Table 4 Results			7.20773	4.54174					

Properties of Alcohol Corrected to 60° F True Proof Rounded to Whole Proof	Proof Gallons per Lb. T4	Wine Gal. per Lb. T4	Lbs. Per Wine Gal Table 5 PF	Lbs. per Proof Gal. Table 5 PF	Parts Alcohol	Parts Water	Total Parts	Specific Gravity in Air	Specific Gravity in
159.0	0.22071	0.138810	7.20417	4.53092	79.50	23.40	102.90	0.86503	0.86518
Quotient Values Using 1/Table 4 Results			7.20409	4.53083					

Alcohol Properties Expanded to 1/10 Proof

What we see is that the true proof is found to 1/100th of a proof (158.73 proof) but for subsequent calculations only the proof rounded to 1/10th proof can be employed.

Table 4 is the only table which was published with 2,000 data points in it. It provides wine gallons per pound and proof gallons per pound at 1/10th-proof increments throughout its range. A proof scale from 0 to 200, with each degree having 10 subdivisions, results in 2,000 data points available. Therefore, the outputs are accurate to 1 part in 2,000, which translates to 5 parts in 10,000 (although there are some duplicate values in it, at low proofs, with respect to Wine Gallons per Pound—but not Proof Gallons per Pound), so one can determine individual mass percentages in air by finding their ratio to each other on a pounds-per-gallon basis. (1/WGPP = lbs. per WG.).

---

[5] I have adopted Table 4 as the standard table for determining pounds per wine gallon since there are slight differences between the Table 5 based values and the quotient values arrived at by using (1/Table 4 gallon per pound values). My experience is that the reciprocal nature of the two Table 4 based values makes them more useful than combining Table 4 values with Table 5 values and, by using one table only, one is putting all one's uncertainty about values into one data set rather than two. Quote from Section § 30.64 § 186.64 "The slight variation between this Table (referring to Table 4) and Tables 2, 3, and 5 on some calculations is due to the dropping or adding of fractions beyond the first decimal in those Tables."

As published by the TTB, Tables 5 and 6 only provide values for whole proofs, providing 200 data points. As we have discussed, using rounded whole proof values can result in errors that exceed the blending tolerance of the batch to be blended. Two hundred data points amounts to an accuracy of 1 part in 200, which translates to 50 parts out of 10,000 (50/10,000). After I had expanded Table 5 from 200 to 2,000 data points, I realized it did not match the Table 4 values and added uncertainty rather than diminishing it.

Tolerance for Blending Operations

We need to determine what blending tolerance to apply to our operations. In my worksheets, the blending tolerance has been calculated using 0.15 percent alcohol by volume which translates into 0.30 proof. This implies that, when blending to 80 proof, customers must receive no less than 79.7-proof alcohol when they purchase a bottle, and that the government will not penalize you if you provide the customer with up to 80.3 proof in the same bottle. Each component tolerance, both for alcohol and water, is independent of the other. One can't add the wrong amount of alcohol and the right amount of water and expect the batch to achieve the target proof and vice versa.

A word about what "0.15 percent alcohol by volume" means. Numerically it is 0.0015 or 15 parts out of 10,000. By multiplying any blending component of a batch by 0.0015, you can obtain the blending tolerance for that component. The published program allows the user to change the tolerance value so one can choose a smaller tolerance, but 0.0015 is the maximum tolerance allowed for unobscured alcohol. We will address how solids like sugar, calcium carbonate in water or other solids can obscure (hide/mask) the true proof later on.[6] Following are several ways in which tolerance can be calculated.

Find Tolerance as a Degree of Proof	
Percentage Scale	100.00
Amount of Scale for Tolerance	0.15%
Scale * % of scale = Fraction of 100	0.15
Divide Fraction of Scale by 100 Scale = Fraction of 1 or Unity	0.0015

Tolerance as Parts	
Tolerance Restated	0.0015
Total Parts to Achieve Unity of Whole	10,000.00
Tolerance * Absolute Scale = Parts per 10,000	15.0

Find Proof Scale Equivalent	
Proof Scale	200
Tolerance fraction of Unity	0.0015
Proof Scale * Fraction = Proof Tolerance	0.30

Quick Way to Find Tolerance as a Degree of Proof	
Tolerance Restated From Input	0.0015
Tolerance * 2 =	0.0030
Tolerance * 2 result * 100 = Degree of Proof	0.30

Batch Plan by Volume

We need to compare the required tolerance to the accuracy of the tables that are provided to conduct our operations. With 2,000 data points available between 0 and 200 proof, and a mass percent running from 0 to 100%, each 0.1 degree of proof represents 1 part out of 2,000. Translating from a 2,000 scale to a 10,000 scale simply involves multiplying each part by five.

---

[6] Alcohol blending tolerance is defined in § 5.37. Alcohol content -(b) Tolerances. "The following tolerances shall be allowed (without affecting the labeled statement of alcohol content) for losses of alcohol content occurring during bottling: (1) Not to exceed 0.25 percent alcohol by volume for spirits containing solids in excess of 600 mg per 100 ml; or (2) Not to exceed 0.25 percent alcohol by volume for any spirits product bottled in 50 or 100 ml size bottles; or (3) Not to exceed 0.15 percent alcohol by volume for all other spirits."

Find .1 proof increment portion of whole	2,000 Scale		10,000 Scale	
Total .1 Proof Increments	2,000		10,000	
Each .1 increment is one part out of 2,000	1.00		5.00	Parts of 10k
Or .05% of the total.	0.05%		0.05%	
Table 4 uncertainty				
Table Parts Increments, Parts of 10,000	5			
Divide by 2 because of rounding	2.5			
Total Parts of Tolerance	15			
Rounding	16.67%			
Parts Scale	10,000			
Tolerance of Parts	15			
One part out of	666.67			

Using rounding to the nearest increment, the uncertainty in obtaining proof is one-half of 5 parts out of 10,000 or 2.5 parts out of 10,000. This uncertainty is itself 2.5/15 = 0.167 or 16.67% of the tolerance allowed. One other way to look at how tolerance is quantified is to take 10,000 and divide by 15 which yields 666.67 (i.e., one part out of 666.67 for the total tolerance allowed for any batch component).

Direct Table 1 Use in Special Case Only

When I began distilling I simply rounded the temperature and proof values of my alcohol to whole numbers and used them along with Table 1 to find the "True Proof." When I gained more experience, and finally understood the tables, regulations, and methods described in them, it became clear to me that the direct use of Table 1 was not the best practice because the direct use of Table 1 is only allowed in the special case where both the apparent proof and temperature are whole numbers. In practice this never occurs. The effect is to make the direct use of Table 1 inappropriate for proofing alcohol. In order to find the amount of error this procedure can induce, we will run several examples: First, we will recreate Example 1 outlined in §186.61 (§30.61 new style) of the regulations showing 190.0 proof as a result.

Table 1 Whole Temp & Whole Proof Method	
Whole Number Temperatrue of Alcohol	72.00
Whole Number Proof of Alcohol	193.00
Table 1 Proof Value Retrieved	190.20
Retrieved Value Rounded to Nearest Whole Number = "True Proof"	190.00

Table 1 Lookup of True Proof

However, if the actual temperature before rounding was 72.6° and the indicated proof was 192.5 and we use those values along with the §186.23-§30.23 procedure for determining true proof, it shows that the true proof is in fact 189.5 rather than 190 proof. Setting the weight so that we come close to a true proof containing 1,000 proof gallons we see that the overestimate of proof caused by using Table 1 directly has overestimated the proof gallons by 3.81 proof gallons and that this amounts to 254% of the tolerance for blending or 154% over the maximum allowed blending tolerance.

Weight of Alcohol (Lbs.)	3,590.00	0.0015	Blending Tolerance
Temperature of Alcohol (F.)	72.60	73.00	Rounded Temp for Table 1 use
Hydrometer Reading of Apparent Proof	192.50	193.0	Rounded Proof For Table 1 use
True Proof @ 60° F. Results from §186.23 -	189.50	190.00	Table 1 Lookup of "True" Proof. §186.61 - §30.61 Procedure
Degrees of Proof Dif. from §186.61 - §30.61 Table 1 Method	-0.50	0.50	Degrees of Proof Dif. from §186.23 - §30.23 Table 1 Method
True Proof Gallons Actually in Batch	1,000.10	1,003.91	Proof Gallons the Distiller will think were added.
Difference in Proof Gallons	-3.81	3.81	Amount More (Less) will be Added using rounding & Table 1
Tax difference for each method	-$51.37	$51.37	Tax Difference
Proof Gallon Tolerance for Batch	1.50		
Absolute Value of Proof Gallon Difference	3.81		
Percentage of Tolerance Used	254%		

Rounding Errors of Table 1

This 0.5 proof difference from 189.5 proof to 190 with rounding is just the error in this case; the errors can be as large as 0.7 proof. We've set up the example to approximate 1,000 proof gallons. Since this is an entirely avoidable error, it is best to use the full §30.23 procedures.

If one used Table 1 directly, they would make a record of and pay excise tax on 1,003.91 proof gallons when in fact they had only included 1,000.10 proof gallons. This shorts the customer because they are getting less alcohol in their blend. The error is compounded by the distiller overpaying the treasury $51.37 in tax for alcohol that didn't exist.[7] Of course the error can also go the other way depending on which way one rounds.

Because the discrepancy can be positive or negative, one batch will be short of proof gallons and the next over-endowed with proof gallons, and you won't be able to control the process or properly account for your inventory: a rounded proofing that goes into or out of the storage account can go any which way and you are trusting to the law of averages to sort things out.

The blending tolerance for a 1,000-proof-gallon batch is 1.5 proof gallons. (1,000 * 0.0015 = 1.5). Following are examples of how the effects of rounding work on an individual basis:

Rounding temperature up results in a higher true proof because more thermal correction is included by Table 1 in order to obtain a 60° result. Accordingly, one adds more proof gallons than they think are in the batch. The customer benefits (which is all well and good) but the distillery fails to pay tax on proof gallons that were actually added to the batch. The final blend will be above the intended target proof.

Weight of Alcohol (Lbs.)	3,590.00	0.0015	Blending Tolerance
Temperature of Alcohol (F.)	72.60	73.00	Rounded Temp for Table 1 use
Hydrometer Reading of Apparent Proof	193.00	193.0	Rounded Proof For Table 1 use
True Proof @ 60° F. Results from §186.23 -	190.10	190.00	Table 1 Lookup of "True" Proof. §186.61 - §30.61 Procedure
Degrees of Proof Dif. from §186.61 - §30.61 Table 1 Method	0.10	-0.10	Degrees of Proof Dif. from §186.23 - §30.23 Table 1 Method
True Proof Gallons Actually in Batch	1,004.73	1,003.91	Proof Gallons the Distiller will think were added.
Difference in Proof Gallons	0.83	-0.83	Amount More (Less) will be Added using rounding & Table 1
Tax difference for each method	$11.15	-$11.15	Tax Difference
Proof Gallon Tolerance for Batch	1.51		
Absolute Value of Proof Gallon Difference	0.83		
Percentage of Tolerance Used	55%		

Rounding Errors of Table 1

---

[7] At $13.50 per proof gallon as of 2016.

Rounding temperature down results in a lower true proof because less correction is required by Table 1 in order to obtain a 60° result. Accordingly, one adds fewer proof gallons than they think are in the batch. The customer loses and the distillery is paying tax on proof gallons that don't exist. The final blend will be below the intended target proof.

Weight of Alcohol (Lbs.)	3,590.00	0.0015	Blending Tolerance
Temperature of Alcohol (F.)	73.40	73.00	Rounded Temp for Table 1 use
Hydrometer Reading of Apparent Proof	193.00	193.0	Rounded Proof For Table 1 use
True Proof @ 60° F. Results from §186.23 -	189.90	190.00	Table 1 Lookup of "True" Proof. §186.61 - §30.61 Procedure
Degrees of Proof Dif. from §186.61 - §30.61 Table 1 Method	-0.10	0.10	Degrees of Proof Dif. from §186.23 - §30.23 Table 1 Method
True Proof Gallons Actually in Batch	1,003.19	1,003.91	Proof Gallons the Distiller will think were added.
Difference in Proof Gallons	-0.72	0.72	Amount More (Less) will be Added using rounding & Table 1
Tax difference for each method	-$9.69	$9.69	Tax Difference
Proof Gallon Tolerance for Batch	1.50		
Absolute Value of Proof Gallon Difference	0.72		
Percentage of Tolerance Used	48%		

Rounding Errors of Table 1

Rounding proof up results in a lower true proof than the rounded result because of the rounding itself. One adds fewer proof gallons than they think are in the batch. The customer loses and the distillery is paying tax on alcohol that is not in the batch. The final blend will be below the intended target proof.

Weight of Alcohol (Lbs.)	3,590.00	0.0015	Blending Tolerance
Temperature of Alcohol (F.)	73.00	73.00	Rounded Temp for Table 1 use
Hydrometer Reading of Apparent Proof	192.50	193.0	Rounded Proof For Table 1 use
True Proof @ 60° F. Results from §186.23 -	189.50	190.00	Table 1 Lookup of "True" Proof. §186.61 - §30.61 Procedure
Degrees of Proof Dif. from §186.61 - §30.61 Table 1 Method	-0.50	0.50	Degrees of Proof Dif. from §186.23 - §30.23 Table 1 Method
True Proof Gallons Actually in Batch	1,000.10	1,003.91	Proof Gallons the Distiller will think were added.
Difference in Proof Gallons	-3.81	3.81	Amount More (Less) will be Added using rounding & Table 1
Tax difference for each method	-$51.37	$51.37	Tax Difference
Proof Gallon Tolerance for Batch	1.50		
Absolute Value of Proof Gallon Difference	3.81		
Percentage of Tolerance Used	254%		

Rounding Errors of Table 1

When rounding proof down, the distiller is deliberately understating the alcohol content of the sample. The true proof is higher than is reported. More proof gallons are added to the batch than reported. The customer benefits, and the distillery fails to pay tax on proof gallons that were actually added to the batch. The final blend will be above the intended target proof.

Together these errors can either offset or amplify each other. Rounding proof down is shown below.

Weight of Alcohol (Lbs.)	3,590.00	0.0015	Blending Tolerance
Temperature of Alcohol (F.)	73.00	73.00	Rounded Temp for Table 1 use
Hydrometer Reading of Apparent Proof	193.40	193.0	Rounded Proof For Table 1 use
True Proof @ 60° F. Results from §186.23 -	190.40	190.00	Table 1 Lookup of "True" Proof. §186.61 - §30.61 Procedure
Degrees of Proof Dif. from §186.61 - §30.61 Table 1 Method	0.40	-0.40	Degrees of Proof Dif. from §186.23 - §30.23 Table 1 Method
True Proof Gallons Actually in Batch	1,007.00	1,003.91	Proof Gallons the Distiller will think were added.
Difference in Proof Gallons	3.09	-3.09	Amount More (Less) will be Added using rounding & Table 1
Tax difference for each method	$41.68	-$41.68	Tax Difference
Proof Gallon Tolerance for Batch	1.51		
Absolute Value of Proof Gallon Difference	3.09		
Percentage of Tolerance Used	204%		

Rounding Errors of Table 1

Worst Case # 1 is rounding proof down and rounding up temperature; this results in the least payment of tax.

Weight of Alcohol (Lbs.)	3,590.00	0.0015	Blending Tolerance
Temperature of Alcohol (F.)	72.50	73.00	Rounded Temp for Table 1 use
Hydrometer Reading of Apparent Proof	193.40	193.0	Rounded Proof For Table 1 use
True Proof @ 60° F. Results from §186.23 -	190.50	190.00	Table 1 Lookup of "True" Proof. §186.61 - §30.61 Procedure
Degrees of Proof Dif. from §186.61 - §30.61 Table 1 Method	0.50	-0.50	Degrees of Proof Dif. from §186.23 - §30.23 Table 1 Method
True Proof Gallons Actually in Batch	1,007.78	1,003.91	Proof Gallons the Distiller will think were added.
Difference in Proof Gallons	3.88	-3.88	Amount More (Less) will be Added using rounding & Table 1
Tax difference for each method	$52.34	-$52.34	Tax Difference
Proof Gallon Tolerance for Batch	1.51		
Absolute Value of Proof Gallon Difference	3.88		
Percentage of Tolerance Used	256%		

Rounding Errors of Table 1

Worst Case # 2 is rounding proof up and rounding down temperature; this results in the most payment of tax without alcohol actually being present.

Weight of Alcohol (Lbs.)	3,590.00	0.0015	Blending Tolerance
Temperature of Alcohol (F.)	73.40	73.00	Rounded Temp for Table 1 use
Hydrometer Reading of Apparent Proof	192.50	193.0	Rounded Proof For Table 1 use
True Proof @ 60° F. Results from §186.23 -	189.30	190.00	Table 1 Lookup of "True" Proof. §186.61 - §30.61 Procedure
Degrees of Proof Dif. from §186.61 - §30.61 Table 1 Method	-0.70	0.70	Degrees of Proof Dif. from §186.23 - §30.23 Table 1 Method
True Proof Gallons Actually in Batch	998.59	1,003.91	Proof Gallons the Distiller will think were added.
Difference in Proof Gallons	-5.31	5.31	Amount More (Less) will be Added using rounding & Table 1
Tax difference for each method	-$71.73	$71.73	Tax Difference
Proof Gallon Tolerance for Batch	1.50		
Absolute Value of Proof Gallon Difference	5.31		
Percentage of Tolerance Used	355%		

Rounding Errors of Table 1

Note that the total outage is 0.7 proof. Also, when entering data into the ABS program one will note that each 0.1°F of temperature change does not automatically result in a change in the true proof result. Table 1 is jumpy like that and sometimes temperatures can vary by one half a degree Fahrenheit before a 0.1 change in proof is returned.

Summary of Rounding:

Rounding Temp Up: you add more true proof gallons than you think and pay less.

Rounding Temp Down: you add fewer true proof gallons than you think and pay more tax - Error of paying tax on proof gallons not in batch & creating gains in Storage Account.

Rounding Proof Up: you add fewer true proof gallons than you think and pay more tax - Error of paying tax on proof gallons not in batch & creating gains in Storage Account.

Rounding Proof Down: you add more true proof gallons than you think and pay less - Error of depleting proof gallons in inventory and not paying tax.

Together these errors can either offset or amplify each other.

Worst Case # 1 is rounding proof down and rounding up temp. Results in least payment of tax.

Worst Case # 2 is rounding proof up and rounding down temp. Results in most payment of tax without proof gallons leaving plant.

The reason that I have taken the time to follow these small discrepancies and to explain them is to inform the user that they exist and to establish that they are part and parcel of the blending regimen prescribed by the TTB. With respect to blending rather than proofing, whole-proof rounding is permitted using the Table 6 procedures and values, so it is a permissible way to blend even though it is not strictly within tolerance.

Batch Blending to Target Proof

Batch Planning by Volume

Before we blend we need to plan our batch and see what components we will need. The *Batch Plan by Volume* worksheet allows the user to determine many of the values required to fill a given number of cases at any target proof including the total weight of the batch and the proof gallons required. One can also fill by weight.

What is the Target Proof of the Batch	80.00		100.00	Planned Cases
How many cases in batch	102.00		2.00%	Processing Margin of Loss
What is the bottle fill amount in milliliters	750.00		2.00	Cases * Margin
How many bottles per case	12.0		102.00	Increased Batch Size
Find Volume Needed for Planned Cases				
Bottle Fill ml * Bottles Per Case * Cases = Tot. ml	918,000.00		1,224.00	Bottles in Batch
Divide by 1,000 = Liters Equivalent	918.00			
Gallons per Liter §19.722	0.2641720		$13.50	Excise Tax Rate
Wine Gallons needed @ Target Proof & 60° F.	242.510		$2,619.10	Excise tax on Gross Batch
Find Batch Total Weight				
Wine Gallon per Lb. @ Target Proof	0.12616		242.510	Batch Gallons Restated
Wine Gal/W.G. per Lb. = Total Lbs. of Batch	1,922.20		2.3775480	Gal/Cases = Gal. per Case
Find Proof Gallons			Find Weight of Alcohol Required	
Proof Gallons per Lb. @ Target Proof	0.10093		194.01	Total Proof Gal. Required
Lbs. * Proof Gal. per Lb. = Proof Gallons	194.01		160.00	True Proof of Alc 1
Proof Gallons per Case	1.9020326		0.222450	Proof Gal. per Lb. of Alc 1
Proof Gallons per Bottle	0.1585027		872.14	PG/PGPP = Lbs. Alc 1
Find Grams of Alcohol per Bottle				
Pounds to Kilograms Factor	0.45359237		2.0	Labels per Bottle
Batch Weight in Kilograms	871.89		2,448.00	Labels Required
Batch Weight in Grams	871,893.83		84.0	Cases Per Pallet
Cases * Bottles per case =Total Bottles	1,224.00		1.21	Number of Pallets Needed
Total Grams/ Total Bottles =Grams per Bottle	712.33156			
Grams to Weight Ounces Converter	0.035273962			
Weight Ounces Per Bottle	25.1268			

Batch Plan by Volume

In planning your bottling runs, you need to account for the inevitable loss of product that occurs during bottling operations. Originally this worksheet included an estimated percentage of loss that increased the batch volumes and proof gallons. However, too much uncertainty was introduced when the numbers were multiplied by the percentage of increase. To ensure the accuracy of the main calculations, I have included a separate module that allows the user to determine the loss rate for the batch and estimate the increase in cases before entering the desired number of cases in the primary calculation that drives the worksheet.

Using the procedure contained in §30.52 for measurement of cased spirits, we determine the conversion from liters to wine gallons with the value of 0.264172 provided by §19.722. The resulting volume calculated is a net value after the change in volume due to blending but that is not an issue, since by using the Table 4 value for wine gallons per pound we can find the total weight of the batch. Then, using the total batch weight, we multiply it by the proof gallons per pound to find the total proof gallons that will be contributed to the batch. From there it is a simple matter to find the remaining values in the worksheet.

Because we are planning the batch based on proof gallons, we do not have to be concerned about the volume reduction that will occur when you blend your components together. You can use proof gallons as an input to plan the batch and blend the batch. Since one is required to maintain alcohol inventory in the form of proof gallons, this allows one to withdraw them in the same unit value in which they are stored and essentially ignore volume for purposes of blending.

I find that different bottle types have different processing losses; for instance, 375-ml bottles run at a higher loss rate than 750-ml because each 375-ml run has twice as many bottles as a 750-ml run. Simply filling more containers with the same volume leads to a larger error. This makes sense because if I was only filling one *big* bottle my loss rate would be essentially zero, except what was left in the filter and had to be disposed of.

Another reason I estimate processing losses is because if I am going to go to the trouble of bottling, the consequence to be guarded against is coming up short on the cases I intended to bottle. In my experience, there is nothing worse than spending a whole day bottling cases and then find out that I have run out of alcohol before achieving the target number of cases bottled. It is better to have two or three more cases than to run short. Because the plan calculates the number of bottles, cases, and labels you will need, it is possible to have the maximum number needed on hand, so those components match the blended volume. If necessary, by playing with the numbers a little, it is possible to use the worksheet to match the batch size to the materials available.

There are many other variables that affect blending and bottling operations that cause operational losses. These variables are extensively covered in the *Bottling Log* sheet outlined later in this manual.

While we are converting from metric to English units, I want to take a moment to examine the accuracy of the converters used to accomplish this task.

The regulations give a conversion value that is valid to three decimal places (3.875)[8]. Using an accepted value that has seven decimal places reduces the conversion error to one case of whiskey out of five million.[9]

Wine Gallons Restated	242.5099	242.5099
Liters per Gal. (Sec 5.11, 3 Places)	3.78500	3.7854118
Net Liters	917.9000	918.000
Net ml	917,899.96	917,999.82
Bottle Size in ML.	750.00	750.00
Total Milliliters/ ML. per Bottle = Bottles in Batch	1,223.87	1,224.00
Bottles in a Case	12.00	12.00
Cases in ideal bottling run 3rd way	101.99	101.9999802
Proof Gal./P.G. Per Case = Ideal Cases from P.G.	102.000	102.000
Case Difference in Methods	0.011116	1.98E-05
Cases/Case Difference in Methods	0.00010898	1.94E-07
Difference Equals One case out of	9,176	5,155,426

Batch Plan by Volume

Applying the same analysis to the proof-gallons-per-case we have calculated and comparing it to the value given in the regulations shows that our methods are essentially sound, with a discrepancy of one proof gallon out of 34,419:

Calculated Proof Gal. per Case	1.902032637
Standard for 750 * 12 80 Proof Case Proof Gal.	1.902061691
Difference as Absolute Value	0.0000291
Percentage Difference	0.0000153
Difference Equals One case out of	65,467.31
Equivalent to one Proof Gallon out of	34,419.66

---

[8] Section 5.11 Meaning of Terms.

[9] For unit conversion I have found that a computer program, *Uconeer*, published by Harvey Wilson and Katmar Software, contains the best and most consistent set of interrelated conversions. The conversions are available to eight decimal places if the particular converter requires them. They are also very closely reciprocal with one another. The version I use is Version 2.4 and the software can be found at http://www.katmarsoftware.com?referrer=Uconeer24A.

Applying the same procedure to the standard wine gallons per case shows an outage of one gallon out of 55,555.13:

Fluid (Wine Gallon) Conversion Test	
Calculated Wine Gallons per Case	2.3775480
Standard for 750*12 100 Proof Case Wine Gal.	2.3775300
Difference as Absolute Value	0.0000180
Percentage Difference	0.0000076
Difference Equals One case out of	132,085.00
Equivalent to one Fluid Gallon out of	55,555.13

Batch Plan by Volume

The only other value we can find that will improve the accuracy of our calculations is to take the three-decimal-place value that is provided by §5.11 (33.814 fluid ounces per liter) and convert it to a liters-per-gallon value side by side with a six-decimal-place value from the *Uconeer* conversion program:

Fluid Ounces to Liters per US Gallon	TTB Value	Uconeer Value
Fluid Ounces per Liter. Section 5.11 Value	33.814	33.814023
Fluid Ounces per Gallon	128	128
Percentage of a Gallon	26.417188%	26.417205%
The Value of One	1	1
One Divided by Gallon Fraction = Liters per Gal.	3.7854143	3.7854118
Comparison Liters per Gal. Value	3.7854118	3.7854118
Difference from Calculated Value	0.0000025	0.0000000
Fractional Difference	0.0000007	0.0000000
Difference Equals One case out of	1,498,919	76,661,792
Equivalent to one Fluid Gallon out of	395,973	20,251,903
Gallons per Liter §19.722	0.2641720	0.2641720
Equivalent to one liter out of	104,605	5,349,986

Batch Plan by Volume

The value that is generated using the TTB value is accurate to one gallon out of 395,973 or one liter out of 104,605. Using the *Uconeer* value extends the accuracy of the conversion to one gallon out of twenty million or one liter out of five million (5,349,986). That's a large difference just for using a few more decimal places.

While we are testing conversion factors, we should look at the ones we use for mass as well. The pounds-to-kilograms and back is very good with a two-way variance of one pound out of one hundred million. If you assume that a one-way conversion would only have half of the error of a two-way conversion, then you can multiply the outage by 2 to find the approximate value of what a one-way conversion costs in terms of uncertainty.

Test Pounds to Kilograms Conversion Factors			Testing Grams to Ounces Conversion Factors	
Pounds	1,922.24		1,000.00	Gram Fill Amount
Lbs. to Kilo's	0.453592370		0.03527396	Grams to Ounces Converter
Kilograms	871.9134		35.274	Ounces Result
Kilograms to Lbs. Avdp	2.20462260		28.34950	Ounces To Grams Converter
Back to Pounds Again	1,922.24		1000.00	Grams Result
Variance	0.0000191		0.000814281	Variance in Grams per Bottle
One Pound out of	100,903,715		1,228,077	One Bottle out of
Times 2 Assuming 1/2 for one conversion	201,807,430		2,456,155	Times 2 Assuming 1/2 for one conversion

Batch Plan by Volume

Proof Gallons to Weight

Now that we have planned our batch and checked our conversion factors, we can find the mass and 60°F volume of our source alcohol that we will need to produce the planned cases. For this blend we will assume that we are using 160-proof alcohol to blend to our target proof of 80 proof. Using the value of 194.01 proof gallons that are required we can determine the weight and volume of the alcohol we will need to remove from our storage account in order to blend the batch.[10]

Number of Proof Gallons to Withdraw	194.01		
Temperature of Alcohol	60.00	0.000	Thermometer Cor.
Apparent Proof of Alcohol	160.00	0.000	Hydrometer Cor.
True Proof Rounded to 1/10 Proof	160.00		
Lbs. per Proof Gallon 1/T4 Value	4.49539	0.22245	Proof Gal. per Lb.
Proof Gal. * Lbs. per = Lbs. of Alcohol	872.15	872.15	Proof Gal./Prof Gal per Lb.
Lbs. Per Wine Gal T4 Quotient	7.19259		
Lbs./ Lbs per WG = 60° F. Vol. of Alcohol	121.257	459.01	Liters
Vol. Cor. Factor for Present Temp	1.0000		
Present Volume of 60° F. Wine Galons	121.26	459.01	Liters
Gallon Difference In Volume	0.00	0.00	Liters

Withdraw Proof Gallons

By using the reciprocal values based on Table 4 one sees that the resulting weight and volume values are identical and that we will need 872.15 pounds, equivalent to 121.257 gallons at 60°F.

Now that we know the weight of alcohol we can apply that value to determine how much water we will need in order to reach our target proof.

Batch Planning by Weight

Before we address the blend itself, I'd like to make some observations on the advantages of blending using the weight of alcohol rather than its volume. The main reason is that the weight of alcohol can be used independently of temperature after true proof has been properly determined. Once true proof at 60°F has been established, all that is required to blend is the correct weight of alcohol and water that will be used. In my practice, being able to ignore temperature and volume after true proof has been determined simplifies matters so much as to be worth the trouble of processing materials by weight. Digital scales that are accurate to 0.2 pounds are very reasonable in price and the only complication is keeping track of the weight of the container itself, the "tare".

Using the *Batch Plan by Weight* worksheet, one can simply enter the target proof, the source proof and the total batch weight in pounds to arrive at essentially the same values as were produced with the *Batch Plan by Volume* worksheet (*Batch Plan Wt.* Worksheet)

---

[10] This procedure is outlined in §30.65 and §186.65 of the regulations and Gauging Manual respectively.

One Alcohol Batch Planning			
Target Proof of the Batch	80.00	0.12616	Wine Gal per Lb. at Target Proof
Total Batch Mass in Lbs.	1,922.20	242.510	Batch Lbs. * WG per Lb. = Total 60° F Gal.
Proof Gallons per Lb. at Target Proof	0.10093	3.7854118	Gallons to Liters (Liters per Gal.)
Mass * Proof Gal per Lb. = Total Proof Gal.	194.01	918.00	Total Liters
Proof of Alcohol	160.00	917,999.8	Total ml.
Proof Gal. per Lb. of Alcohol	0.222450	750.00	What is the bottle fill amount in milliliters
Proof Gal/Prof Gal. per Lb. = Lbs. Alc.	872.14	12.0	How many bottles per case
Total Batch Lbs. - Alc. Lbs. = H2O Lbs.	1,050.06	9,000.00	Total ml per case
Lbs. to Kg. Converter	0.45359237	102.000	Total Cases
Batch Kilograms	871.89	1,224.00	Total Bottles
		712.3317	Grams of Alcohol per Bottle
Excise Tax Rate	$13.50	0.035274	Grams to Weight Ounces Converter
Excise tax on Gross Batch	$2,619.10	25.1268	Weight Ounces Per Bottle
Labels per Bottle	2.0	242.510	Wine Gallons Restated
Labels Required	2,448.0	2.37755	Batch Gal/Cases = Wine Gal. per Case
Cases Per Pallet	84.0		
Number of Pallets Needed	1.21		

It is also a simple matter to plan a batch using two source alcohols if one assigns a percentage of proof gallons to be used by one of the source alcohols (in this case 16% for alcohol 1 and the remainder to alcohol 2).

Two Alcohol Batch Planning			
Target Proof of the Batch	80.00	0.45359237	Lbs. to Kg. Converter
Total Batch Mass in Lbs.	1,922.22	871.91	Batch Kilograms
Proof Gallons per Lb. at Target Proof	0.10093		
Total Proof Gal. Required	194.01	0.12616	Wine Gal per Lb. at Target Proof
Alcohol 1 %	16.00%	242.513	Batch Lbs. * WG per Lb. = Total 60° F Gal.
Alc 1 Proof Gal.	31.04	3.7854118	Gallons to Liters (Liters per Gal.)
True Proof of Alc 1	160.00	918.01	Total Liters
Proof Gal. per Lb. of Alc 1	0.222450	918,012.41	Total ml.
PG/PGPP = Lbs. Alc 1	139.54	750.00	What is the bottle fill amount in milliliters
Alc 2 Proof Gal.	162.97	12.0	How many bottles per case
Proof of Alc 2	190.00	9,000.00	Total ml per case
Proof Gal. per Lb.	0.279640	102.001	Total Cases
Proof Gal/Prof Gal. per Lb. = Lbs. Alc.	582.78	1,224.02	Total Bottles
Total Lbs. of Alcohol 1 and 2	722.32	712.3317	Grams of Alcohol per Bottle
Total Batch Lbs. minus Alc. Lbs. = H2O Lbs.	1,199.90	0.035274	Grams to Weight Ounces Converter
Excise Tax Rate	$13.50	25.1268	Weight Ounces Per Bottle
Excise tax on Gross Batch	$2,619.13	2.37755	Batch Gal/Cases = Wine Gal. per Case

Batch Plan by Weight

The tax payable is within 3 cents of what the one alcohol blend determined and validates the procedure.

I have found that metering volume is a difficult task when the density and temperature of the medium being metered varies from batch to batch based on the temperature and proof. Mechanical meters are simply not up to the task. Sight glass tanks rarely will provide adequately precise measurements and also require volume correction to be calculated before a useful 60°F volume can be determined. Blending by weight saves time, since it is less likely that the final blend will have to be adjusted with the addition of more alcohol or water to achieve the target proof.

One should re-gauge their alcohol just prior to blending to see if the alcohol content has changed though evaporation. If the proof has changed, it is a simple matter to recalculate the amount of water necessary to reach the target proof. If you do find that the true proof has changed while spirits have been in the storage account, the loss generated can be reported on line 22 of the storage report before the spirits are transferred to processing. This avoids having storage losses show up as if they were processing losses.

Blend One Alcohol by Weight

Returning to our single alcohol blend using the batch plan value of 872.15 pounds of 160-proof alcohol that we will need to blend to 80 proof we will enter it into the *Blend by Weight* worksheet as the amount of alcohol to be used. This spreadsheet follows the §186.64 - §30.64 procedures for blending one alcohol with water to obtain a target proof.[11]

Blend One Alcohol by Weight using Table 4 then add components by weight or volume

Temp of Alcohol	60.00	0.000	Thermometer Cor.
Apparent Proof of Alcohol	160.00	0.000	Hydrometer Cor.
True Proof Result	160.00	0.00150	Blending Tolerance
Target Proof	80.00	0.300	Tolerance as ° of Proof
Weight of Alcohol	872.15	395.60	Kilograms (Air)
Volume of Alcohol at 60° F.	121.257	459.01	Liters
Table 7 Volume Correction Factor to be applied	1.0000		
Volume of Alcohol corrected to Present Volume	121.257	$13.50	Tax per Proof Gallon
Present Volume Difference from 60° F. Volume	0.00	750.00	ML per Bottle
60° F. Volume compared to present volume	0.00	12.00	Bottles per Case

Results			
Proof Gallons of Alcohol	194.01	$2,619.11	Tax
Gallons of Water @ 60° F. to Add	126.09	97.00	Absolute Alcohol Gallons = P.G./2
Gross Volume of Batch @60°F.	247.34	242.51	Net Blended Wine Gallons
Volume Reduction due to Blending	-4.83	39.999%	Absolute Alc. Gal./Net. Gal. = % Alc
Actual "net" 60° F. Volume available for Bottling	242.51	-18.28	Liters
Weight of Water to Add	1,050.07	-24.37	# of 750 ML bottles "Short"
Weight of Alcohol	872.15	1.95%	Vol Difference/Gross Gallons
Total Weight of Batch	1,922.22	1.99%	(Gross Gal/Net Gal)-1

Blend One Alcohol by Weight

There is more information in the worksheet than we need at the moment but we will refer to this diagram several more times. For now, we will look at the methodology of calculating the blend to 80 proof.

Using the batch plan value for the weight of alcohol and multiplying it by the proof gallons per pound for 160-proof alcohol we find the proof gallons that will be required:

Find Proof Gallons from Weight	
Pounds of Alcohol Restated	872.15
Proof Gallons Per Lb. at True Proof of Alcohol Used	0.22245
Alcohol Pounds * Proof Gal. Per Lb. = Proof Gallons in Batch	194.01

This value of 194.01 proof gallons also matches the one obtained in the *Batch Planning* worksheet. We can then take the proof gallons required and divide them by the proof gallons per pound at the target proof of 80 proof; this gives the total batch weight as the result.

Find Weight of Entire Blended Batch at the Target Proof	
Proof Gallons Restated	194.01
P.G. Per Lb @ target proof	0.10093
Proof Gal. /Proof Gal. per Lb. @ Target Proof = Weight	1,922.22

This value of 1,922.22 lbs. is also the same value as we determined in our batch plan worksheet. Then we subtract the weight of the alcohol from the total batch weight and this gives the weight of water to be used in the blend.

---

[11] §186.64 "This Table (referring to Table 4) may also be used for ascertaining the quantity of water required to reduce to a given proof. To do this, divide the proof gallons of spirits to be reduced by the fractional part of a proof gallon per pound of spirits at the proof to which the spirits are to be reduced, and subtract from the quotient the net weight of the spirits before reduction. The remainder will be the pounds of water needed to reduce the spirits to the desired proof."

Total Weight Restated	1,922.22
Weight of Alcohol in Pounds Restated	872.15
Total Weight - Alcohol Weight = Water Weight	1,050.07

Blend One Alcohol by Weight

By using the regulation-prescribed value for the weight of a wine gallon of water at 60°F we can also calculate the gallons of water needed at that temperature and the total gallons in the batch:[12]

Find 60° F. Gallons of Water to add to Batch	
Water Weight Restated	1,050.07
Weight Per Gallon of Water @ 60° F.	8.328230
Water Weight / Weight per Water Gal. = 60° F. Gallons	126.09

Find Total 60° F. Wine Gallons	
Water Wine Gal. @60° F. in Batch Restated	126.09
Alcohol Wine Gal. @60° F. in Batch Restated	121.26
Water Gal. + Alcohol Gal. = Total Gal. Deposited	247.34

Blend One Alcohol by Weight

Finding the total volume contributed is a simple matter of adding the two known volume values.

Lastly, we will determine the volume change (usually a reduction in volume) that occurs when alcohol and water are blended together by multiplying the total batch weight times the wine gallons per pound value at the target proof to establish 60°F volume:[13]

Finding Corrections to Volume as a result of Blending - Volume Change as a Result of Combining Alcohol and Water		
Total Weight of Batch Restated	1,922.22	
Wine Gal. Per Lb @ Target Proof	0.12616	
Wt. of Batch * W.G. per Lb @ Target = Net Vol.	242.51	Blended Volume of Batch
Total Gallons Deposited To Batch Restated	247.34	
Blended Gallons - Deposited Gallons = Vol Change	-4.83	60°F. Wine Gallon Change in Volume as a result of Blending
Same result stated as a positive number	4.83	
% Vol. Difference "Change in Vol."	1.99%	(Deposited Vol. /Blended Vol.) -1

Blend One Alcohol by Weight

Note that the net blended wine gallons are the same value (242.51) as in our batch planning worksheet so we can have some assurance our batch is properly sized for the job.

Volume Change Due to Blending

I want to provide at least an introduction to the topic of why the process of blending alcohol and water results in a volume that is not the sum of the component volumes.

Water, consisting of two hydrogen atoms and one atom of oxygen, is a very stable molecule and not highly reactive but it is also the universal solvent. The reason for this is that water is a bi-polar molecule with respect to electric charge. Even though it is electrically neutral as a whole, there is a small amount of electric charge that is exposed given the structure of the molecule itself. The nature of the charge is that there is negative charge peeking out from the two white colored hydrogen atoms and positive charge associated with the oxygen atom's lobe. This structure means that water is *amphoteric* in that it is both an acid and a base; it produces H+ and OH− ions by self-ionization. This is why things dissolve in water so easily.

---

[12] There are actually several different values in the regulations and tables. §30.41 Wine Gal. per Lb. = 0.120074 and 1/0.120074= 8.32820 Lbs. per Gallon.at 60°F. Table 5 gives a value of 8.32823 and the reciprocal of Table 4 value of 0.12007 Wine Gal. per Lb. is 8.32848. Also Interpolated values from HCP 89th Value based 60°F 8.32834, HCP 72nd Lbs. per Gal. Air at 15.556 8.32826.

[13] The volume change is usually a reduction but, when blending to low proofs, there can be a small increase in volume. See Volume Change section infra.

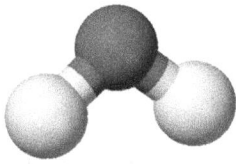

Water Molecule Ball & Stick[14]

Looking at the molecular structure of alcohol (ethanol) shows that it also has the hydrogen atoms sticking out that provide negative charge lobes which can form bonds with the water atoms.

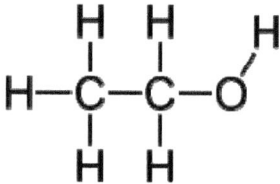

Ethanol Molecule Diagram[15]   *Ethanol Molecule Ball & Stick*[16]

When water and alcohol are combined, they form chemical bonds, or, if not actual bonds, affinities for locating themselves around certain parts of the other molecules. As a result of this, the molecules of each substance sometimes act like sand filling in the spaces between gravel resulting in a reduction in volume. However, when combining lower proof alcohol and water, one will find that the components can act like sticks building chains that increase the volume over the total of the components contributed. Looking at two blends by volume will illustrate this point. In the first case, we will blend 100 gallons of 190-proof alcohol down to 10 proof and note that there is a 4.8-gallon reduction in volume. The second example shows 100 gallons of 100-proof alcohol blended down to 10 proof and that blend shows a 0.59-gallon increase in volume.

Temp of Alcohol	60.00		Temp of Alcohol	60.00
Apparent Proof of Alcohol	190.00		Apparent Proof of Alcohol	100.00
True Proof Result	190.00		True Proof Result	100.00
Target Proof	10.00		Target Proof	10.00
Volume of Alcohol at Present Temp to be used	100.00		Volume of Alcohol at Present Temp to be used	100.00
Table 7 Volume Correction Factor to be applied	1.0000		Table 7 Volume Correction Factor to be applied	1.0000
Volume of Alcohol corrected to 60° F.	100.00		Volume of Alcohol corrected to 60° F.	100.00
Present Volume difference from 60° F. Volume	0.00		Present Volume difference from 60° F. Volume	0.00
60° F. Volume compared to present volume	0.00		60° F. Volume compared to present volume	0.00
Results			Results	
Proof Gallons of Alcohol	190.00		Proof Gallons of Alcohol	100.00
Gallons of Water @ 60° F. to Add	1,805.42		Gallons of Water @ 60° F. to Add	899.74
Gross Volume of Batch @60°F.	1,905.42		Gross Volume of Batch @60°F.	999.74
Gallon Volume Change Due to Blending	-4.80		Gallon Volume Change Due to Blending	0.59
Actual "net" 60° F. Volume available for Bottling	1,900.62		Actual "net" 60° F. Volume available for Bottling	1,000.33
Vol Difference/Gross Gallons = % Change	0.25%		Vol Difference/Gross Gallons = % Change	0.06%

Blend One Alcohol by Volume

[14] Water Molecule Ball & Stick: Benjah-bmm27, own work, public domain, https://commons.wikimedia.org/w/index.php ?curid=1997535

[15] Ethanol Diagram: Benjah-bmm27, own work, public domain, https://commons.wikimedia.org/w/index.php?curid=2067606

[16] Ethanol Molecule Ball & Stick: public domain, https://commons.wikimedia.org/w/index.php?curid=1587150

Typically, the distiller will be taking relatively high-proof spirits and blending them down to some commercially standard proof. In the following case, we will blend our 190-proof spirits down to 80 proof and see that the volume change is 2.85 % of the volume employed.

Temp of Alcohol	60.00
Apparent Proof of Alcohol	190.00
True Proof Result	190.00
Target Proof	80.00
Volume of Alcohol at Present Temp to be used	100.00
Table 7 Volume Correction Factor to be applied	1.0000
Volume of Alcohol corrected to 60° F.	100.00
Present Volume difference from 60° F. Volume	0.00
60° F. Volume compared to present volume	0.00

Results	
Proof Gallons of Alcohol	190.00
Gallons of Water @ 60° F. to Add	144.45
Gross Volume of Batch @60°F.	244.45
Gallon Volume Change Due to Blending	-6.96
Actual "net" 60° F. Volume available for Bottling	237.49
Vol Difference/Gross Gallons = % Change	2.85%

Any change in volume resulting from blending does not affect the proof gallons employed. The same amount of alcohol is present in the mixture; it is just confined in a different volume than the sum of the two input volumes. This change does not affect the tax payable and does not cause a loss in processing because you store your spirits in the quantity of proof gallons and withdraw them in proof gallons as well.

The only way the change in volume due to blending can affect your operations is if you plan your batch to produce the gross volume of each component that is added.

Exothermic Reactions Due to Blending Alcohol and Water

The process of blending alcohol and water together results in an exothermic reaction that liberates substantial amounts of heat. Following are the results of a blending operation conducted at my distillery that shows just how much energy is released. All the components of the batch were the same temperature to begin with because they were all in the same shop and been there long enough to acquire the same temperature as the ground temperature of the shop.

Diagram follows on the next page.

While the temperature change was only 5.3°F the large thermal mass of the batch itself means that this small difference amounted to enough energy released to power a 100-watt bulb for a full day. (100 watts * 24 = 2.4 kilowatt-hours).

Exothermic Reaction Resulting From Blending Alcohol and Water				
Target Proof	100.00			
Temp. in ° F. of Components	58.20			
Blended Temperature of Components	63.50			
Temp change	5.30			
Btu's per ° of temp rise per pound	1.00			
Total Btu's per Lb.	5.30			
Weight of Alcohol in Blend	805.23			
Weight of Water in Blend	859.59			
Total Weight of Components in Pounds	1,664.82			
Temp Change times Total Pounds	8,823.55			
Specific Gravity of Blend @ Target Proof	0.934180			
BTU's Times Specific Gravity = Net BTU's	8,242.78			
		Constants		
Kilowatt Hours	2.41	0.0002929	1 BTU to KW Hours	
Horse Power Hours	3.24	0.0003930	1 BTU to HP hours	
Calories	2,077,181	252.00	1 BTUu to Calories	
Foot Pounds	6,409,990	777.65	1 BTU to Foot Pounds	
Joules	8,690,775	1,054.35	1 BTU to Joules	
Watt Seconds	8,690,775	1,054.35	1 BTU to Watt Seconds	

Distilling Ops Workbook: Exothermic Reactions

As components are mixed together, the initial blended temperature of a batch will always be above the temperature of each component part used. However, if you let the batch sit for a while, it will assume the ambient temperature of the distillery, which may be above or below 60°F. Above 60°F, volume expansion of the batch will cause more bottles to come off the line than the 60°F blended volume would indicate. This results in a gain in the processing account. This gain will offset any losses in processing during bottling. Below 60°F, volume contraction of the batch will cause fewer bottles to come off the line than the 60°F blended volume would indicate. Below 60°F, the loss in the processing account is in addition to other losses that inevitably occur during bottling operations.

Once the batch is blended and in the processing account, the only way out of the processing account is by bottling it with the resulting loss usually associated with that operation. The temperature of the batch at the time of bottling will either offset (above 60°F) or amplify (below 60°F) those losses. The regulations allow for this discrepancy in bottling temperature from the 60°F standard in a catch-all section that requires sound practices and also recognizes that evaporation does occur:

§5.47a (1) *Discrepancies due to errors in measuring which occur in filling conducted in compliance with good commercial practice.* [Author Note: i.e., at a comfortable working temperature.] ... *And (3) Discrepancies in measure due to differences in atmospheric conditions in various places and which unavoidably result from the ordinary and customary exposure of alcoholic beverages in bottles to evaporation.*

I always want my alcohol and water as cold as possible when I blend. This is in part to account for the exothermic effect, but also in part because the components just seem to work better together the cooler they are. Hot alcohol has a real tendency to develop colloidal opacity because of the corn oil, calcium carbonate, or other congeners that are in the alcohol. The same is true of water: in particular the dissolved minerals in water can combine with oils in the alcohol to form colloidal substances. The danger is that the batch can become cloudy. This is a real problem when trying to blend clear spirits, as you can imagine.[17]

---

[17] B. G. Kyle, *Chemical and Process Thermodynamics*, Third Edition. Prentice Hall, 1999, p. 157. ISBN 978-0130874115.

Volume Correction

Before we move on to blending by volume we need to take a look at the volume correction algorithm prescribed by the TTB regulations. Table 7 provides the values required in order to properly calculate the volume of alcohol or water required for our batch.

First, we will reprint our *Blend One Alcohol by Weight* example so we will have the 60°F values close at hand, as well as the tolerance values:

Temp of Alcohol	60.00
Apparent Proof of Alcohol	160.00
True Proof Result	160.00
Target Proof	80.00
Weight of Alcohol	872.15
Volume of Alcohol at 60° F.	121.26
Table 7 Volume Correction Factor to be applied	1.0000
Volume of Alcohol corrected to Present Volume	121.255
Present Volume Difference from 60° F. Volume	0.00
60° F. Volume compared to present volume	0.00

Results	
Proof Gallons of Alcohol	194.01
Gallons of Water @ 60° F. to Add	126.09
Gross Volume of Batch @60° F.	247.34
Volume Reduction due to Blending	-4.83
Actual "net" 60° F. Volume available for Bottling	242.51
Weight of Water to Add	1,050.07
Weight of Alcohol	872.15
Total Weight of Batch	1,922.22

Tolerance Values	
Pounds of Alcohol Tolerance	1.31
Pounds of Water Tolerance	1.58
Total Weight of Tolerance	2.88
Proof Gallon Blending Tolerance	0.29
Volume of Alcohol Tolerance	0.18
60° F. Water Gallon Tolerance	0.19
Total Tolerance Volume	0.37

We know we need 121.26 gallons of alcohol at 60°F and 126.09 gallons of water at the same temperature. Looking first at the alcohol component we will assume that we're doing this in the summer and the temperature of the alcohol is 75°F:

What is the True Proof @ 60°F. Already Determined.	160.00	Must know	True Proof	0.0015	Tolerance
What is the Volume @ 60° F. Already determined	121.26	459.02	Liters	0.300	Tolerance as ° of Proof
What is the present Temperature of the Alcohol	75.00	0.0000	Therm Cor.	0.18	60° F. Wine Gal. Tolerance
Correction Factor to be applied for this Temp.	1.0085			567%	% of Tolerance
Present Volume of 60° F. Wine Galons	122.29	462.92	Liters		
Wine Gallon Difference	1.031	3.90	Liters		

Volume Correction from 60° Vol to Present Vol.

The increase in temperature over the 60°F standard has resulted in an increase in volume amounting to 1.031 gallons or 3.90 liters. Accordingly, one should meter 122.29 gallons of alcohol into the blending tank at the present temperature, not 121.26 60°F gallons. Failure to do this will result in the batch being short by 1.65 proof gallons. With respect to tolerance, the alcohol volume tolerance is 0.18 gallons, so the batch volume will be off by 567% ($1.031/0.18 \approx 5.67$).

Following is the methodology for finding the correct volume correction factor to be applied:

Volume Correction from True Proof and 60° Volume to Present Volume

Find Table 7 Entry Temp

Temperature of Alcohol	75.00
Correction Factor for this Thermometer	0.00
Temperature Corrected	75.00
Temperaure Rounded for Table 7 Entry	75.0

Find Table 7 Entry Proof

True Proof @60° F. Restated	160.00
True Proof Rounded to whole proof for Table 7 Entry	160.00

If Temp is Odd Then Lookup 1 Temp Lower	0.9920	At	74°
If Temp is Odd Then Lookup 1 Temp Higher	0.9910	At	76°
Subtract the smaller value from the larger	0.001		
Divide the Result by 2	0.0005		
Temperature Corrected Value Only	0.9915		

Second Stage of Procedure to Determine Proof Correction Factor if Needed for Table 7 Intermediate Proof

Restated True Proof Rounded for Table 7 Entry	160.00		
Temp Adjusted to Nearest Lower Temp if Odd Temp	74.00°		
Look up Closest Higher Proof Match	0.9920	165	Closest Higher Match to Table
Look up Closest Lower Proof Match	0.9920	160	Closest Lower Match to Table
Difference between Factors	0.0000		
Divide Factor Difference by 5	0.0000		Difference per degree of proof
Find difference between Lower Value & Actual Proof	0.00		
Multiply Per Proof difference by actual Difference	0.0000		Add this amount to the Temp Dif. Factor
Temperature Correction Factor Restated	0.9915		
Add in Proof Correction Factor	0.0000		
Correction for Odd Temp and intermediate Proof	0.9915	Result	
True Proof @ 60° F. Volume Restated	121.26		
Volume if Correction factor was applied directly	120.23		
Adjustment needed to correct table 7 to read the other way	1.0000		
Difference of Correction Factor from 1	0.0085		
Reversed Correction Factor	1.0085		
Present Volume of 60° F. Wine Galons	122.29		
Difference In Volume	1.03	Gallons	

The algorithm has two components—the first for temperature correction and the second for proof. In this case because we have an odd temperature we must adjust the temperature correction factor by half the difference between the 74° and 76° values. For this example, the second stage of the procedure is not required because there is no difference between the Table 7 values for 160 and 165 proof. The table is designed primarily to correct a known present volume to its 60°F volume so applying the volume correction factor generated (0.9915) directly will find the 60° volume of a known volume.

In this case we are going from a known 60°F volume to unknown present volume of the alcohol. In order to accomplish this, we must make the table work the opposite way to find the present volume. We find the difference of the table value from the value of one (1.0) and add that value to 1.0 to obtain a "reversed" correction factor of 1.0085 and multiply the 60° volume by that value.

Water does not expand nearly as fast as alcohol does with increasing temperature. In our blend by weight example the volume of water required is 126.09 gallons and, assuming the same 75°F temperature, the difference in water volume is only 0.252 gallons; yet this is still 133% of the tolerance for the blend:

What is the True Proof @ 60°F. Already Determined.	0.00		Must know True Proof	0.0015	Tolerance
What is the Volume @ 60° F. Already determined	126.09	477.30	Liters	0.300	Tolerance as ° of Proof
What is the present Temperature of the Alcohol	75.00	0.0000	Therm Cor.	0.189	60° F. Wine Gal. Tolerance
Correction Factor to be applied for this Temp.	1.0020			133%	% of Tolerance
Present Volume of 60° F. Wine Galons	126.34	478.26	Liters		
Wine Gallon Difference	0.252	0.95	Liters		

Note that for the same temperature difference and similar volumes of alcohol and water (126.09 for water and 121.26 for alcohol), the 160-proof alcohol expanded by over one gallon (1.03 gal.) while the water only expanded by 0.252 gallons, indicating that 160-proof alcohol expands four times as much as the water does.

Table 7 procedures and examples are reproduced below.[18]

---

[18]§ 30.67 § 186.67 "Table 7, for correction of volume of spirituous liquors to 60 degrees Fahrenheit. This Table is prescribed for use in correcting spirits to volume at 60 degrees Fahrenheit. To do this, multiply the wine gallons of spirits which it is desired to correct to volume at 60 degrees Fahrenheit by the factor shown in the Table at the percent of proof and temperature of the spirits. The product will be the corrected gallonage at 60 degrees Fahrenheit. This Table is also prescribed for use in ascertaining the true capacity of containers where the wine gallon contents at 60 degrees Fahrenheit have been determined by weight in accordance with Tables 2, 3, 4, or 5. This is accomplished by dividing the wine gallons at 60 degrees Fahrenheit by the factor shown in the Table at the percent of proof and temperature of the spirits. The quotient will be the true capacity of the container.

"*Example*. It is desired to ascertain the volume at 60 degrees Fahrenheit of 1,000 wine gallons of 190 proof spirits at 76 degrees Fahrenheit: 1,000×0.991 equals 991 wine gallons, the corrected gallonage at 60 degrees Fahrenheit.

"*Example*. It is desired to ascertain the capacity of a container of 190 proof spirits at 76 degrees Fahrenheit, shown by Table 2 to contain 55.1 wine gallons at 60 degrees Fahrenheit: 55.1 divided by 0.991 equals 55.6 wine gallons, the true capacity of the container when filled with spirits of 60 degrees temperature.

It will be noted the Table is prepared in multiples of 5 percent of proof and 2 degrees temperature. Where the spirits to be corrected are of an odd temperature, one-half of the difference, if any, between the factors for the next higher and lower temperature, should be added to the factor for the next higher temperature.

"*Example*. It is desired to correct spirits of 180 proof at 51 degrees temperature: 1.006 (50°)−1.005 (52°)=0.001 divided by 2=0.0005 0.0005+1.005=1.0055 correction factor at 51°F.

"*Example*. It is desired to correct spirits of 180 proof at 53 degrees temperature: 1.005 (52°)−1.003 (54°)=0.002 divided by 2=0.001 0.001+1.003=1.004 correction factor at 53°F.

"Where the percent of proof is other than a multiple of five, the difference, if any, between the factors for the next higher and lower proofs should be divided by five and multiplied by the degrees of proof beyond the next lower proof, and the fractional product so obtained should be added to the factor for the next lower proof (if the temperature is above 60 degrees Fahrenheit, the fractional product so obtained must be subtracted from the factor for next lower proof), or if it is also necessary to correct the factor because of odd temperature, to the temperature corrected factor for the next lower proof.

"*Example*. It is desired to ascertain the correction factor for spirits of 112 proof at 47 degrees temperature:1.006 (46°)−1.005 (48°)=0.001 divided by 2=0.0005 0.0005+1.005=1.0055 corrected factor at 47°F.1.007 (115 proof)−1.006 (110 proof)=0.001 0.001 divided by 5=0.0002 (for each percent of proof)×2 (for 112 proof)=0.0001 0.0004=1.0055 (corrected factor at 47°F.)=1.0059 correction factor to be used for 112 proof at 47°F.

"*Example*. It is desired to ascertain the correction factor for spirits of 97 proof at 93 degrees temperature: 0.986 (92°)−0.985 (94°)=0.001 divided by 2=0.0005 0.0005+0.985=0.9855 corrected factor at 93°F. 0.986 (95 proof)−0.985 (100 proof)=0.001 0.001 divided by 5=0.0002 (for each percent of proof)×2 (for 97 proof)=0.0004 0.9855 (corrected factor at 93°F.)=0.0005=0.9851 correction factor to be used for 97 proof at 93°F."

For completeness the following reproduces the §186.67-§30.67 regulation for using Table 7 for correction of volume of spirituous liquors to 60°F, and works a full example given in the regulations that includes a proof interpolation in addition to the temperature one.[19]

Volume Correction of True Proof Alcohol to 60° F. Volume

Known Present Volume	1,000.00	3,785.41	Liters
What is the Present Temperature of the Alcohol	47.00	0.0000	Therm Cor.
What is the True proof @ 60° F of the Alcohol	112.00	You must know true proof already	
Correction Factor to be applied	1.0059		
60° F. Wine Gallons in Container	1,005.90	3,807.75	Liters
Diffrence between Present Vol. and 60° F. Vol.	-5.90	-22.33	Liters
Proof/100 = Proof Gallon Conversion Factor	1.1200		
60° F. Wine Gallons * Proof / 100 = Proof Gallons	1,126.61		

Volume Correction Using Table 7 Using §186.67 - §30.67 Procedure

Temperature of Alcohol	47.00		
Correction Factor for this Thermometer	0.000		
Temperature Corrected	47.00		
Temperature Rounded to whole Temp for Table 7 Entry	47.00		
True Proof Restated	112.00		
True Proof Rounded to whole proof for Table 7 Entry	112.00		
If Temp is Odd Then Lookup 1 Temp Lower	1.006	At	46°
If Temp is Odd Then Lookup 1 Temp Higher	1.005	At	48°
Subtract the smaller value from the larger	0.001		
Divide the Result by 2	0.0005		
Temperature Corrected Value Only	1.0055		

Second Stage of Procedure to Determine Proof Correction Factor if Needed for Table Intermediate Proof

If Odd Temp above Modify to Nearest Even Lower Temp	46°		
Look up Closest Higher Proof Match	1.007	115	Closest Higher Match to Table
Look up Closest Lower Proof Match	1.006	110	Closest Lower Match to Table
Difference between Factors	0.001		
Divide Factor Difference by 5	0.0002	Difference per degree of proof	
Find difference between Lower Value & Actual Proof	2.00000		
Multiply Per Proof difference by actual Difference	0.0004	Add this amount to the Temp Dif.	
Temperature Correction Factor Restated	1.0055		
Add in Proof Correction Factor	0.0004		
Add for Factor for Odd Temp and intermediate Proof	1.0059	Table 7 Correction Factor to be applied	
Volume at Present Temp Restated	1,000.00		
Volume * Factor = Corrected Gallons to 60° F.	1,005.90	Wine Gal. * Factor – Gallons @ 60°	
Present Gal. Dif. from 60° F. Volume	-5.90		

Vol Cor. Present Vol. to 60 F. Vol.

[19] "*Example.* It is desired to ascertain the correction factor for spirits of 112 proof at 47 degrees temperature:1.006 (46°)−1.005 (48°)=0.001 divided by 2=0.0005 0.0005+1.005=1.0055 corrected factor at 47°F 1.007 (115 proof)−1.006 (110 proof)=0.001 0.001 divided by 5=0.0002 (for each percent of proof)×2 (for 112 proof)=0.0001 0.0004=1.0055 (corrected factor at 47°F)=1.0059 correction factor to be used for 112 proof at 47°F."

Blending One Alcohol by Volume

We can now turn our attention to blending alcohol by volume. Let's use the same values as we did in our prior blending project but use volume instead of weight to calculate the components to be added to the blend. Because we are including a temperature for the alcohol other than 60°F, I have indicated an apparent proof of 164.9, which in conjunction with the 75°F alcohol temperature will result in a true proof of 160. Similarly, we are using the 75°F volume (122.29 gallons) of the alcohol we calculated in our volume correction algorithm above.

Blend One Alcohol by Volume using Table 4 then add components by weight or volume

Temp of Alcohol	75.00	0.000	Thermometer Cor.	$13.50	Tax per Proof Gallon
Apparent Proof of Alcohol	164.90	0.000	Hydrometer Cor.	750	ML per Bottle
True Proof Result	160.00	0.0015	Blending Tolerance	12.00	Bottles per Case
Target Proof	80.00	0.300	Tolerance as a ° of Proof	2.377530	Wine Gal. per Case
Volume of Alcohol at Present Temp to be used	122.290			1.902062	Proof Gallons Per Case
Table 7 Volume Correction Factor to be applied	0.9915			101.99	Proof Gal. Ideal Output
Volume of Alcohol corrected to 60° F.	121.25			0.158505	Proof Gallons Per Bottle
Present Volume difference from 60° F. Volume	1.04		Positive = Present Volume is Greater than 60° F. Vol.		
60° F. Volume compared to present volume	-1.04		Negative = 60° F. Gal. are less than present Vol.		

Results

Proof Gallons of Alcohol	194.00		$2,618.97	Tax
Gallons of Water @ 60° F. to Add	126.08		97.00	Absolute Alcohol Gallons = P.G./2
Gross Volume of Batch @60°F.	247.33		242.50	Net Blended Wine Gallons
Gallon Volume Change Due to Blending	-4.83		39.999%	Absolute Alc. Gal./Net Gal. = % Alc. By Vol.
Actual "net" 60° F. Volume available for Bottling	242.50		-18.28	Liters of Change due to Blending
Vol Difference/Gross Gallons = % Change	1.95%		-24.37	# of 750 ML bottles "Short"
Weight of Water to Add	1,050.02		1.95%	Vol Difference/Gross Gall [0.00%]
Weight of Alcohol	872.11		1.99%	(Gross Gal/Net Gal)-1
Total Weight of Batch	1,922.12			

Blend One Alcohol by Volume

We see that the weight of our alcohol is within 0.03 lbs. of that calculated using weight as the input (872.15 v. 872.12 lbs.). Other batch values are similarly aligned to values well within tolerance.

The first task in this blending procedure is to find the 60°F volume of the spirits (which we have already seen above); this time we don't use the reversed correction factor but the present volume to 60°F factor of 0.9915:

First Stage - Methodology For Temp adjustment for table 7 Entry

True Proof Restated	160.00		
True Proof Rounded to whole proof for Table 7 E	160.00		
Temperature Rounded to whole Temp for Table	75.00		
If Temp is Odd Then Lookup 1 Temp Lower	0.992	At	74°
If Temp is Odd Then Lookup 1 Temp Higher	0.991	At	76°
Subtract the smaller value from the larger	0.001		
Divide the Result by 2	0.0005		
Temperature Corrected Value Only	0.9915		

Blend One Alcohol by Volume

Since we already know that there is no proof-based difference, there is no reason to reproduce that part of the method.

After we have found the 60°F volume, we can find the weight of the alcohol using the Table 4 wine gallons per pound value:

60° F. Alcohol Volume Restated	121.25	
Wine Gallons Per Lb. of Alcohol @ True Proof.	0.13903	From Table 4
60° Vol. /W. Gal. per Lb. @ 60° F. True Proof = Wt.	872.12	Weight of Alcohol

Having found the weight, we can then find proof gallons by multiplying the weight by the proof gallons per pound value for the true proof of the alcohol.

Find Proof Gallons from Weight		
Pounds of Alcohol Restated	872.12	
Proof Gallons Per Lb. at True Proof of Alcohol	0.22245	Table 4 Value
Alcohol Pounds * Proof Gal. Per Lb. = Proof Gallons	194.00	

With these two Table 4 transformations we are able to calculate the pounds of alcohol required in order to use the Table 4 blending regimen. The remaining steps are the same as we applied in the *Blend One Alcohol by Weight* section and there is no need to reproduce them here.

Blend Two Alcohols

Now let's look at using two alcohols of different proof in our batch and blend them together with water in order to achieve a planned target proof. In my distillery, I ferment and distill corn whiskey. By definition, corn whiskey cannot be taken at more than 160 proof and still be called by that name. To make my final product, I then blend the roughly 160-proof corn whiskey in a ratio with 190-proof grain neutral spirits (GNS) that I purchase in bulk to make my *Mountain Moonshine®* product.

You may recall from the batch planning worksheet that we needed 194.01 proof gallons to create our 102-case batch. My blend requires that I use, in this case, 16% on a proof-gallon basis to achieve the taste profile of my product. That value fluctuates slightly from year to year depending on the intensity of the corn flavor in my alcohol for that year's production. Planning by percentage of proof gallons employed makes it very easy to work with my inventory on hand. For tax and regulatory purposes, from my storage reports I already know that I have—say, 500 proof gallons of corn whiskey in the shop and 1,000 proof gallons of 190 proof. Using somewhat idealized whole numbers for the two alcohol proofs, we see that sixteen percent of 194.01 is 31.04 proof gallons of corn whiskey that are required with the remaining 162.97 proof gallons coming from the 190-proof Grain Neutral Spirits (GNS):

For Two Alcohol Blend Find Weight of Alcohol Required	
Total Proof Gal. Required	194.01
Proof Gal. % of Alcohol 1	16.00%
Alc 1 Proof Gal.	31.04
True Proof of Alc 1	160.00
Proof Gal. per Lb. of Alc 1	0.222450
PG/PGPP = Lbs. Alc 1	139.54
Alc 2 Proof Gal.	162.97
Proof of Alc 2	190.00
Proof Gal. per Lb.	0.279640
Proof Gal/Prof Gal. per Lb. = Lbs. Alc.	582.77
Total Lbs. of Alcohol 1 and 2	722.31
Total Batch Lbs. minus Alc. Lbs. = H2O Lbs.	1,199.88

Batch Plan by Volume

Knowing the proof gallons of each component, it is an easy matter to divide the proof gallons required by the proof-gallons-per-pound and obtain the weight of each alcohol. Of course, if one applied the wine-gallon-per pound value, the answer in 60°F wine gallons would be the result and one could blend by volume.

Knowing the weight of each alcohol we can enter that into the *Blend Two Alcohols by Weight* worksheet and see how well the batch matches our planned values.

Blend Two Alcohols by Weight then using Table 4 add components by weight or volume

What is the Target Proof of the Batch	80.00		0.0015	Tolerance
What is the Temperature of Alcohol 1	60.00	True Proof	0.000	Thermometer Cor.
What is the Apparent Proof of Alcohol 1	160.00	160.0	0.000	Hydrometer Cor.
Weight Alcohol 1 at Present Temp to be used	139.54		0.000	Thermometer Cor.
			0.000	Hydrometer Cor.
What is the Temperature of Alcohol 2	60.00		$13.50	Excise Tax Rate
What is the Apparent Proof of Alcohol 2	190.00	190.0	750.00	Bottle fill in milliliters
Weight of Alcohol 2 to use	582.77		12.00	How many bottles per case

		Vol. as %	P.G. As %.	Wt as %	
Proof Gallons of Alcohol 1	31.04	18.45%	16.00%	27.46%	
Proof Gallons of Alcohol 2	162.97	81.55%	84.00%	80.68%	
Total Proof Gallons	194.01	4.421	5.250	4.176	Ratios
60° F. Wine Gallons of Water to add to Alcohol 1	20.17		4.176	Weight Ratio of A1 to A2	
60° F. Wine Gallons of Water to add to Alcohol 2	123.90		4.421	Present Volume Ratio of A1 to A2	
Total Gallons of water to add	144.07		4.421	60° F Vol. Ratio of Alc 1 to Alc 2	
Gross Volume of Batch	249.25		5.250	Proof Gallon Ratio	
Total Volume Reduction due to Blending	-6.74				
Actual "net" Volume Available for Bottling	242.51		25.53	Volume Difference in Liters	
Weight Both Alcohols	722.31		34.04	# of 750 ML bottles "Short"	
Weight of Water to add to Alcohol 1	168.01		-2.71%	Vol Difference/Gross Gallons	
Weight of Water to add to Alcohol 2	1,031.87		2.78%	(Gross Gal/Net Gal)-1	
Total Weight of Water to Add	1,199.88		97.00	Absolute Alcohol Gallons	
Total Weight of Batch	1,922.19		40.000%	% Alcohol By Volume	

Blend Two Alcohols by Weight

The total weight of the batch is calculated as 1,922.19 pounds and from our batch planning worksheet it was calculated as 1,922.20 pounds, so we are well within tolerance. The weight of water is very different (1,199.98 v. 1,050.06 lbs.) but the total batch mass comes out the same.

What this spreadsheet demonstrates is that you can independently determine, as if for two separate batches, the amount of alcohol and water necessary to blend to the same target proof and then combine each of those individual batch results into one larger batch and the larger batch will also reach the target proof. This procedure can be used for any number of alcohols that you desire, and, so long as each batch is planned to blend to the same target proof, they will successfully do so when combined.

Of course, it is also possible to do the same two alcohol blend by volume using the methods described above and simply blending each alcohol to the same target proof and summing the combined values. The wine gallons are found from mass using the wine gallons per pound or pounds per wine gall on values for the true proof of the alcohols involved:

Blend Two Alcohols by Volume then using Table 4 add components by weight or volume

What is the Target Proof of the Batch	80.00		0.0015	Tolerance
What is the Temperature of Alcohol 1	60.00	True Proof	0.000	Thermometer Cor.
What is the Apparent Proof of Alcohol 1	160.00	160.0	0.000	Hydrometer Cor.
What is the Temperature of Alcohol 2	60.00		0.000	Thermometer Cor.
What is the Apparent Proof of Alcohol 2	190.00	190.0	0.000	Hydrometer Cor.
Volume of Alc. 1 at Present Temp to be used	19.40		$13.50	Excise Tax Rate per Proof Gallon
Volume Correction Factor applied	1.0000		750.00	Bottle fill amount in ml
Volume of Alcohol 1 corrected to 60° F.	19.40		12.00	How many bottles per case
Present Volume Difference from 60°F of Alc. 1	0.00			
Present Volume of Alcohol 2 to use	85.78	105.18	Total Present Volume	
Volume Correction Factor applied	1.0000			
Volume of Alcohol 2 corrected to 60° F. Volume	85.78	105.18	Total 60° F. Volume	
Present Volume Difference from 60°F of Alc. 2	0.00			

		Vol. as %	P.G. % of Total P.G.		
Proof Gallons of Alcohol 1	31.04	18.45%	16.00%	$419.05	Tax on Alcohol 1
Proof Gallons of Alcohol 2	162.98	81.55%	84.00%	$2,200.20	Tax on Alcohol 2
Total Proof Gallons	194.02			$2,619.25	Total Tax
Total Gallons of water to add	144.09		6.74	Volume Difference in Gallons	
Gross Volume of Batch	249.27		25.52	Liters	
Total Volume Reduction due to Blending	-6.74		34.03	# of 750 ML bottles "Short"	
Actual "net" Volume Available for Bottling	242.53		2.71%	Vol Difference/Gross Gallons	
Weight of Alcohol 1	139.54		2.78%	(Gross Gal/Net Gal)-1	
Weight of Alcohol 2	582.77		97.01	Absolute Alcohol Gallons	
Total Weight of Water to Add	1,200.01		39.999%	% Alcohol By Volume	
Total weight of Batch	1,922.32				

Proof Gallon Tolerance A1	0.05		0.21	Alcohol Weight Tolerance A1
Proof Gallon Tolerance A2	0.24		0.87	Alcohol Weight Tolerance A2
Total Proof Gallon Tolerance	0.29		1.08	Total Alcohol Weight Tolerance

Water Gallon Tolerance A1	0.03		0.25	Water Weight Tolerance A1
Water Gallon Tolerance A2	0.19		1.55	Water Weight Tolerance A2
Total Water Gallon Tolerance	0.22		1.80	Total Water Weight Tolerance

Present Volume Ratio of A1 to A2	4.421
60° F Volume Ratio of Alcohol 1 to Alcohol 2	4.421
Proof Gallon Ratio	5.251

Blend Two Alcohols by Vol. in a Ratio

And finally, by taking the ratio found in the above example (4.421), one can perform the blend by the ratio of volume.

Blend Two Alcohols by Volume in a ratio to one another then using Table 4 add components by weight or volume

What is the Target Proof of the Batch	80.00		0.0015	Tolerance
What is the Temperature of Alcohol 1	60.00	True Proof	0.000	Thermometer Cor.
What is the Apparent Proof of Alcohol 1	160.00	160.0	0.000	Hydrometer Cor.
What is the Temperature of Alcohol 2	60.00		0.000	Thermometer Cor.
What is the Apparent Proof of Alcohol 2	190.00	190.0	0.000	Hydrometer Cor.
Volume of Alcohol 1 at Present Temp to be used	19.40		$13.50	Excise Tax Rate per Proof Gallon
Correction Factor applied	1.0000		750.00	What is the bottle fill amount in ml
Volume of Alcohol 1 corrected to 60° F.	19.40		12.00	How many bottles per case
Present Volume Difference from 60°F of Alc. 1	0.00			
Ratio of 60° F. Alcohol 1 To Alcohol 2	4.421			
60° Vol. of Alcohol 2 to use	85.78	105.18	Total 60° F. Volume	
Volume of Alcohol 2 at Present Temp to be used				
Correction Factor applied	1.0000			
Volume of Alcohol 2 corrected to Present Volume	85.78			
Present Volume Difference from 60°F of Alc. 2	0.00	105.18	Total Present Volume	

		Vol. as %	P.G. % of Total P.G.		
Proof Gallons of Alcohol 1	31.04	18.45%	16.00%	$419.1	Tax on Alcohol 1
Proof Gallons of Alcohol 2	162.98	81.55%	84.00%	$2,200.3	Tax on Alcohol 2
Total Proof Gallons	194.02			$2,619.3	Total Tax
Total Gallons of water to add	144.09		6.74	Volume Difference in Gallons	
Gross Volume of Batch	249.28		25.52	Liters	
Total Volume Reduction due to Blending	-6.74		34.03	# of 750 ML bottles "Short"	
Actual "net" Volume Available for Bottling	242.53		2.71%	Vol Difference/Gross Gallons	
Weight of Alcohol 1	139.54		2.78%	(Gross Gal/Net Gal)-1	
Weight of Alcohol 2	582.79		97.01	Absolute Alcohol Gallons	
Total Weight of Water to Add	1,200.04		39.999%	% Alcohol By Volume	
Total weight of Batch	1,922.37				

Blend Two Alcohols by Vol. in a Ratio

Table 6 Blend to Target

Before we move on to increasing the proof of an alcoholic blend, I'd like to introduce Table 6 which is discussed in §186.66-§30.66.[20] The easiest way to do this is to conduct our standard blend using Table 6 by volume to a target proof and along the way we can compare the results to a Table 4 blend and see how well the two methods match up.

Table 6 uses the parts of alcohol and parts of water, at given proof, to accomplish its blend.[21] As you see below the parts alcohol are related to the proof but the parts of water vary according to the proof. We will manipulate these relationships to accomplish our blend:

Alcohol Proof	Parts Alcohol	Parts Water	Total Parts
160.0	80.00	22.87	102.87

Target Proof	Parts Alcohol	Parts Water	Total Parts
80.0	40.00	63.42	103.42

Using the Table 6 method results in a very close correspondence with the Table 4 method resulting in 0.12 lbs. difference in the total batch weight.

Blend one Alcohol with water to a target proof using Table 6 With T4 Blend for comparison

What is the Temperature of the Alcohol	60.00		0.000	Therm Cor.		
What is the Apparent Proof of the Alcohol	160.00		0.000	Hydr. Cor.	1.31	Pounds of Alcohol Tolerance
What is the Target Proof of the Batch	80.00		0.0015	Tolerance	0.18	Gallon Vol. of Alcohol Tolerance
Volume of Alcohol at Present Temp to be used	121.26		0.300	Tol. as ° of Proof	0.29	Batch Proof Gallon Tolerance
Correction Factor applied	1.0000		$13.50	Excise Tax Rate	1.57	Pounds of Water Tolerance
Volume of that Alcohol corrected to 60° F.	121.26		$2,619.11	Tax on Batch	0.19	60° F. Water Gallon Tolerance
Present Gal. Vol. Difference from 60°F Alcohol	0.00	Gallons				

	Table 6	Table 4		Dif.	% of Tolerance
True Proof@ 60° F.					
Proof Gallons of Alcohol	194.01	194.01		0.00	0.31%
Gallons of Water to Add	126.07	126.08		-0.02	-8.02%
Gross Gallon Volume of Batch	247.32	247.34		-0.02	
Volume Reduction due to Blending	-4.83	-4.83		0.00	
Actual Volume available for Bottling	242.49	242.51		-0.02	
Weight of Water to Add	1,049.93	1,050.06		-0.13	-8.02%
Weight of Alcohol	872.14	872.14		0.00	0.00%
Total Weight of Batch	1,922.07	1,922.19		-0.13	

Blend 1 Alc T6 & T4

Knowing that the Table 6 blending algorithm can give this good an account of itself, it is worth seeing how it is accomplished. Table 6 blending is based on a scale of 100 fluid gallons but can be scaled up to any batch size as we will see below. But first let's run through the 100-gallon scale procedure. First, we find the ratio of the alcohol parts to each other. Then we multiply that ratio by the ending parts water to find what I call the AW ratio for (alcohol/water). Then by a nifty piece of legerdemain the AW ratio minus the starting parts water results in the total gallons of water required to accomplish the blend. From there it is a simple matter to convert to mass at the scale of 100 gallons:

---

[20] §30.66 §186.66 "Table 6, showing respective volumes of alcohol and water and the specific gravity in both air and vacuum of spirituous liquor. This Table provides an alternate method for use in ascertaining the quantity of water needed to reduce the strength of distilled spirits by a definite amount. To do this, divide the alcohol in the given strength by the alcohol in the required strength, multiply the quotient by the water in the required strength, and subtract the water in the given strength from the product. The remainder is the number of gallons of water to be added to 100 gallons of spirits of the given strength to produce a spirit of a required strength."

[21] I expanded table 6 by the inflationary interpolation method described infra from 200 data points to 2,000 so that 1/10th proof accuracy could be achieved.

Table 6 Blending	
Starting Parts Alcohol	80.00
Ending Parts Alcohol at Target Proof	40.00
Starting Parts /Ending Parts = Alcohol Parts Ratio	2.00000
Ending Parts Water @ Target Proof	63.420
Alc. Parts Ratio * Ending Parts Water = AW Ratio	126.840
Starting Parts Water	22.870
AW Ratio - Starting Parts Water = Gallons H2O	103.970
100 + Gal H20 = Tot. Gal. at 100 Scale	203.97
Convert Alcohol to Mass @ 100 Scale	
True Proof Lbs. Per Wine Gallon	7.192594
100 * Lbs. per Wine Gal = Mass of Alc @ 100 Scale	719.259

Blend 1 Alc T6 & T4

In order to scale up the batch to the size we want, one simply creates a scaling ratio between 100 gallons and the gallons actually employed by the batch and then multiplies the 100-scale volume and mass units by that factor to obtain the scaled-up values:

Scaling Up Table 6 by Volume	
Total 60° F. Wine Gallons of Alcohol to be used	121.26
Standard Table 6 Batch Size	100.00
Total Gal./Factor = Scaling Factor	1.2126
Gal. of Water Per 100 Gal. Alcohol Restated	103.97
Scaling Factor * Gal per 100 = Total Gal. of Water	126.07
Alc. Gal. + Water Gal = Tot 60° F. Gal.	247.324

Scale Up Mass of T6 Blend	
Weight of Alcohol * Scaling Factor = Tot. Alc. Wt.	872.14
Weight of Water * Scaling Factor = Mass of H20	1,049.93
Total Scaled up Weight	1,922.07
Wine Gal. Per Lb @ Target Proof	0.126163
Batch Lbs. * W.G. per Lb @ Target = Blended Vol.	242.494
Total Gallons Deposited To Batch	247.324
Deposited Gal. - Blended Vol. - Change in Vol.	-4.83
% Vol. Difference	-1.953%

When using the Table 6 algorithm, it is important to disregard the advice to truncate the alcohol parts ratio as they suggested in the regulations. Just like determining pounds per gallon using 1/Wine Gallons per Pound by the quotient method, the same principal applies to the ratios required here. The most accurate results are obtained without truncating or rounding.

Second Method of Parts Alcohol Blending

There is a second method by which the "parts alcohol and parts water" table can be applied to blend to a target proof.[22] Let's run through a blend of 190-proof alcohol down to 80 proof as our example and compare the TTB method with the alternative "Second Method" in the adjacent column:

Blend one Alcohol with water to a target proof using Table 6   With T4 Blend for comparison

What is the Temperature of the Alcohol	60.00		0.000	Therm Cor.	
What is the Apparent Proof of the Alcohol	190.00		0.000	Hydr. Cor.	1.02 Pounds of Alcohol Tolerance
What is the Target Proof of the Batch	80.00		0.0015	Tolerance	0.15 Gallon Vol. of Alcohol Tolerance
Volume of Alcohol at Present Temp to be used	100.00		0.300	Tol. as ° of Proof	0.29 Batch Proof Gallon Tolerance
Correction Factor applied	1.0000		$13.50	Excise Tax Rate	1.80 Pounds of Water Tolerance
Volume of that Alcohol corrected to 60° F.	100.00		$2,565.00	Tax on Batch	0.22 60° F. Water Gallon Tolerance
Present Gal. Vol. Difference from 60°F Alcohol	0.00	Gallons			

	Table 6	Table 4	Dif.	% of Tolerance
True Proof@ 60° F.				
Proof Gallons of Alcohol	190.00	190.00	0.00	0.69%
Gallons of Water to Add	144.46	144.46	0.00	-2.09%
Gross Gallon Volume of Batch	244.46	244.46	0.00	
Volume Reduction due to Blending	-6.96	-6.96	0.00	
Actual Volume available for Bottling	237.50	237.50	0.00	
Weight of Water to Add	1,203.06	1,203.10	-0.04	-2.09%
Weight of Alcohol	679.39	679.39	0.00	0.00%
Total Weight of Batch	1,882.45	1,882.49	-0.04	

Table 6 Blending		Second Method	
Starting Parts Alcohol	95.00	95.00	Starting Parts Alcohol
Ending Parts Alcohol at Target Proof	40.00	100.00	Volume Scale
Starting Parts /Ending Parts = Alcohol Parts Ratio	2.37500	9,500.00	Parts Alc * Volume Scale
		40.00	Ending Parts Alcohol
Ending Parts Water @ Target Proof	63.430	237.50	Division = Blended Volume
Alc. Parts Ratio * Ending Parts Water = AW Ratio	150.646		
Starting Parts Water	6.190	95.00	Starting Parts Alcohol
AW Ratio - Starting Parts Water = Gallons H2O	144.456	63.43	Ending Parts Water
100 + Gal H20 = Tot. Gal. at 100 Scale	244.46	6,025.85	Multiplication Result
Convert Alcohol to Mass @ 100 Scale		40.00	Ending Parts Alcohol
True Proof Lbs. Per Wine Gallon	6.793894	150.65	Division Result = Gross Water Vol.
100 * Lbs. per Wine Gal = Mass of Alc @ 100 Scale	679.389	6.19	Starting Parts Water
		144.46	Subtraction Result = Water to Add
		244.46	Alc Vol. + H20 = Total Vol.

Blend 1 Alc T6 & T4

At a scale of 100 we know that the 190-proof spirits contain 95 parts alcohol by volume. Intermixed with this alcohol is 6.18 parts water which can be thought of as 95 gallons of alcohol and 6.18 gallons of water for a total of 101.18 gallons in our 100-gallon volume.

When this is diluted to 80 proof the resulting alcohol will have 40 parts alcohol and 63.42 parts of water for a total of 103.42 gallons "comingled" within each 100 gallons of product.

The 40 parts of alcohol in the final blend are the same 95 parts that we start with, just diluted to the target proof value. The dilution to the blended volume is simply starting parts per gallon times the 100 scale we are working with divided by ending parts per gallon (or 95 * 100 / 40) for 237.5 gallons as the net blended volume, accounting for the change in volume due to blending.

The water in the 80-proof blend can be calculated knowing that it is 63.42 parts relative to the 40 parts of alcohol, so multiplying 95 times 63.42 yields 6,024.90; when divided by 40, this results in 150.65 gallons of water.

---

[22] This method is credited to Harvey Wilson and published on his website at Katmar Software: http://www.katmarsoftware.com http://www.katmarsoftware.com/articles/alcohol-dilution-calculator.html

But the original 190-proof alcohol also contained 6.19 gallons of water per 100 gallons so we must subtract that value from the amount of water that we need to add, so 150.65 - 6.19 = 144.46; this result matches the TTB method. It is also evident that the total volume contributed is the same in both methods: 100 + 144.46 = 244.46. The benefit of this method is that it does not require the ratios that are attendant to the first method.

Specific Gravity Blending

The TTB tables, when referring to proof, assume atmospheric density for all of the components. Most scientific work is done on a vacuum basis.

Proof hydrometers are calibrated under atmospheric conditions, or at least they should be if the manufacturers are doing their job correctly. In any event only the part of the hydrometer that is emergent from the liquid is subject to atmospheric pressure, but even that small stem volume is squeezed by the air, more on high-pressure, clear, and sunny days and less on cloudy and rainy, low-pressure ones. On high-pressure days, the stem is lifted very slightly out of the fluid it is measuring so that more of the hydrometer stem is exposed, indicating a lower proof than it would if it was in a low-pressure chamber with very little atmosphere surrounding it. Obviously in a true vacuum the fluid would boil away like it does in outer space.[23]

The same principal holds true for a tank of fluid sitting on a scale. If it indicated a weight of 1,000 pounds in air sitting in a pressure or altitude chamber and we pumped the air out as best we could, the scale would read approximately 1,001 pounds because the atmosphere had been buoying up the tank by that amount. That would be close to the vacuum mass of the alcohol.

Density, in terms of kilograms per meter cubed or any other unit is also usually expressed on a vacuum basis. Apart from proof hydrometers, almost all other hydrometers are calibrated on a vacuum or absolute density basis. We have to be careful about what this means. It means that the demarcations of density units themselves, as inscribed on the instrument or on the paper inside the hydrometer stem, reflect the density as if it was in a vacuum. The same holds true for relative density hydrometers that are marked in units of specific gravity; they are almost always demarcated in units of SG vacuum rather than SG air.

The hydrometer itself may be designed to be used under atmospheric conditions, with only the stem exposed to the atmosphere, but the readings one will obtain from it are on a vacuum or absolute density basis.[24] Because only the stem of the hydrometer is in the atmosphere and therefore only a small amount of the hydrometer itself is supported by air pressure it is possible to determine density in air but report it in vacuum.

Because Table 6 provides specific gravity values that match proof values at 60°F, I have included worksheets that allow one to determine alcoholic content using specific gravity in both air and vacuum regimens. These tables have been expanded from 200 to 2,000 places in the same manner as I used for expanding the part alcohol and water tables and Table 5.

---

[23] The surface tension associated with alcohol and water mixtures changes over the range of density from water to absolute alcohol and that is a factor in determining accurate alcohol content of a sample. This topic is addressed in Volume 2 of this work.

[24] Apart from the proof scale of measuring density the only other density regimen that I've seen calibrated on an air basis is the Baumé scale. One sees air and vacuum Baumé tables and you have to know which is which. We will also cover Baumé in Volume 2.

We will look at our standard batch blended using SG Vac instead of proof. The method is simple: using the Table 6 values that are equated to proof, one finds the equivalent proof value and then performs the operation as if one was using a proof hydrometer:

Specific Gravity Vac Blend one Alcohol to Target SG by Weight					
Temperature of Alcohol in ° F	60.00	15.56	Celsius	0.000	Thermometer Cor.
Apparent Specific Gravity of Alcohol (Vac)	0.863810	160.00	= Proof	0.00000	Hydrometer Cor.
Weight of Alcohol In Pounds	872.15	395.60	Kilograms	0.0015	Tolerance
				$13.50	Excise Tax Rate
Target Specific Gravity in 60/60 ° F. (Vac)	0.95180	80.00	Target Proof	0.300	Tolerance as ° of Proof
				$2,619.13	Tax on Proof Gallons
True Specific Gravity @ 60° to 1/10th proof	0.86364	160.00	True Proof Rounded to 1/10 Proof		
Results					
Proof Gallons of Alcohol being used	194.01		121.26	Volume of Alcohol @ 60° F.	
Gallons of Water to Add to reach target proof	126.09		121.26	Present Volume	
Gross Volume of Batch	247.34		0.00	Volume Difference in Gallons	
Volume Reduction due to Blending	-4.83				
Actual "net" Volume available for Bottling	242.51	@ 60° F	18.28	Liters	
Weight of Water to Add	1,050.07		24.38	# of 750 ML bottles "Short"	
Weight of Alcohol Restated	872.15		0.00	Vol Difference/Gross Gallons	
Total Weight of Batch	1,922.22		0.02	(Gross Gal/Net Gal)-1	
Blending Tolerance					
Batch Proof Gallon Tolerance	0.29		0.19	Batch Water Gallon Tolerance	
Pounds of Alcohol Tolerance	1.31		-1.31	Pounds of Water Tolerance	
Gallon Volume of Alcohol Tolerance	0.18		0.69	Liters of Water Tolerance	
Alcohol Temperature Results			97.00	Absolute Alcohol Gallons	
Corrected Temperature Rounded to 1/10th° F.	60.00		242.51	Net Blended Wine Gallons	
			40.00%	% Alcohol By Volume	

Blend SG 60/60 Vac by Weight

The only interesting part of how this works is the mechanism for arriving at what proof the indicated SG is closest to:

SG Vac @ 60/60 Inputs to Apparent Proof	Alcohol
SG Vac @ 60/60	0.86386
Match Descending Col. Greater Than (-1 type Match)	1600
Index SG Vac	0.86395
Subtract Entered Den from Greater Than Lookup	0.00009
Index Proof	159.90
Add 1 Row # to Match Above	1601
Index SG Vac	0.86381
Compare	-0.00005
Absolute Value	0.00005
Index Proof	160.00
Choose "Winner" Closest 1/10th Proof Increment	160.00

Blend SG 60/60 Vac by Weight

In this case, SG is descending as proof increases, which makes designing the means of matching a little more complicated. The first command looks in the column of SG to find the last SG value in the column that is greater than what the user entered. Then we index that column location to find the equivalent SG. Then we subtract the entered value from the greater than value to obtain the difference between them. In this case a greater SG indicates a lower proof. It is not necessary to have an absolute value of this number because a greater than match will always result in a positive number when our apparent SG value is subtracted. Then we add one row number to our lookup and index the next SG down the column. We see that this value is less than our apparent SG by -0.00005. It is possible to design lookups where both numbers come out positive so that no absolute value command is required but I like to ensure that I've bracketed the number I'm looking for by seeing a positive number on one side and a negative number on the other. The choice for the "Winner" proof closest to the entered value is 160 proof (because 0.00005 is less than 0.00009). The instruction to do this is a simple IF statement where those two numbers are compared. The last question is what to do when one is equally far away from either 1/10th proof increment of the tables. My inclination is that, if the chances of something are 50/50, then it is not advisable to count on it, particularly when dealing with alcohol.

We won't spend too much time on this since it is unlikely that many people have SG hydrometers calibrated at 60°F in either air or vacuum but many hydrometers are calibrated at 15° Celsius (which is 59°F) and the thermal correction factor to allow them to operate with the 60°F. TTB tables is fairly minor and easy—we will cover that later. We will also look at how to adapt a hydrometer calibrated at 20° Celsius in vacuum to use the TTB tables in air as well.

Blend one Alcohol by Total Batch Weight

By using the target proof value in units of proof gallons per pound in conjunction with the proof gallons per pound of the source alcohol one can first find the total proof gallons required for a batch based upon the total batch mass and then find the mass of concentrate alcohol to use to achieve that blend. One can also substitute grams per bottle for the input and then convert to pounds:

Blend One Alcohol by Grams per Bottle	
Grams per Bottle	712.33385
How many bottles per case	12.0
Total Cases	102.00
Total Bottles	1,224.00
Total Grams Required	871,896.64
Kilograms (Air)	871.90
Total Batch Lbs. Weight	1,922.20

Blend One Alcohol by Total Batch Weight			Blend One Alcohol by Grams per Bottle con't	
Temp of Alcohol	60.00			
Apparent Proof of Alcohol	160.00			
True Proof Result	160.00		Temp of Alcohol	60.00
Target Proof	80.00		Apparent Proof of Alcohol	160.00
			True Proof Result	160.00
Total Batch Mass in Lbs.	1,922.22		Target Proof	80.00
Proof Gallons per Lb. at Target Proof	0.10093		Proof Gallons per Lb. at Target Proof	0.10093
Mass * Proof Gal per Lb. = Total Proof Gal.	194.010		Mass * Proof Gal per Lb. = Total Proof Gal.	194.008
Proof Gal. per Lb. of Source Alcohol	0.222450		Proof Gal. per Lb. of Alcohol	0.222450
Proof Gal/Prof Gal. per Lb. = Lbs. Alc.	872.151		Proof Gal/Prof Gal. per Lb. = Lbs. Alc.	872.142
Gallons of Water @ 60° F. to Add	126.09		Gallons of Water @ 60° F. to Add	126.08
Gross Volume of Batch @60°F.	247.34		Gross Volume of Batch @60°F.	247.34
Gal. Vol. Change due to Blending	-4.83		Gal. Vol. Change due to Blending	-4.83
Blended 60° F. Volume	242.51		Blended 60° F. Volume	242.51
Lbs. Weight of Water to Add	1,050.07		Lbs. Weight of Water to Add	1,050.06

Blend One Alcohol by Total Weight   *Blend One Alcohol by Grams per Bottle*

Blend by Total Batch Volume

Of course, it is also possible to blend by total batch volume as long as one uses the net volume and works backwards to find the gross volume to contribute. In this case we are using pounds per gallon in conjunction with proof gallons per pound to find the mass of alcohol to be used and then converting it to a 60°F volume:

Blend One Alcohol by Total Batch Volume					
Temp of Alcohol	60.00	0.000	Therm. Cor.	0.00150	Blending Tolerance
Apparent Proof of Alcohol	160.00	0.000	Hyd. Cor.	0.300	Tolerance as a Degree of Proof
True Proof Result	160.00			$13.50	Tax per Proof Gallon
Target Proof	80.00			$2,619.11	Tax
Total Batch Blended Volume at 60° F.	242.5132	918.01	Liters		
Lbs. per Gal. at Target Proof	7.92625				
Gallons * Lbs. per Gal. = Total Lbs.	1,922.22	871.90	Kilograms	121.257	Gal. Volume of Alcohol at 60° F.
Proof Gallons per Lb. at Target Proof	0.10093			459.01	Liters
Mass * Proof Gal per Lb. = Total Proof Gal.	194.008			1.0000	Table 7 Volume Correction Factor to be applied
Proof Gal. per Lb. of Alcohol	0.222448			121.257	Gal Vol. of Alcohol corrected to Present Volume
Proof Gal/Prof Gal. per Lb. = Lbs. Alc.	872.149			0.00	Present Gal. Vol. Difference from 60° F. Volume
Gallons of Water @ 60° F. to Add	126.09			97.00	Absolute Alcohol Gallons = P.G./2
Gross Volume of Batch @ 60°F.	247.34			242.51	Net Blended Wine Gallons
Gal. Vol. Change due to Blending	-4.83			40.00%	Absolute Alc. Gal./Net. Alc. Gal. = % Alc. By Vol.
Blended 60° F. Volume	242.51			-18.28	Liters Change in Vol.
Lbs. Weight of Water to Add	1,050.07			1.95%	Vol Difference/Gross Gallons
				1.99%	(Gross Gal/Net Gal)-1

Blend One Alcohol by Total Volume

Finding volume from weight involves simply multiplying the mass times the wine gallons per pound or dividing by the pounds per wine gallon. In this case we are multiplying by wine gallons per pound:

Find 60° F. Volume of Alcohol	
Weight Restated	872.15
Wine Gallons per Pound @ True Proof	0.13903
Weight * W.G. per Lb. = 60° F. Alc. Vol.	121.26

Finding the weight and volume of water is straightforward—by taking the total weight and subtracting the weight of alcohol. From that net weight of water, one can use the pounds per gallon of water and by division find the 60°F gallons of water to use:

Find Water Weight needed to blend to Target Proof		
Total Weight Restated	1,922.22	
Weight of Alcohol in Pounds Restated	872.15	
Total Weight - Alcohol Weight = Water Weight	1,050.07	Weight of Water to Add to Batch

Find 60° F. Gallons of Water to add to Batch		
Water Weight Restated	1,050.07	
Weight Per Gallon of Water @ 60° F.	8.328230	60° F. Water Weight Per Gallon
Water Weight / Wt. per Water Gal. = 60° F. Gal.	126.09	60° F. Gallons of Water to Add

Plan by Volume and Blend

Now let's expand our batch plan in order to plan a batch and blend it in one operation.

We can simply enter the number of cases and the bottle fill amount in milliliters to find the total milliliters required and, by translating that into a gallon value, continue the blend using TTB values and procedures:

Plan by Volume & Blend One Alcohol							
Total Cases	102.00			0.00150	Blending Tolerance		
Bottle fill amount in milliliters	750.00			0.300	Tolerance as a Degree of Proof		
How many bottles per case	12.0			$13.50	Tax per Proof Gallon		
				$2,619.07	Tax		
Temp of Alcohol	60.00	0.000	Therm. Cor.				
Apparent Proof of Alcohol	160.00	0.000	Hyd. Cor.				
True Proof Result	160.00						
Target Proof	80.00						
Total ml per case	9,000.00						
Total ml Required	918,000.0	918.00	Blended Liters				
Batch Blended US Gal. Vol. at 60° F.	242.510						
Lbs. per Gal. at Target Proof	7.92625						
Gallons * Lbs. per Gal. = Total Lbs.	1,922.20	871.89	Kilograms	121.255	Gal. Volume of Alcohol at 60° F.		
Proof Gallons per Lb. at Target Proof	0.10093			459.00	Liters		
Mass * Proof Gal per Lb. = Total Proof Gal.	194.005			1.0000	Table 7 Volume Correction Factor to be applied		
Proof Gal. per Lb. of Alcohol	0.222448			121.255	Gal Vol. of Alcohol corrected to Present Volume		
Proof Gal/Prof Gal. per Lb. = Lbs. Alc.	872.138			0.00	Present Gal. Vol. Difference from 60° F. Volume		
Gallons of Water @ 60° F. to Add	126.08			97.00	Absolute Alcohol Gallons = P.G./2		
Gross Volume of Batch @ 60°F.	247.34			242.51	Net Blended Wine Gallons		
Gal. Vol. Change due to Blending	-4.83			40.00%	Absolute Alc. Gal./Net. Alc. Gal. = % Alc. By Vol.		
Blended 60° F. Volume	242.51			-18.28	Liters Change in Vol.		
Lbs. Weight of Water to Add	1,050.06			1.95%	Vol Difference/Gross Gallons		
				1.99%	(Gross Gal/Net Gal)-1		

Plan and Blend One Alcohol by Total Volume

Just always keep in mind that these are 60°F gallons and that a volume correction will be necessary to use them at other temperatures.

Notes

Other Blending Formulas

Pearson's Square Methods of Dilution

Pearson's Square is a simplified method of conducting dilution calculations and has traditionally been used to fortify wine with brandy in order to make port. The method does not account for the change in volume due to blending but is otherwise serviceable when greater precision is not required or when dealing with substances that do not change volume when combined. First, we will reproduce our standard blend for comparison:

Temp of Alcohol	60.00
Apparent Proof of Alcohol	160.00
True Proof Result	160.00
Target Proof	80.00
Volume of Alcohol at Present Temp to be used	121.255
Table 7 Volume Correction Factor to be applied	1.0000
Volume of Alcohol corrected to 60° F.	121.26
Present Volume difference from 60° F. Volume	0.00
60° F. Volume compared to present volume	0.00

Results	
Proof Gallons of Alcohol	194.01
Gallons of Water @ 60° F. to Add	126.09
Gross Volume of Batch @60°F.	247.34
Gallon Volume Change Due to Blending	-4.83
Actual "net" 60° F. Volume available for Bottling	242.51
Vol Difference/Gross Gallons = % Change	1.95%

Blend One Alcohol by Volume

Now we will compare that blend to Pearson's Square that has been set up to blend to a total known volume of our blended volume of 242.51:

Pearson's Square Method to Dilute			
Proof of Alcohol 1	160.00	80.00	% Alcohol By Volume
Proof of Alcohol 2 or Water	0.00	0.00	% Alcohol By Volume
Target Proof of Blend	80.00	40.00	% Alcohol By Volume
Number of Gallons Needed	242.51		
Results			
Gallons Alcohol 1	121.26		
Gallons of Alcohol 2	121.26		

Method Demonstrated				Parts	Proportion	Gallons	
Alcohol 2 or Water Vol %	0.00	Target %	40.00	0.500	121.26	Gal of A2
		40.00				
Alcohol 1 Vol %	80.00		40.00	0.500	121.26	Gal of A1
		Total Parts	80.00			

Pearson's Sq. Blend Total Vol.

To work the method, one subtracts the *Vol %* values from left to right diagonally across the diagram to find the *Parts* value. The Alcohol 2 top side is 40 - 0 = 40 to obtain that result but the Alcohol 1 bottom to top subtraction runs 80 - 40 = 40. Then divide each *Parts* result by the *Total Parts* to find the proportion. Then multiply the proportion by the total gallons needed to find the gallons of each component in the mixture. Without accounting for the change in volume due to blending, the result is short by 126.09 - 121.26 = 4.83 gallons of water in this case. Accordingly, the proof will always come out to be greater than anticipated and more alcohol and less total volume will leave the plant than intended.

It is also possible to use Pearson's square to dilute a known quantity of high-strength alcohol. This is the same as our *Blend One Alcohol by Volume*, but without accounting for the change in volume. It will, however, work for other fluids that do not change in volume due to blending.

Pearson's Square Method to Dilute Known Volume of Concentrate			
Proof of Alcohol 1	160.00	80.00	% Alcohol By Volume
Known Vol. of Alcohol 1	121.26		
Proof of Alcohol 2 or Water	0.00	0.00	% Alcohol By Volume
Target Proof of Blend	80.00	40.00	% Alcohol By Volume
Results			
Gallons of Alcohol 2 or Water to Use	121.26		
Total Volume	242.52		

Method Demonstrated			Parts	Proportion	Gallons	
Alcohol 2 or Water Vol %	0.00	Target %	40.00	2.000	242.52	Total Vol.
		40.00				
Alcohol 1 Vol %	80.00		40.00	0.500	121.26	Gal of A2
		Total Parts	80.00			

Pearson's Sq. Dilute Known Concentrate

In this case the net volume comes out properly, but again the amount of water is off by the same 4.83 gallons (126.09 - 121.26 = 4.83). The equation has been changed in that the top Alcohol 2 value for the parts of the total volume is divided into the total parts. That proportion is then multiplied by the known volume of concentrate to find the total volume. The parts of Alcohol 1 are divided by the total parts to find the proportion. The proportion is then multiplied by the total volume to find the gallons of Alcohol 1.

While I have used gallons for these examples this method is actually dimensionless as to volume. One can substitute liters or any other volume unit and as long as you are dealing only in that unit; then the ratios and calculations will work fine.

Algebraic Methods of Dilution

Apart from Pearson's Square, there is also an algebraic method for diluting to a predicted concentration. It also does not account for change in volume (and as such is not an approved method), but for completeness I will include the formula and an example and we can compare it to our standard batch and see how it fares.

Formula 1: Find the amount of water to add to reduce one alcohol to bottling proof:

$$\frac{(Volume_{Spirits}*True\ Proof)-(Volume_{Spirits}*Desired\ Proof)}{Desired\ Proof}$$

What is the proof of Alcohol	160.00
What is the Volume of Alcohol	121.26
What is the target Proof	80.00
Stage 1	
Volume of Alcohol	121.26
Proof of Alcohol	160.00
Proof * Volume = Factor 1	19,401.60
Stage 2	
Volume of Alcohol	121.26
Target Proof	80.00
Volume * Target Proof = Factor 2	9,700.80
Factor 1 minus Factor 2 = Factor 3	9,700.80
Stage 3	
Factor 3 divided by Desired Proof = Volume of Water	121.26

Algebraic Formulas

We see that the answer is exactly the same as that provided by Pearson's Square (121.26 gallons of water to add, rather than the 126.09 actually required). It is dimensionless as to volume and will work with any unit (gallons, liters, etc.). Because we have added less water than necessary the proof will be too high.

There is one other algebraic formula that I'd like to note. Formula 2 allows one to blend to an unknown destination with known input quantities and will estimate the proof of a blend of two alcohols:

$$\frac{(Volume_{Spirit1} * Proof_{Spirit1}) - (Volume_{Spirit2} * Proof_{Spirit2})}{Volume_{Spirit1} + Volume_{Spirit2}}$$

Blending our two equal volumes of 121.26 does indeed produce a resulting proof of 80 proof, but it is again a wrong answer because we are dealing with alcohol.

Proof of Alcohol 1	160.00
Volume of Alcohol 1	121.26
Volume of Alcohol 2	121.26
Proof of Alcohol 2	0.00
Proof of Resulting Blend	80.00

Stage 1 - Proof Factor	
A1 Vol * Proof	19,401.60
A2 Vol * Proof	0.00
Total of both values	19,401.60
Stage 2 - Volume Factor	
Volume of Alcohol 1	121.26
Volume of Alcohol 2	121.26
Total Both Values	242.52
Stage 3	
Divide Proof Factor by Volume Factor = Proof of Blend	80.00

Algebraic Formulas

The defect in this method is that the result will be off by the value of the volume change due to blending alcohol and water. It will work best with blends of two high-proof alcohols where there is little water to cause volume change—and probably blends of two low-proof alcohols where mostly water is involved. The result should not be considered accurate enough to use in blending in a distilled spirits plant except for adding very small quantities of spirits to a larger batch to move by 1 or 2 proof.

There is a way to quantify what this blend of 121.26 gallons of each component would actually come out to: that is to use the *Blend Two Alcohols to Unknown Target Proof by Volume* spreadsheet. The result is 81.6 proof. We will cover this method shortly. But it also serves to compare it to the algebraic Formula 2. Using 126.09 gallons of water indicates that the target proof would be 78.44, which subtracted from 80 = 1.56 proof; this is essentially the same difference in proof, but in the opposite direction.

Blend to Unknown Target by Volume	
What is the Temperature of Alcohol 1	60.00
What is the Apparent Proof of Alcohol 1	160.00
Volume of Alcohol 1 at Present Temp to be used	121.26
True Proof Result	160.00
Correction Factor applied	1.0000
Volume of Alcohol 1 corrected to 60° F.	121.26
Present Volume Difference from 60°F of Alc. 1	0.00

What is the Temperature of Alcohol 2	60.00
What is the Apparent Proof of Alcohol 2	0.00
Volume of Alcohol 2 at Present Temp to be used	121.26
True Proof Result	0.00
Correction Factor applied	1.0000
Volume of Alcohol 2 corrected to 60° F.	121.26
Present Volume Difference from 60°F of Alc. 1	0.00

Total Alcohol Volume at 60° F.	242.52
Change in Volume due to blending (of 60° F. Gal.)	-4.761
Net Volume after blending at 60° F.	237.76
Blended Proof of to nearest 1/10 Proof	81.60

Blend Unknown Target by Volume

Proof of Alcohol 1	160.00
Volume of Alcohol 1	121.26
Volume of Alcohol 2	126.09
Proof of Alcohol 2	0.00
Proof of Resulting Blend	78.44

Stage 1 - Proof Factor	
A1 Vol * Proof	19,401.60
A2 Vol * Proof	0.00
Total of both values	19,401.60
Stage 2 - Volume Factor	
Volume of Alcohol 1	121.26
Volume of Alcohol 2	126.09
Total Both Values	247.35
Stage 3	
Divide Proof Factor by Volume	
Factor = Proof of Blend	78.44

Algebraic Formulas

Increase Proof

Let's use our standard batch and assume that a mistake in blending was made and that the blended proof has come out to 77 True Proof. Usually when I miss the mark, it is at 79 or 81 proof, but we will exaggerate the norm slightly so that we can see the effect of increasing the proof more clearly. It's impossible to know which component was put in incorrectly, but by knowing the true proof of the batch is only 77, we know that there is extra water in the batch.

In order to use this procedure, we will have to know the mass or volume of material we are dealing with. Since most people don't have their blending tanks situated on load cells, chances are we are stuck with volume. If you don't have a sight glass on your tank—and most us won't—we will have to calculate the volume we are dealing with. It doesn't matter which component was added incorrectly; we need to get a fix on the volume and proof for the *Increase Proof* algorithm to work. To find the volume of our batch, we can co-opt a formula that calculates the size of a round vessel to help us out. Measure the depth of the material in the tank with a dip stick and apply the measured value to the *Size of Vessel* formula as follows:

Calculate Size of Round Vessel	
Diameter of Vessel in Inches	38.00
Radius = Diameter/2	19.00
Radius Squared	361.00
Pi	3.14159
Radius Sq. * Pi = Area Sq. In.	1,134.11
Height of Vessel inches	51.32
Sq. in. * H = Cubic Inch Volume	58,202.78
Cubic Inches per Cubic Foot	1,728.00
Total Cubic In./Cubic in per cubic foot =Cubic Feet	33.68
Cubic Inches per Gallon	231.00
US Gallon Capacity of Vessel	251.96

Size of Vessel

Now, most tanks have a dish bottom and it would be helpful if you knew how many gallons were in the "dish" at the bottom and then use a height above it to measure the remaining quantity in the tank. It will also be helpful if you run a volume correction and find 60°F volume in the tank as well. In this case, assuming an ideal cylinder we have 251.96 gallons in the tank; let's assume that they are in fact 60°F gallons.

Entering the actual true proof of 77 and the 60°F volume into a *Blend by Total Volume* worksheet tells us that there are in fact 194.01 proof gallons in the tank. This will allow us to correct the batch by adding more alcohol to it. Now that we know the proof gallons are correct, we can see the other side of the problem by running a blend to target using the right amount of alcohol and identify that we actually added 135.65 gallons of water to the mixture:

Blend One Alcohol by Total Batch Volume	
Temp of Alcohol	60.00
Apparent Proof of Alcohol	160.00
True Proof Result	160.00
Target Proof	77.00
Total Batch Blended Volume at 60° F.	251.9600
Lbs. per Gal. at Target Proof	7.94534
Gallons * Lbs. per Gal. = Total Lbs.	2,001.91
Proof Gallons per Lb. at Target Proof	0.09691
Mass * Proof Gal per Lb. = Total Proof Gal.	194.011
Proof Gal. per Lb. of Alcohol	0.222448
Proof Gal/Prof Gal. per Lb. = Lbs. Alc.	872.162
Gallons of Water @ 60° F. to Add	135.65
Gross Volume of Batch @ 60°F.	256.91
Gal. Vol. Change due to Blending	-4.95
Blended 60° F. Volume	251.96
Lbs. Weight of Water to Add	1,129.74

Temp of Alcohol	60.00
Apparent Proof of Alcohol	160.00
True Proof Result	160.00
Target Proof	77.00
Volume of Alcohol at Present Temp to be used	121.255
Table 7 Volume Correction Factor to be applied	1.0000
Volume of Alcohol corrected to 60° F.	121.26
Present Volume difference from 60° F. Volume	0.00
60° F. Volume compared to present volume	0.00
Results	
Proof Gallons of Alcohol	194.01
Gallons of Water @ 60° F. to Add	135.65
Gross Volume of Batch @60°F.	256.90
Gallon Volume Change Due to Blending	-4.95
Actual "net" 60° F. Volume available for Bottling	251.95
Vol Difference/Gross Gallons = % Change	1.93%
Weight of Water to Add	1,129.71
Weight of Alcohol	872.14
Total Weight of Batch	2,001.85

Blend One Alcohol by Total Volume   *Blend One Alcohol by Volume*

Following are the results of the *Increase Proof by Volume* calculations to bring the proof from 77 up to 80. Then we will examine the formula we've used:

Temperature of Alcohol 1	60.00	0.000	Thermometer Cor.
Apparent Proof of Alcohol 1	77.00	0.000	Hydrometer Cor.
True Proof of Alcohol 1	77.00		

	$13.50	Excise Tax Rate
	0.0015	Tolerance
	0.300	Tolerance as ° of Proof

Present Volume of Alcohol 1	251.96	
Correction Factor applied to find 60° F. Vol.	1.0000	
Volume of that Alcohol corrected to 60° F.	251.96	60° F. Volume

What is the Target Proof	80.0	3.0	Proof Difference

Properties of Alcohol A2 to be used to reach target proof

Temperature of Alcohol 2	60.00	0.000	Thermometer Cor.
Apparent Proof of Alcohol 2	160.0	0.000	Hydrometer Cor.
True Proof of Alcohol 2	160.0		

Results			
Proof Gallons of Alcohol 1 Already in Batch	194.01		
Proof Gallons of Alcohol 2 to be Used	14.73		
Total Proof Gal. in Batch after raising proof	208.74		
Weight of Alc. 2 to be Added to Alc. 1	66.21		
60° F. Wine Gal. of Alc. 2 to Add to Alc 1	9.20	@ 60° F.	
Total Wine Gallons in Blended Batch	261.16		
Volume Reduction as a result of Blending	-0.25		
60° F. Blended Batch Vol. at Target Proof	260.92		

$2,619.15	Tax on Proof Gal. already in Batch
$198.82	Tax on Proof Ga. Added
$2,817.96	Total Tax in Blended Batch
0.25	Volume Change Due to Blending
0.93	Volume Reduction in Liters
1.24	# of 750 ML bottles "Short"
0.09%	Vol Difference/Gross Gallons
0.09%	(Gross Gal/Net Gal)-1

Using the same 160-proof base stock we can raise the proof of the batch from 77 to 80 by adding 66.21 pounds or 9.20 gallons of alcohol to the batch. The formula that accomplishes this blend uses Table 6 values for the respective alcohols and is applied as follows:

Find Volume and Mass of Alcohol 2 to use	
60° F. Wine Gallons Restated	251.96
Target Proof Parts Alcohol	40.00
Target Proof Parts Water	63.42
Target Parts Alc./Target Parts H2O = Target Parts	0.63072
Alcohol Parts of Alcohol 1	38.50
Water Parts of Alcohol 1	64.84
A 1 Parts Alcohol/A1 Parts Water = A1 Ratio	0.59377
Alcohol Parts of Alcohol 2	80.000
Water Parts of Alcohol 2	22.870
A2 Parts Alcohol/A2 Parts Water = A2 Ratio	3.49803
Target Parts Ratio * A2 Parts Water	14.424
Subtract Result from Alcohol Parts of A2	65.576
Wine Gallons * Target Ratio * Water Parts A1	10,304.06
Wine Gallons * Alcohol Parts A1	9,700.46
Subtract Result 2 from Result 1 =	603.60
Divide by	65.58
Result = 60° F. Wine Gallons of A2 to be added to A1	9.20
Find Weight of A2 and P.G. of A2 to Use	
Wine Gal Per Pound of Alcohol 2	0.13903
Wine Gallons A2/ Wine Gal per Lb. A2 = Lbs. A2	66.21

Increase Proof by Volume

One can get a fix on the proof gallons if one knows the total mass in the tank by blending to 77 proof and then using the *Increase Proof by Weight* worksheet:

Temperature of Alcohol 1	60.00
Apparent Proof of Alcohol 1	77.00
True Proof of Alcohol 1	77.00
Weight of Alcohol 1	2,001.96
Wine Gallons per Pound	0.12586
Wt. * W.G. per Lb. = Wine Gal. A1	251.97
What is the Target Proof	80.0

Blend One Alcohol by Total Batch Weight		Properties of Alcohol A2 to be used to reach target proof	
Temp of Alcohol	60.00		
Apparent Proof of Alcohol	160.00	Temperature of Alcohol 2	60.00
True Proof Result	160.00	Apparent Proof of Alcohol 2	160.0
Target Proof	77.00	True Proof of Alcohol 2	160.0
Total Batch Mass in Lbs.	2,001.96	Results	
Proof Gallons per Lb. at Target Proof	0.09691	Proof Gallons of Alcohol 1 Already in Batch	194.02
Mass * Proof Gal per Lb. = Total Proof Gal.	194.016	Proof Gallons of Alcohol 2 to be Used	14.728
Proof Gal. per Lb. of Source Alcohol	0.222448	Total Proof Gallons in Batch after raising proof	208.74
Proof Gal/Prof Gal. per Lb. = Lbs. Alc.	872.186	Weight of Alcohol 2 to be Added to Alcohol 1	66.21

Blend One Alcohol by Total Weight　　　　*Increase Proof by Weight*

Pearson's Square Method to Increase Proof

It is also possible to use Pearson's Square to increase the proof of a mixture. Because this method is often applied to use brandy to make port from wine we will stick with those labels for the components. One drawback of using this method directly is that you must play with batch size in order to fit the components available to the batch size and it also does not account for change in volume due to blending:

Pearson's Square Method to Increase Proof of a Mixture

Proof of "Wine" to be Increased	77.00	38.50	% Alcohol By Volume
Proof of "Brandy" to be used	160.00	80.00	% Alcohol By Volume
Target Proof of "Port" Blend	80.00	40.00	% Alcohol By Volume
Number of Gallons Needed	251.60		

Results

Gallons of Wine To Use	242.51	
Gallons of Brandy to Use	9.09	
Total Gal. after Brandy Addition	251.60	

Method Demonstrated

				Parts	Proportion	Gallons	
Wine % Alcohol	38.50	Target %		40.00	0.964	242.51	Gallons of Wine To Use
		40.00					
Brandy % Alcohol	80.00			1.50	0.036	9.09	Gallons of Brandy to Use
		Total Parts		41.50		251.60	Total Gallons

Pearson's Square Increase Proof

In this case we have manually adjusted the total number of gallons needed in order to match the 242.51 gallons that were planned for the net volume of the original batch that was mis-blended. We obtain an approximately accurate value of 9.09 rather than the 9.21 gallons of 160-proof alcohol to add to the batch to bring the proof up to 80. The formula is the same as the Pearson's dilution formula. The reason the formula is working "better" in this example is because there is much less change in volume to account for and thus the error is less.

What this points out is that, with respect to alcohol, Pearson's square works best when one is using similar density alcohol as in blending low-proof alcohol with other low-proof alcohol or high-proof alcohol with high-proof alcohol (where the change in volume is not a large factor).

Batch Proof Too High

I blend by weight and so I keep track of my batch by the weight of alcohol and water added. I also plan my blends in two stages. I will "short" the blend by an amount of water that I think will end up three (3.0) proof higher than my target proof. Then I let the batch sit for an hour or so and proof it. If I'm correct then I can simply add the amount of water that I've reserved to blend to the target proof.

To quantify this, one can take a batch that has been shorted of water. It is an easy matter to determine how much water to add to reduce it to the desired bottling proof. I've run our standard batch to 83 proof for starters. All the alcohol is in the batch and we are just adding less water to achieve 83 proof.

83 Proof Blend to Target:

Temp of Alcohol	60.00
Apparent Proof of Alcohol	160.00
True Proof Result	160.00
Target Proof	83.00
Weight of Alcohol	872.15
Volume of Alcohol at 60° F.	121.2552
Table 7 Volume Correction Factor to be applied	1.0000
Volume of Alcohol corrected to Present Volume	121.255
Present Volume Difference from 60° F. Volume	0.00
60° F. Volume compared to present volume	0.00
Results	
Proof Gallons of Alcohol	194.01
Gallons of Water @ 60° F. to Add	117.18
Gross Volume of Batch @60°F.	238.44
Volume Reduction due to Blending	-4.69
Actual "net" 60° F. Volume available for Bottling	233.74
Weight of Water to Add	975.92
Weight of Alcohol	872.15
Total Weight of Batch	1,848.07

Blend One Alcohol by Weight

83 to 80 Proof Blend to Target:

Temp of Alcohol	60.00
Apparent Proof of Alcohol	83.00
True Proof Result	83.00
Target Proof	80.00
Volume of Alcohol at Present Temp to be used	233.740
Table 7 Volume Correction Factor to be applied	1.0000
Volume of Alcohol corrected to 60° F.	233.74
Present Volume difference from 60° F. Volume	0.00
60° F. Volume compared to present volume	0.00
Results	
Proof Gallons of Alcohol	194.01
Gallons of Water @ 60° F. to Add	8.90
Gross Volume of Batch @60°F.	242.64
Gallon Volume Change Due to Blending	-0.14
Actual "net" 60° F. Volume available for Bottling	242.50
Vol Difference/Gross Gallons = % Change	0.06%
Weight of Water to Add	74.16
Weight of Alcohol	1,848.04
Total Weight of Batch	1,922.20

Blend One Alcohol by Volume

By knowing the 60°F volume is 233.74 gallons at 83 proof, we can run a blend by volume from 83 proof down to 80 and find that we need 8.90 gallons or 74.16 pounds of water to blend to the target proof (for a total of 126.08 gallons). I've even blended by 0.1 proof to find the fraction of a gallon/weight of water I need to move that 0.1 proof. As I've said repeatedly, my practice is to prefer using weight as the means by which alcohol is accounted for. I find it results in significant savings in time and provides increased accountability in alcohol processing operations.

Blend to Unknown Target by Volume

It is possible to use the parts alcohol and water values contained in Table 6 to develop an equation to blend to an unknown target proof. This can be useful when you want to combine two alcohols of known quantity and quality but predict what the result will be before accomplishing the blend. We are keeping this simple and just using our same standard batch numbers to verify that the formula does arrive at the correct "unknown" proof of 80. This sheet results in a volume of 242.52 gallons which is the same as our standard blend result of 242.52 by mass input and slightly less than the 251.95 gallons when using the *Blend by Volume* worksheet:

Blend to Unknown Target by Volume			
What is the Temperature of Alcohol 1	60.00	0.000	Thermometer Cor.
What is the Apparent Proof of Alcohol 1	160.00	0.000	Hydrometer Cor.
Volume of Alcohol 1 at Present Temp to be used	121.255		
True Proof Result	160.00		
Correction Factor applied	1.0000		
Volume of Alcohol 1 corrected to 60° F.	121.26		
Present Volume Difference from 60°F of Alc. 1	0.00		

What is the Temperature of Alcohol 2	60.00	0.000		Thermometer Cor.
What is the Apparent Proof of Alcohol 2	0.00	0.000		Hydrometer Cor.
Volume of Alcohol 2 at Present Temp to be used	126.09		0.0015	Tolerance
True Proof Result	0.00		0.300	Tolerance as ° of Proof
Correction Factor applied	1.0000		1.31	Pounds of A2 Tolerance
Volume of Alcohol 2 corrected to 60° F.	126.09		0.13	Proof Gal. A2 Tolerance
Present Volume Difference from 60°F of Alc. 1	0.00		0.189	Wine Gal. A2 Tolerance

Total Alcohol Volume at 60° F.	247.35
Change in Volume due to blending (of 60° F. Gal.)	-4.829
Net Volume after blending at 60° F.	242.52
Blended Proof of to nearest 1/10 Proof	80.00

Blend Unknown Target by Volume

The way Table 6 is used to accomplish this blend is show as follows:

Find Alcohol to Water Ratio A1	
Alcohol 1 Volume Restated	121.26
Volume/100 =	1.2126
Parts Alcohol of A1	80.00
Volume/100 * Alcohol Parts of A1 = Alc. Parts Ratio A1	97.00

Volume/100 Restated	1.2126
Parts Water of A1	22.88
Volume/100 * Alcohol Parts of A2 = Water Parts Ratio A1	27.74

Find Alcohol to Water Ratio A2	
Alcohol 2 Volume Restated	126.09
A2 Volume/100 =	1.2609
Parts Alcohol of A2	0.00
Volume/100 * Alcohol Parts of A1 = Alc. Parts Ratio A2	0.00

A2 Volume/100 Restated	1.2609
Parts Water of A2	100.00
Volume/100 * Alcohol Parts of A2 = Water Parts Ratio A2	126.09

A1 Alc. Parts Result + A2 Alc. Parts Result = Total Alc Parts	97.00
Alc 1 H2O parts Result + Alc 2 H2O parts result = Tot. Parts	153.83
Tot. Alc. Parts / Tot. Water Parts = Alc. to Water Ratio	0.63058

Blend Unknown Target by Volume

The final alcohol-to-water ratio is then looked up in Table 6 to find the proof of the destination density. It requires that one create an *Alcohol Water Ratio* term in the table as a separate column of calculations to be indexed for each individual proof.

Determining Proof of High Proof Blend Using A/W Ratio Results	
A/W Ratio	0.63058
Match Less than or Equal Row	800
Index Ratio	0.62943
Add 1 for Next Row Higher	801
Index Ratio	0.63072
Winner Ratio	0.63072
Winner Row Exact Match	801
Index Proof	80.00

Blend Unknown Target by Volume

The formula to find the volume to blend to an unknown target proof using the traditional terms of wine and brandy to fortify wine to port (using V_1 = Volume Wine, V_2 = Volume Brandy, Pw_1 = Parts Water of Wine, Pw_2 = Parts Water of Brandy, Pa_1 = Parts Alcohol Wine, Pa_2 = Parts Alcohol Brandy) follows:

$$V_2 = \frac{R(V_1*Pw_1)-(V_1*Pa_2)}{Pa_2-(R*Pw_2)}$$

Blend to Unknown Target by Weight

I would like to include one example of the *Blend to An Unknown Target Proof by Weight* algorithm. The formula to accomplish this blend is the same as for volume and this requires that we calculate the 60°F volume from the weight before we can use the formula. Using 1,000 lbs. of alcohol for each component with respective proofs of 190 and 125 shows that the resulting blend will arrive at 160 proof. I have also included a summary of the component alcohols' properties, the most useful of which is the proof gallon proportion of the blend:

What is the Temperature of Alcohol 1	60.00	0.000	Thermometer Correction Alcohol 1
What is the Apparent Proof of Alcohol 1	190.00	0.000	Hydrometer Correction Alcohol 1
Weight of Alcohol 1 To Be Used in Batch	1,000.00	190.00	True Proof of Alcohol 1
What is the Temperature of Alcohol 2	60.00	0.000	Thermometer Correction Alcohol 2
What is the Apparent Proof of Alcohol 2	125.00	0.000	Hydrometer Correction Alcohol 2
Weight of Alcohol 2 To Be Used in Batch	1,000.00	125.00	True Proof of Alcohol 2
Total Alcohol Weight	2,000.00		
Total Alcohol Volume	279.47	@ 60° F.	Tolerance
Reduction in Volume due to blending	-1.41	0.00150	0.300	Tolerance as ° of Proof
Net Volume after blending	278.06	1.500	Pounds of A1 Tolerance
Blended Proof of to nearest 1/10 Proof	160.00	1.500	Pounds of A2 Tolerance
		3.00	Total Pounds Tolerance as ° of Proof
Absolute Alcohol Gallons	222.50	0.42	Wine Gallon Volume of Alcohol Tolerance
% Alcohol By Volume	80.0%	0.67	Batch Proof Gallon Tolerance

Rounded to Nearest 1/10th Proof Method	Weight Restated	Wine Gallons Per Pound	Volume = Weight * W.G. Per Lb.	P.G. Per Lb.	Gallons = Lb's * P.G. Per Lb.	% by Vol. of Gross Gallons	P.G. as % of total P.G.	% by Weight of each
Weight of Alcohol 1	1,000.00	0.14718	147.18	0.27964	279.64	52.66%	62.84%	50.00%
Weight of Alcohol 2	1,000.00	0.13229	132.29	0.16536	165.36	47.34%	37.16%	50.00%
Total Weight of Alcohol	2,000.00	Total Vol.	279.47	Total P.G.	445.00	100.00%	100.00%	100.00%

Blend Unknown Target by Weight

Table 7 Volume Correction

Volume Correction Present to 60°F

We've covered volume correction previously when we discussed our standard batch as a *Blend by Volume* batch; but there are several other applications for this procedure that need to be addressed. Volume correction is essential in any larger-scale facility where it is not possible to have a scale or load cells with sufficient capacity to permit using weight as the means of accounting for alcohol. Even in smaller operations, some operators are more comfortable dealing with volumes; but because alcohol and water mixtures change volume with temperature, it is necessary to run volume correction procedures to determine the correct volume either presently or at 60°F. The importance of volume correction is noted since the regulations give no fewer than six examples of how to use Table 7 to accomplish the procedure. We will start with the most basic volume correction algorithm explained in §186.67-§30.67 and use Example 6 as our guide in order to show the full extent of the procedure:[25]

Present Volume	1,000.00	3,785.41	Liters
What is the Present Temperature of the Alcohol	93.00	0.0000	Therm Cor.
What is the True proof @ 60° F of the Alcohol	97.00	You must know true proof already	
Correction Factor to be applied	0.9851		
60° F. Wine Gallons in Container	985.10	3,729.01	Liters
Diffrence between Present Vol. and 60° F. Vol.	14.90	56.40	Liters
Proof/100 = Proof Gallon Conversion Factor	0.9700		
60° F. Wine Gallons * Proof / 100 = Proof Gallons	955.55		

Volume Correction Using Table 7 Using §186.67 - §30.67 Procedure			
Temperature of Alcohol	93.00		
Correction Factor for this Thermometer	0.000		
Temperature Corrected	93.00		
Temperature Rounded to whole Temp for Table 7 Entry	93.00		
True Proof Restated	97.00		
True Proof Rounded to whole proof for Table 7 Entry	97.00		
If Temp is Odd Then Lookup 1 Temp Lower	0.986	At	92°
If Temp is Odd Then Lookup 1 Temp Higher	0.985	At	94°
Subtract the smaller value from the larger	0.001		
Divide the Result by 2	0.0005		
Temperature Corrected Value Only	0.9855		
Second Stage of Procedure to Determine Proof Correction Factor if Needed for Table Intermediate Proof			
If Odd Temp above Modify to Nearest Even Lower Temp	92°		
Look up Closest Higher Proof Match	0.985	100	Closest Higher Match to Table
Look up Closest Lower Proof Match	0.986	95	Closest Lower Match to Table
Difference between Factors	-0.001		
Divide Factor Difference by 5	-0.0002	Difference per degree of proof	
Find difference between Lower Value & Actual Proof	2.0		
Multiply Per Proof difference by actual Difference	-0.0004	Add this amount to the Temp Difference Factor	
Temperature Correction Factor Restated	0.9855		
Add in Proof Correction Factor	-0.0004		
Add for Factor for Odd Temp and intermediate Proof	0.9851	Table 7 Correction Factor to be applied	
Volume at Present Temp Restated	1,000.00		
Volume * Factor = Corrected Gallons to 60° F.	985.10	Wine Gallons * Correction factor = Gallons @ 60°	
Present Gal. Dif. from 60° F. Volume	14.90		

Vol Cor. Present Vol. to 60 F. Vol.

---

[25] "*Example.* It is desired to ascertain the correction factor for spirits of 97 proof at 93 degrees temperature: 0.986 (92°) - 0.985 (94°) = 0.001 divided by 2 = 0.0005. 0.0005 + 0.985 = 0.9855 corrected factor at 93°F. 0.986 (95 proof) - 0.985 (100 proof) = 0.001. 0.001 divided by 5 = 0.0002 (for each percent of gauging) ×2 (for 97 proof) = 0.0004. 0.9855 (corrected factor at 93°F) - 0.0005 = 0.9851 correction factor to be used for 97 proof at 93°F."

We see that the correction factor of 0.9851 has been solved for and that the net gallons at 60°F will be 985.10.

Where the *Present Gallons* difference is a positive value this indicates that the gallons being measured will shrink to reach the 60°F volume. Conversely, when the present gallons are a negative value, it means that the gallons being measured will increase in volume to reach their 60°F volume. What I found in constructing this worksheet is that it was important to maintain the negative values generated and not use absolute values in order for the algorithm to reproduce the TTB example results.

By including a true proof correction, it is possible to go directly from apparent proof and temperature to true proof and then perform a volume correction. This example shows the difference in proof gallons engendered by using the two different approved methods of determining proof gallons, which in this case turns out as −2.25% of the tolerance—which it pretty good:

Known Present Volume	1,000.00	3,785.41	Liters
What is the present Temperature of the Alcohol	93.00	0.0000	Thermometer Correction
What is the Apparent Proof of the Alcohol	110.50	0.0000	Hydrometer Correction
True Proof Result	97.00		
Correction Factor to be applied for this Temp.	0.9851		
60° F. Gallon Vol.	985.10	3,729.01	Liters
Present Volume Difference from 60° F. Wine Gallons	14.900	56.40	Liters

If Negative Wine Gallons will "Grow" to reach 60° F If Positive Wine Gallons will Shrink to reach 60°F

Find Proof Gallons Method 1 - Table 4 Standard Transformation - 60° F. Vol. to Weight		
Wine Gallons Per Pound @ True Proof	0.12814	
60° F. Volume/W.G. Per Lb. = Lbs.	7,687.69	
P.G. Per Lb. value	0.12430	
Proof Gallons per Lb. * Lbs. = Total Proof Gallons	955.58	Proof Gallons

Find Proof Gallons Method 2	
True Proof / 100	0.970
60° F. Wine Gallons * (Proof / 100) = Proof Gallons	955.55
Proof Gallon Difference in Methods	-0.03
Difference as % of Tolerance	-2.25%

Volume Correction Apparent to 60° F. Vol.

What the regulations do not go to great lengths to explain is that the Table 7 procedures can be adapted to take 60°F volumes and predict the present volume of alcohol. This is useful when a blend has been prepared and the 60°F volumes have already been determined but now one must go out into the real world and determine how many gallons at present temperature to withdraw in order to transfer the correct 60°F volume into the blending tank. To accomplish this, we need to make the correction value go the other way or to reverse it.

Suppose a distiller has prepared a blend that requires 985.1 gallons at 60°. Going to the tank yard, the distiller finds that the temperature of the tank is 93°F and it is already known that this tank contains 97-proof alcohol. The following worksheet solves for how many gallons of alcohol to withdraw in order to accomplish the blend:

What is the True Proof @ 60°F. Already Determined.	97.00	Must know True Proof	
What is the Volume @ 60° F. Already determined	985.10	3,729.01	Liters
What is the present Temperature of the Alcohol	93.00	0.0000	Therm Cor.
Correction Factor to be applied for this Temp.	1.0149		
Present Volume of 60° F. Wine Galons	999.78	3,784.57	Liters
Wine Gallon Difference	14.678	55.56	Liters

True Proof divided by 100	0.9700
60° F. Wine Gal. * (Proof / 100) = Proof Gallons	955.55
Proof Gallons Obscured by temp. Dif. from 60 °F.	-14.24

Volume Correction from 60° Vol to Present Vol.

The answer turns out to be just about converse of the previous example indicating that one should remove 999.78 US Gallons from the storage tank at this temperature in order to obtain 985.1 gallons at 60°F. Table 7 is not a perfect volume correction algorithm and is off by 0.22 fluid gallons in this example, but it is eminently serviceable for commercial purposes.

The procedure to do this is to calculate the factor as if one was determining the 60°F volume (already determined to be 985.1 US gallons). Subtract the Table 7 correction factor (0.9851) from one; add this difference (0.0149) to one; the result (1.0149) will be the factor required: 985.10 * 1.0149 = 999.78 US gallons:

Correction for Odd Temp and intermediate Proof	0.9851
True Proof @ 60° F. Volume Restated	985.10
Adjustment needed to correct table 7 to read the other way	1.0000
Difference of Correction Factor from 1	0.0149
Reversed Correction Factor	1.0149
Present Volume of 60° F. Wine Galons	999.78

Including a true-proof correction and a second volume correction in a worksheet allows one to go to the tank yard, simultaneously obtain a true proof and, by adjusting the present volume required, obtain the correct amount to meter out of the tank in order to obtain the correct volume required for the blend at the present temperature:

Present Measured Wine Gallons in Tank	1,000.00	3,785.41	Liters
Present Temperature of the Alcohol	93.00	0.0000	Thermometer Cor.
Apparent Proof of the Alcohol	110.50	0.0000	Hydrometer Cor.
True Proof @ 60°F. of the Alcohol	97.00	955.55	Proof Gallons
Temperature at which it is desired to know Volume Temp. that Volume is to be corrected to in ° F.	60.00		
Factor from Present Vol. to Desired Temp/Volume	0.9851		
Present Wine Gallons Corrected to Desired Temp	985.10	3,729.01	Liters
Desired Temp Wine Gallon Dif. from Present Volume	14.90	56.40	Liters

Volume Correction Any Temperature

In order to determine tank capacity room must be left for expansion. In bottling this is known as *headspace*, the proportion above the fill line in the bottle to account for expansion and contraction.

Suppose we have a 10,000-gallon tank and we want to load it with 190-proof alcohol. We need to calculate what will happen when the temperature goes up to 100°F on a summer day.

Present Measured Wine Gallons in Tank	10,000.00	37,854.12	Liters
Present Temperature of the Alcohol	60.00	0.0000	Thermometer Cor.
Apparent Proof of the Alcohol	190.00	0.0000	Hydrometer Cor.
True Proof @ 60°F. of the Alcohol	190.00	19,000.00	Proof Gallons
Temperature at which it is desired to know Volume Temp. that Volume is to be corrected to in ° F.	100.00		
Factor from Present Vol. to Desired Temp/Volume	1.0240		
Present Wine Gallons Corrected to Desired Temp	10,240.00	38,762.62	Liters
Present Temp Dif. From Desired Temp Wine Gallons	-240.00	-908.50	Liters

Volume Correction Any Temperature

If that were to happen, and the tank was full to begin with, the tank would overflow by 240 gallons, so it would be important to leave 250 gallons of headspace in the tank.

One last use of the *Volume Correction to Any Temperature* procedure is to find out how much space can be saved in packing alcohol into a container. If one had a tank car that held 100,000 volumetric gallons and we wanted to load the car with as many gallons of 190-proof alcohol as possible, we would determine how much space could be saved by cooling the alcohol down. We would adjust the present measured wine gallons upwards and the temperature downwards from 60°F to determine how much space could be saved. We will assume it is still 93°F in the tank yard but that the alcohol when proofed is 190 proof:

Present Measured Wine Gallons in Tank	103,250.00	390,843.77	Liters
Present Temperature of the Alcohol	93.00	0.0000	Thermometer Cor.
Apparent Proof of the Alcohol	197.45	0.0000	Hydrometer Cor.
True Proof @ 60°F. of the Alcohol	190.00	192,349.59	Proof Gallons
Temperature at which it is desired to know Volume Temp. that Volume is to be corrected to in ° F.	40.00		
Factor from Present Vol. to Desired Temp/Volume	0.9685		
Present Wine Gallons Corrected to Desired Temp	99,997.63	378,532.19	Liters
Desired Temp Wine Gallon Dif. from Present Volume	3,252.38	12,311.58	Liters

Volume Correction Any Temperature

If we had a chiller for our 93°F alcohol, we could load an extra 3,252 gallons into the tank if they were chilled to 40°F. If tax was paid by the volumetric gallon as it left the plant, there would be a strong incentive to invest in tank chillers during transport. You'd still have to leave some headspace in the tank, but the principal is sound.

Volume Correction with Sight Glass

Some distilleries are fortunate in having sight glasses that show the level of alcohol in the tank itself. By incorporating the height in inches and the wine gallons per inch in the worksheet one can determine how many proof gallons are in the tank at any given time. This worksheet tracks the §186.51-§30.51 Example 1: [26]

What is the Temperature of the Alcohol	72.00	0.000	Thermometer Cor.
What is the Apparent Proof of the Alcohol	92.00	0.000	Hydrometer Cor.
True Proof @ 60°F. of the Alcohol	86.80		
Height of Fluid in Tank in Inches	88.00		
Wine Gallons per inch	48.96		
Inches * W.G. per inch = Present Gallons in Tank	4,308.48	16,309.37	Liters
Volume Correction Factor	0.9950		
Wine Gallons * Correction factor = 60° F. Gal.	4,286.94	16,227.82	Liters
60° F. Wine Gallon Dif. from Present Volume	21.54	81.55	Liters
True Proof divided by 100 value	0.8680		
Corrected Gallons * Proof / 100 = Proof Gallons	3,721.06		
Proof Gallons Rounded to 1/10 proof	3,721.10		

Vol. to P.G. T7 Sight Glass

This function assumes that the distiller has a sight glass associated with the tank they are using and that the tank has been calibrated in wine gallons per inch of height in the tank.

---

[26] §30.51 "Procedures for measurement of bulk spirits. §186.51 Where the quantity of spirits (including denatured spirits) in bulk is to be determined by volume as authorized by this chapter, the measurement shall be made in tanks, by meters as provided in 27 CFR part 19, or by other devices or methods authorized by the appropriate TTB officer, or as otherwise provided in this chapter, or such measurement may be made in tank cars or tank trucks if calibration charts for such conveyances are provided and such charts have been accurately prepared, and certified as accurate, by engineers or other persons qualified to calibrate such conveyances. Volumetric measurements in tanks shall be made only in accurately calibrated tanks equipped with suitable measuring devices, whereby the actual contents can be correctly ascertained. If the temperature of spirits (including denatured spirits) is other than the standard of 60 degrees Fahrenheit, gallonage determined by volumetric measurements shall be corrected to the standard temperature by means of Table 7. In the case of denatured spirits, the temperature-correction factor for the proof of the spirits used in denaturation will give sufficiently accurate results, except that the temperature-correction factor used for specially denatured spirits, Formula No. 18, should be that given in Table 7 for 100 proof spirits. When the quantity of spirits, in wine gallons, has been determined by volumetric measurement, the number of proof gallons shall be obtained by multiplying the wine gallons by the proof of the spirits as determined under §30.31. *Example* Gauge glass reading inches 88. Wine gallons per inch: 48.96. Temperature °F: 72. Proof of spirits: 86.8. Temperature correction factor (Table 7): 0.995. 48.96 W.G.×88=4308.48 wine gallons. 4308.48 W.G.×0.995=4286.94 wine gallons. 4286.94 W.G.×0.868=3721.06392=3721.1 proof gallons."

For those who do not have such a tank, it is possible to find the wine gallons per inch and then use a dip stick to substitute for the sight glass reading. You can then measure tank volumes using the level on the dip stick if one knows the volume per inch in the tank.

The stick itself will expand and contract with temperature and you might want to find out the coefficient of expansion for the material used. I think wood has a low coefficient of linear expansion with temperature change and that is why it is commonly used. Stainless steel may not be the best choice from a linear expansion accuracy standpoint.

Volume Correction of Blend Water

While water expands and contracts far less than high-proof alcohol it is important to perform a volume correction on water that is used in a batch since the effect can exceed the tolerance for the batch:

What is the Temperature in ° F of the Water to be Used	75.00		0.0015	Tolerance
Wine Gallons @ 60° F. Required for Batch	1,000.00		1.500	Gallon Tolerance
Factor to Obtain 60° Gallons	0.9980			
Amount Present Temp Gal. Differ From 60° F. Gal.	998.00			
Extra (Shortage) of Gallons to meter into batch. If Positive Wine Gallons will Shrink to reach 60°F If Negative Wine Gallons will "Grow" to reach 60° F	2.00			Meter More or (Less) to achieve Target Value
Net Gallons to Meter At current Temp	1,002.00			Gallons to Meter at the Current Water Temperature

Volume Correction for Blend Water

Volume Corrections When Bottling

Another volume correction problem that occurs is during bottling operations. After blending we must leave the ideal 60°F world of the tables and go back to the real world. The batch was blended either by weight or volume to a volume as if all components were 60°F after they were combined. We've discussed the exothermic reaction that increases temperature when combining alcohol and water and in many cases the bottling temperature will be above the 60°F standard simply because working conditions require a comfortable temperature in the shop (although I've bottled on cold days when the tank was below 60°F, as well).

Let's assume that the blended temperature is 75°F at the time of bottling for our standard batch of 102 cases containing 1,224 bottles at 750 ml each:

What is the present Temperature of the Batch	75.00	0.0000	Thermometer Correction		
True Proof of Blend	80.00				
Blended 60° F. Volume of Batch	242.51				
Correction Factor to be applied for this Temp.	1.0055			0.00150	Tolerance
Present Gallon Volume Differential from 60° F. Vol.	1.33			1.13	ml Tolerance
Total Present Volume	243.84				
Proof Gallons	194.01				
Other Results					
Present # of Liters Difference from 60° F. Liters	5.05			102.00	Cases Planned
Bottle Size in milliliters	750.00			12.00	Bottles per Case
Present # of bottles or Excess or (Short)	6.73			1,224.00	Total Bottles
% of Bottling Run	0.55%			5.50	ml per Bottle Off

Volume Correction for Bottling

The result is that if we fill to a line that indicates "750 ml" on each bottle, each bottle will contain 5.5 milliliters less alcohol than we intended when assuming that bottle was 60°F. Cumulatively that will amount to 6.73 more bottles coming off the line than planned.

When our bottles sit on the shelf in the store at 60°F (or 68°F for that matter), their fills will contract compared to the bottling temperature and will look low. Customers will not purchase them because there is too much headspace above the fluid in the bottle. In this case we should set our bottling machine so that all the "fills" are high by 5.5 milliliters to achieve the desired 60°F volume in each bottle. Even though this is a positive number, since we are measuring present volume compared to intended volume, it is in fact negative with respect to what we are trying to achieve.

This effect is hard to see, except on the bottle itself, because usually this temperature-induced gain in processing is completely swamped by the general loss in processing that occurs through rejected bottles or leakage, spillage and the alcohol that accumulates in filters and hoses and is not recovered.

Fortunately, as I mentioned before §5.47a of the regulations recognizes that bottling is a commercial process and "good commercial practice" defines what is reasonable.[27]

Volume from Apparent Proof and Weight

There is a clever way to determine present volume without having to perform a Table 7 volume correction by using the pounds per wine gallon value from Table 4. The drawback is that one must know the weight of the alcohol involved.

Traditionally Table 4 has been used to calculate the 60°F volume of alcohol with the following formula:

Weight * Wine Gallons per pound at True Proof (60°F) = equals the 60°F volume.

In measuring present volume, it is possible to measure the apparent proof and use the corresponding wine gallons per pound to obtain the present volume with the following formula:

Weight * Wine Gallons per pound at the Apparent Proof = Present Volume.

Starting from a known weight, the only value that will fluctuate is the volume. The only variable that will affect the volume is the temperature.[28] Temperature affects density and since our hydrometer reads density it is the instrument to use. After finding the present density we can use Table 4 or Table 5 to tell us how much volume a pound of that density material occupies at the present temperature. What we cannot learn from this procedure is the alcohol content of the mixture. To do that we would need to fix both temperature and density and then use §183.23 to find the true proof at 60°F.

For just finding volume without regard to alcohol content we can take the apparent proof and make it an input rather than obtaining it as a result and go from that value through Table 4 or Table 5 to obtain the volume per pound of the apparent density. By eliminating alcohol content as a variable to be fixed and only focusing on weight and present density one can find the "wine gallons, at the prevailing temperature, of most liquids within the range of the Tables" according to the regulations.[29, 30]

---

[27] §5.47a "(1) Discrepancies due to errors in measuring which occur in filling conducted in compliance with good commercial practice. [Author Note: i.e. at a comfortable working temperature.] And (3) Discrepancies in measure due to differences in atmospheric conditions in various places and which unavoidably result from the ordinary and customary exposure of alcoholic beverages in bottles to evaporation."

[28] Other variables that can affect the measurement are the thermal expansion/contraction of the hydrometer, the atmospheric pressure on the hydrometer stem, the surface tension of the alcohol clinging to the stem, and perhaps local gravity differences from Standard *g*.

[29] §30.62 §186.62 "Table 2, showing wine gallons and proof gallons by weight. ... In addition, this Table may be used to obtain the wine gallons, at the prevailing temperature, of most liquids within the range of the Table, from the weight of the liquid and the uncorrected reading of the hydrometer stem. An application of this would be in determining the capacity of a package." [Author note: ABS uses Table 4 Quotient values rather than Table 2 to perform calculations Because Table 2 is in one-fourth pound increments it is cumbersome to use and difficult to program.]

[30] This method is also referenced in §30.65 §186.65 dealing with Table 5: "This Table also shows the weight per wine gallon (at the prevailing temperature) corresponding to each uncorrected reading of a proof hydrometer."

Applying this method of finding volume from apparent proof to our standard batch at the time of bottling shows several interesting features. Because the bottling temperature is 75°F we have to change the apparent proof to 86.6 to obtain a true proof of 80 for our batch. Using this apparent proof of the batch and the known weight of the batch yields a present volume of 243.85 via Table 7 and 243.89 using the apparent wine gallons per pound value from Table 4 to find the present volume:

Using Apparent Proof and Weight to find Present Volume.

Temp of Alcohol	75.00	0.000		Thermometer Cor.
Apparent Proof of Alcohol	86.60	0.000		Hydrometer Cor.
True Proof Result	80.0		0.0015	Tolerance
Weight in Pounds	1,922.22		0.300	Tolerance as ° of Proof
Weight of Container (Tare) if needed	0.00		2.88	Batch Weight Tolerance
Net Weight for Calculation	1,922.22		0.364	60° F. Alcohol Vol. Tolerance
Find Present Volume from Apparent Proof and Weight				
Wine Gallons Per Pound @ apparent proof	0.12688		7.88161	1/WGPP = Lbs. per Wine Gal
Lbs. * W.G. per Lb. @ Apparent Proof = Present Wine Gal.	243.89		243.89	Lbs. / Lbs. per WG = Present Gal. Vol.
Compare to Table 7 Volume Correction				
Wine Gallons Per Pound at True Proof	0.12616		7.92626	Lbs. per Wine Gal. at True Proof
Weight * Wine Gallons per Lb. = 60° F. Wine Gallons	242.513		242.513	Wt./ Wine Gal. per Lb. = 60° F. Gal.
Table 7 Correction factor from 60°F. Vol to Present Vol.	1.0055			
Factor * 60° F. Wine Gallons = Present Wine Gallons	243.85		1.33	Present Vol. Dif. From 60° F. Vol.
Difference in Gallons	0.040			
Percent of Tolerence	11.04%			

Apparent Proof and Weight to Present Vol.

In this case there is very close agreement (0.04 gal.) between the method of finding true proof and doing a Table 7 volume correction and using the apparent proof and the wine gallons per pound at that apparent proof. The true proof correction is only needed for the Table 7 comparison and not for the §186.62-§30.62 procedure.

From this we can determine that in general a volume calculated from weight and wine gallons per pound at the apparent proof is as good a predictor of present volume as a Table 7 volume correction procedure applied to weight and true proof.

§30.41 Bulk spirits. §186.41 Determination of Quantity by Weight

This regulation contains several useful procedures. Part (a) describes a method of finding present volume and part (b) describes a method of determining wine gallons per pound from specific gravity readings. While the regulation specifies that the procedures are to be used with obscured alcohol above 600 milligrams per 100 milliliters, the procedures also work well for unobscured alcohol.

Part (a) takes the apparent proof and the wine gallons per pound at the apparent proof and combines them with a known mass to arrive at the present volume of the alcohol just as we have discussed above using the §30.62 procedure.

The second procedure (b) describes how to use the true proof SG and the wine gallons per pound of water to find the wine gallons per pound of alcohol. Notice that I have calculated the wine gallons per pound in vacuum at 60°F to find the vacuum density value as well.

	Air	Vac
Find Pounds per Wine Gal. per Lb. using §30.41		
SG 60/60 F.	0.95174	0.95180
Wine Gal. per Pound of Water	0.120074	0.119946
SG/WG per Lb. H2O = Lbs. per Wine Gal.	7.92628	7.935229
Lbs. per Wine Gal. Difference from T4 Value	0.0000	
Find Wine Gal. per Lb. using §30.41		
SG 60/60 F.	0.95174	0.95180
Wine Gal. per Pound of Water	0.120074	0.119946
WG per Lb. H2O/SG = Lbs. per Wine Gal.	0.126163	0.126020

Apparent Proof and Weight to Present Vol.

This demonstrates that one can use the density of water and specific gravity to translate into wine gallons per pound and thence into pounds per wine gallon accurately.[31]

---

[31] §30.41 Bulk spirits. §186.41 "When spirits (including denatured spirits) are to be gauged by weight in bulk quantities, the weight shall be determined by means of weighing tanks, mounted on accurate scales. .... From the weight and the proof thus ascertained, the quantity of the spirits in proof gallons shall be determined by reference to Table 4........ by: (a) Use of a precision hydrometer and thermometer, in accordance with the provisions of §30.23, to determine the apparent proof of the spirits (if specific gravity at the temperature of the spirits is not more than 1.0) and reference to Table 4 for the wine gallons per pound, or (b) Use of a specific gravity hydrometer, in accordance with the provisions of §30.25, to determine the specific gravity of the spirits (if the specific gravity at the temperature of the spirits is more than 1.0) and dividing that specific gravity (corrected to 60 degrees Fahrenheit) into the factor 0.120074 (the wine gallons per pound for water at 60 degrees Fahrenheit)."

Operational Procedures

Gauging Drums of Alcohol

Using the §30.63 and §30.64 procedures it is possible to make up a worksheet that will assist you in gauging alcohol. When I receive alcohol at my distillery it arrives in 55-gallon drums. When I started out in 1998, they were sometimes metal drums lined with Teflon or some alcohol-resistant substance; subsequently all the grain neutral spirits suppliers began to use plastic drums. When the drums arrive, I have to put the alcohol, in the form of proof gallons, on my books. If I get 10 drums of alcohol and have to proof it all, that can take most of the afternoon. My procedure is to look at the lot numbers of the drums. If they come from different lots, then each lot will have to be checked. Even if the lot numbers are the same, I will check all the drums to make sure they actually contain alcohol but only proof three of the 10 drums to obtain a true proof. The supplier will not tell you how many proof gallons they are sending you, only that you are receiving 10 drums of 190-proof minimum USP beverage grade alcohol. Some suppliers are now actually delivering almost absolute alcohol near to 200 proof to save shipping the small amount of water contained in 190 proof. This is surprising because one cannot directly distill alcohol above 194.37 proof; at that proof the mixture becomes an azeotrope (a mixture of two or more liquids that when boiled give off vapors of exactly the same composition as the liquid) and further separation is not possible. In order to create pure absolute alcohol, one must perform an azeotropic distillation that uses a third chemical such as benzene or trichloroethylene to bond with the remaining water so pure alcohol can be drawn off separately. Since these chemicals are quite toxic, I'm concerned about what process is used to remove them prior to certifying the alcohol as beverage grade.

At a minimum each full drum will have to be weighed and a best estimate of the tare made. If you can arrange to empty one drum by blending or bottling close to the time it is received you will at least have one actual tare to apply to all the drums received. For establishing the correct tare, both drum caps should be left on the drum.

With those limitations in mind, following is the *Gauging Drums* worksheet that you can use to proof your alcohol and measure the proof gallons. The example is taken from two drums I received:

	Drum 1	Drum 2	Drum 3	Drum 4
Drum Gross Weight	165.00	475.20	475.00	474.60
Weight of Drum Empty ("Tare")	21.40	21.40	21.40	23.40
Fluid Weight in Drum	143.60	453.80	453.60	451.20
Thermometer Cor. Factor	0.000	0.000	0.000	0.000
Hydrometer Cor. Factor	0.000	0.000	0.000	0.000
Temperature of Alcohol 1	56.70	57.40	56.40	57.20
Apparent Proof of Alcohol 1	189.20	189.20	189.30	189.10
True Proof of Alcohol 1	190.00	189.80	190.20	189.80
Proof Gallons in Drum	40.16	126.72	127.05	125.99
Wine Gallons in Drum @ 60° F.	21.14	66.76	66.80	66.38

Multiple Drum Results				
Number of Drums	1.00	1.00	1.00	1.00
Total Proof Gallons	40.16	126.72	127.05	125.99

Single Drum Results	Fluid Weight Restated	Proof Gallons Per Lb.	Lb's * P.G. Per Lb. = Proof Gal.	Wine Gallons Per Pound	Weight * W.G. Per Lb. = 60° F. Volume	% of W.G. to total Vol	% of P.G. to total P.G.
Container of Alcohol 1	143.60	0.279660	40.16	0.14719	21.14	9.56%	9.56%
Container of Alcohol 2	453.80	0.279234	126.72	0.14712	66.76	30.21%	30.18%
Container of Alcohol 3	453.60	0.280087	127.05	0.14726	66.80	30.20%	30.26%
Container of Alcohol 4	451.20	0.279234	125.99	0.14712	66.38	30.04%	30.00%
Total Weight of Alcohol	1,502.20	Total PG	419.91	Total Vol.	221.08	100.00%	100.00%

Gauging Drums of Alcohol

If you have multiple drums that all weigh the same (say within 0.2 or 0.3 lbs.), then one can simply multiply the proof gallons of one drum by the number being received into storage. The *Storage Report* worksheet has such a procedure built into it.

I've found that grain neutral spirits do not keep more than a year in a metal drum, even lined, without picking up some taste from the drum; plastic drums will also impart taste to the alcohol if left in too long. It is best to have your GNS supply arrive and be used within six months or so of arrival. I wouldn't suggest buying a two-year supply because if it starts to acquire an off taste there is no way to remove it.

One nice aspect of obtaining GNS is that I've always received higher proof alcohol than the 190 minimum the supplier guarantees. Usually the true proof is between 190.5 and 191.5. The problem with gauging the drums of alcohol received is that you don't know the weight of the drums themselves. Also, drums vary in weight from drum to drum. Of course, once a drum is emptied it is possible to weigh the drum and use its weight as a reference (if the others are the same type). Plastic drums usually don't vary by more the 0.4 pounds between them. If you use the same supplier over and over, you will usually get the same drums as well and you can use an average drum weight as the tare to gauge the alcohol. What one does not want to do is to underestimate the weight of the drum and put proof gallons on your books that don't exist. Nor do you want to assign too high a proof to the shipment since that will do the same thing. Accordingly, I adopt a conservative approach to proofing and drum weights when receiving proof gallons into my bonded premises.

Withdrawing Proof Gallons by Weight

There are several other operational procedures similar to gauging drums that we can cover in short order.

Since we are planning our batches by the number of proof gallons and moving them around for purposes of blending, it is helpful to have a means of determining the mass or volume required.

Let's look at the *Weight to Proof Gallon with SG* worksheet using the first example given in §186.64-§30.64, which envisions a tank car containing 81,000 lbs. of alcohol at 190 proof, and obtain the proof gallons in the tank. "*Example.* It is desired to ascertain the wine gallons and proof gallons of a tank of 190-proof spirits weighing 81,000 pounds." 81,000 × 0.14718 = 11,921.58 = 11,921.6 wine gallons. 81,000 × 0.27964 = 22,650.84 = 22,650.8 proof gallons.

Using Weight to find Proof Gallons and Wine Gallons							
Temperature of Alcohol	60.00	0.0000	Thermometer Cor.		0.0015	Tolerance	
Apparent Proof of Alcohol	190.00	0.0000	Hydrometer Cor.		33.98	Proof Gal Tol.	
True Proof Result	190.00				17.88	Vol. Tolerance	
Weight in Lbs.	81,000.00						
Weight of Container (Tare) if needed	0.00						
Net Weight for Calculation	81,000.00		Second Method				
			8.328201	Weight Constant for 1 Gal. of Water			
			0.81577	Specific Gravity in Air of Alcohol			
Wine Gal. per Lb.	0.14719		6.79388	Water Wt. * SG of True Proof = Lbs. per Gal.			
Weight * Wine Gal. per Lb. = Wine Gal.	11,922.50		11,922.50	Wt. * Lbs. per Gal = 60° F. Wine Gal. Result			
Proof Gal. per Lb.	0.27964		1.900	True Proof/100			
Weight * Proof Gal. per Pound = Proof Gal.	22,650.84		22,652.75	Times Wine Gallons = Proof Gallons			
Correction Factor to TTB Gal.	0.9999230						
TTB US Fluid Gallons	11,921.58						

Weight to Proof Gallons & Wine Gal. SG

The proof gallons come out almost exactly because we are using weight and the original TTB proof-gallon-per-pound values. This is important for taxation that we retain these values as close as possible.

A good deal of this book will be devoted to how and why I have adjusted the density values at 15.556°C (60°F) in air to conform to the modern definition of a liter and consequently a US gallon. We note in this case that the gallon value calculated by the modified Table 4 value is 11,922.50 using a value of 0.14719, where the traditional TTB Table 4 value is 0.14718 resulting in 11,921.58 gallons for a difference of essentially one gallon out of ~12,000. This is well within the tolerance of 17.88 US gal. (5% of tolerance) and the benefits of being able to work conveniently in vacuum densities that convert easily and reciprocally to air as well as relating to modern published densities of both water and of alcohol and water mixtures offset the small discrepancies with the original Table 4 values.

For some reason one is not able to correct this issue using 0.999973 cm^3 for the definition of a liter in effect pre-1964 because it is at 4°C and it would not work. By the same token one cannot use 1.000027 to correct the other way. Also, we are dealing with air densities rather than vacuum ones, which further complicates the issue. I tried using the Table 37 density correction procedure to change the density from 4°C to 15.556°C and conversely to see if either parameter would affect the correction. Neither was effective. Finally, I hand-tuned a correction factor of 0.999923 that will correct to the TTB US fluid wine gallons. Since we are using mass in this example this correction is not strictly necessary but can be applied elsewhere to greater effect when we are using volume.

I later learned that the definition of the inch had changed in 1930 when the Industrial Inch was adopted. This changed the cubic inches per gallon for any container manufactured after the widespread adoption of the Industrial Inch (which is now ubiquitous). The effects of this change are more fully explained in the Gallon Discrepancy section.

This worksheet also shows a secondary method of finding proof gallons using SG air and the density of water in pounds per gallon. There are also differences caused by using Table 5 values rather than Table 4 of about the same magnitude as for my modified Table 4 values.

Now we can apply the procedure from §186.65-§30.65 to reverse the calculation to obtain the weight of alcohol to withdraw when we want to employ it for blending:

Withdrawing Proof Gallons by Weight

Number of Proof Gallons to Withdraw	22,650.84			0.0015	Tolerance
Temperature of Alcohol	60.00	0.000	Thermometer Cor.	0.300	Tolerance as ° of Proof
Apparent Proof of Alcohol	190.00	0.000	Hydrometer Cor.	$13.50	Excise Tax Rate
True Proof Rounded to 1/10 Proof	190.00			$305,786.34	Excise Tax Payable
Lbs. per Proof Gallon 1/T4 Value	3.57603	0.27964	Proof Gal. per Lb.	17.88	Gallons of Tolerance
Proof Gal. * Lbs. per = Lbs. of Alcohol	81,000.00	81,000.00	Proof Gal./Prof Gal per Lb.	121.50	Lbs. of Tolerance
				33.98	PG of Tolerance

Lbs. Per Wine Gal T4 Quotient	6.79388		
Lbs./ Lbs per WG = 60° F. Vol. of Alcohol	11,922.50	45,131.58	Liters
Vol. Cor. Factor for Present Temp	1.0000		
Present Volume of 60° F. Wine Galons	11,922.50	45,131.58	Liters
Gallon Difference In Volume	0.00	0.00	Liters

Correction Factor to TTB Gal.	0.9999230
TTB US Fluid Gallons	11,921.58

Withdraw Proof Gallons.

The weight comes out just fine but we will in fact be taking out one more gallon of ~12,000 because the definition of a liter (dm^3) has changed and the inch has also changed since the tables and volume values were created. With 17.88 US gallons of tolerance in this worksheet the effect is about 5.6% of the tolerance when dealing with volume.

Wine Gallons to Weight & Proof Gallons

Similarly, we can use the volume of alcohol, find the true proof and then determine the mass and proof gallons. This procedure is set forth in §186.65-§30.65:

Using Wine Gallons and Proof to find Weight and Proof Gallons of Alcohol					
Temperature of Alcohol	60.00	0.0000	Thermometer Cor.		
Apparent Proof of Alcohol	190.00	0.0000	Hydrometer Cor.		
True Proof of Alcohol	190.00				
Volume of Alcohol at Present Temp	11,922.50		0.0015	Tolerance	
Correction Factor applied to Volume	1.0000		0.300	Tolerance as ° of Proof	
Volume of Alcohol corrected to 60° F.	11,922.50		17.88	Gallons of Tolerance	
Present Volume Difference from 60°F of Alcohol	0.00		121.50	Lbs. of Tolerance	
			33.98	PG of Tolerance	
Pounds Per Wine Gallon @ 1/10 th proof	6.79388		$13.50	Excise Tax Rate	
60° F. Vol. * Lbs. per Wine Gal. = Lbs.	81,000.00		$305,786.34	Excise Tax Payable	
Proof Gal. per Lb. at True Proof	0.27964				
PG per Lb. * Lbs. = Proof Gallons	22,650.84				

Wine Gallons to Weight & Proof Gallons

We see that the operation is reciprocal to the one just above in that we can take the volume of alcohol and by using the pound per wine gallon derive the same total pounds under consideration and then using the proof gallons per pound arrive at the same number of proof gallons.

Fill by Weight

§186.44-§30.44[32] explains how to fill a container by the weight of material needed. To fully exploit the procedure, it is helpful to add a volume correction algorithm to ensure that the container is not overfilled. In this case we have scaled up the procedure to 100 gallons and changed the temperature to see the effect. The result is that if we just add the weight indicated at the present temperature that we will overfill the container by 1.2 gallons. Accordingly, I have included a margin of error module that will allow for reserving a portion of the intended fill and the margin can be adjusted to exceed the overfill:

Gallon Size of Container	100.00	Gallons
Temp of Alcohol	80.00	0.0000 Therm. Cor.
Apparent Proof of Alcohol	198.20	0.0000 Hyd. Cor.
True Proof Of Alcohol	194.00	
Wine Gal Per Lb	0.14867	
Pounds of Fluid Needed	672.63	Gallons/W.G. Per Lb.
Find Pounds and Ounces		
Fractional Pound	0.63	
Ounces Per Pound	16.00	
Fractional Pounds * Ounces	10.02	Ounces
Round Up or down to whole ounce	10.00	
Total to Fill in Pounds	672.00	Pounds
Plus Ounces	10.00	Ounces
Pounds Avdp to Kilograms	0.45359237	
Kilograms of Alcohol Needed to Fill.	305.098	

Volume Correction for this Temperatrue		Alcohol in this Capacity Container	Find Grams of fill for Bottles	
Present Volume of Alcohol	101.20		Size of Container in liters	0.7500
Correction Factor applied	0.9880		Gallons per Liter §19.722	0.2641720
Present Volume corrected to 60° F.	98.80		Size of Container in US Gal.	0.1981290
Present Volume Difference from 60°F	1.20		True Proof Of Alcohol	80.00
			Wine Gal Per Lb	0.12616
Optional Margin of Error to reserve in case of overfill			Gallons/W.G. Per Lb. = Lbs.	1.570421
Margin of Error %	2.00%	Percent	Pounds Avdp to Kilograms	0.453592370
Pounds of Margin of Error	13.44	Reserve this Wt.	Result= Kilograms of Alcohol Fill.	0.712331
Pounds minus margin of error	659.19	First Fill Wt.	Grams of Alcohol (Air)	712.331
Fill this 60° F. Volume First	98.00		Margin of Error %	1.00%
Reserve this 60° F. Volume	2.00		Reserve this Wt. for "Topping Off"	7.12331
Total 60° F. Volume Cross check	100.00		Fill This Weight First	705.20759

Fill by Weight

Also demonstrated is the method to calculate the mass of alcohol in air that one can use to fill any container using liters and grams along with a margin of error in grams.

What this sheet demonstrates first is that the present volume is greater than the capacity of the container and will always be so as long as the temperature is above 60°F. Therefore, we need to reserve a certain volume to account for this effect. I really haven't examined the effects of filling with cold alcohol and then having it try to

---

[32] § 30.44 Weighing containers. §186.44 "(c) Containers of other proofs or sizes. Where containers of proofs or sizes not shown above are to be filled, the following rule may be used for ascertaining the weight of the spirits to be placed in the container: Divide the number of gallons representing the quantity of spirits to be placed in the container by the fractional part of a gallon equivalent to 1 pound, to obtain the weight of the spirits in pounds and fractions of a pound to two decimal places. Reduce the decimal fraction of a pound to ounces by multiplying by 16, calling any fraction of an ounce a whole ounce. The pounds and ounces thus obtained will determine the point to which the spirits must be weighed to produce the results desired. If the weight must be marked on the container in pounds and decimal fractions of a pound, it will be necessary to convert the ounces to hundredths of a pound. The fraction of a gallon equivalent to 1 pound at any given proof shall be ascertained by reference to Table 4. However, if the spirits contain solids in excess of 600 milligrams per 100 milliliters, the fraction of a gallon equivalent to 1 pound shall be determined as prescribed for such spirits in §30.41. Example. It is desired to fill a 1-gallon can with precisely 1 wine gallon of 194 proof spirits:1.00 divided by 0.14866=6.73 pounds. 0.73 multiplied by 16=11.68 ounces, rounded to 12 ounces. Weight of spirits-6 pounds, 12 ounces. Weight, if required, to be marked on can-6.75 pounds."

expand inside the container. This would create pressure that might exceed the capacity of the container to contain it.

Capacity of Container

§186.67-§30.67 This example uses a Table 7 volume correction to find the capacity of a container at 60°F. I've scaled up the example by a factor of 10 from 55.1 gallons to 551 gallons in order to see the effect more clearly. Using the TTB *Example 2*: "It is desired to ascertain the capacity of a container of 190-proof spirits at 76 degrees Fahrenheit, shown by Table 2 to contain 55.1 wine gallons at 60 degrees Fahrenheit: 55.1 divided by 0.991 equals 55.6 wine gallons, the true capacity of the container when filled with spirits of 60 degrees temperature."

Capacity of Container when True Proof & 60° F. Volume are Known					
60° F. Volume	551.00	2,085.76	Liters		
What is the Present Temperature of the Alcohol	76.00	0.0000	Therm. Cor.		
What is the True proof @ 60° F of the Alcohol	190.00	Known True Proof			
Correction Factor to be applied	0.9910				
Capacity of Container in wine gallons when filled with 60° F. Spirits	556.00	60°F. Volume/Correction Factor			
Gallon Difference In Volume	5.00	18.94	Liters		
Find Proof Gallons					
Proof divided by 100	1.9000				
60° F. Wine Gallons * Proof / 100 = Proof Gallons	1,046.90				
Second Method to find Present Volume					
Volume at 60° F. & True Proof Restated	551.00		0.0015	Tolerance	
Adjustment needed to correct table 7 correction factor to read the other way	1.000		0.83	Gallons of Tolerance	
Difference of Correction Factor from # 1	0.0090		3.13	Liters of Tolerance	
Reversed Correction Factor	1.0090				
60° F. Wine Gallons * reversed correction factor	555.96				
Gallon Difference In Volume	4.96	18.77	Liters		
If Negative Wine Gallons will "Grow" to reach 60° F If Positive Wine Gallons will Shrink to reach 60°F					

Capacity of Container

What this shows is that dividing the 60°F volume by the correction factor produces the present volume which is also the "capacity of the container." The second method of reversing the correction factor induces a small difference in volume generated. However, the difference is well within tolerance.

One can also find the capacity of a full container from the apparent proof and weight using the §186.62-§30.62 method. This is the same procedure we discussed in the section on volume correction.

Accounting and Reports

Storage Report

The easiest way to track alcohol through the blending process is to start with the storage account. Using our standard two-alcohol-blend batch plan and blend we can withdraw those calculated proof gallon amounts from the account in order to keep track of the tax payable.

We will assume that we have on the bonded premises and in storage 300 proof gallons of corn whiskey and 900 proof gallons of grain neutral spirits (GNS). The corn whiskey comprises ~16% of the proof gallons required; the remainder will be the GNS at 190 proof:

Storage Report	Alcohol 1	Alcohol 2	Totals	
Beginning Proof Gallons	300.00	900.00	1,200.00	Line 1
Received during month	0.00	0.00	0.00	Line 2
Sub total	300.00	900.00	1,200.00	Line 6
Transferred to Processing	31.04	162.97	194.01	Line 17
Ending	268.96	737.03	1,005.99	Line 23
Total	300.00	900.00	1,200.00	Line 24
Cross Check	0.00	0.00	0.00	

Storage Report

Withdrawing the 31.04 proof gallons of corn whiskey and 162.97 gallons of GNS for a total of 194.01 proof gallons begins the process. The indicated line numbers are the lines of the TTB Storage Report. At the end of the month one can total all the withdrawals from storage, offset it with any receipts and file the report.

Bottling Log & Processing Report

Following our standard batch though a bottling run we can see the results of our efforts by looking at the bottling log.

We had planned to bottle 102 cases but, in this case, we have actually bottled 99 cases and 4 bottles for a gross operational loss of 2.61%. This is par for the course as far as my facility goes and occasionally I will have batches that are in the 3% to 5% loss range, which can attract the attention of regulators if it keeps up that way. I have found that average losses of 3.5% for bottling runs are apparently acceptable because many of my monthly reports end up at that level.

Proof Gallons of Alcohol 1 in Batch	31.04		80.00	What is the Blended Proof of the Batch
Proof Gallons of Alcohol 2 In Batch	162.97		12.0	How many bottles per case
Total Proof Gallons Deposited to Batch	194.01		750.00	What is the bottle fill amount in milliliters
Number of Saleable Cases Produced	99.00		9.000	Liters per Case
Proof Gallons Per Case	1.902076091		0.2641720	Liter to Gallon Converter
Proof Gallons in Saleable Cases	188.31		2.377548	60° F. Gallons Per Case
Remnant Good Bottles Produced	4.00		0.126160	Wine Gallon per Lb. @ Target Proof
Proof Gallons in Bottles	0.634		18.85	Wine Gal/W.G. per Lb. = Total Lbs.
Proof Gallons Cased Line 9 Bottled or packed	188.94		0.10093	Proof Gallons per Lb. @ Target Proof
Gross Proof Gallon (Loss) or gain in Processing	(5.070)		1.9020761	Lbs. * Proof Gal. per Lb. = Proof Gal. per Case
Gross Loss Percentage	2.61%		0.158506	Proof Gallons per Bottle

Bottling Log

The good news is that some of the 2.61% loss can be accounted for and the operational losses are less than the gross loss. My practice is to always set aside the first case that comes off the bottling line for close examination. Generally, the first alcohol out of the bottling machine does not meet the quality standards necessary for my product. Even with good cleaning and a good, work-hardened filter, the spirits that first flow out of the bottling line have an off taste. One has to get things moving through the system, and quickly, before the taste becomes acceptable. I also sample the second case off the line, only one small taste of one bottle, to see if that off taste has dissipated. If it has not, then we keep rejecting cases until it has gone away. The rejects are recycled back to the production account/report into the unfinished spirits or heads and tails account under "Part V - Used in Redistillation" section. The distiller can create several accounts, one for Heads and Tails and one for "Bottling Run Recycled". Up to eight accounts can be created and used in this part of the form. They will be added to the next double run for re-distillation:

Accounting For Loss that is still on premises		
Rejected and Recycled Cases	1.00	Reject Cases To Production Act.
Proof Gallons In rejected Cases	1.90208	
Rejected Bottles	2.00	Floaters or chips, flaws in glass - Bad Labels
Proof Gallons in rejected Bottles	0.32	
Total Rejected Cases and Bottle Proof Gallons	2.22	
Line 17 Total Transfer to Production for redistillation as unfinished spirits - Heads & Tails.		
Total Accounted For P.G.	191.16	
P.G. Loss restated as a positive number	5.070	
P.G. Loss - Rejects = Net Loss	2.85	Line 24 Losses in Part I Bulk Ingredients
Operational Loss For this Batch	1.47%	

In this run we only had to reject one case and two bottles and by recovering these proof gallons we have reduced our effective loss rate 1.47%.

We can also account for losses that are not recoverable but know what they consist of:

Broken Bottles	1.00
P.G. In Broken Bottles	0.159
Other Accounted for Losses in P.G.	0.00
P.G. In Filter and lines Lost/Destroyed	0.50
Accounted for P.G. Loss	0.66
Total Transferred or accounted for loss	2.88
P.G. Loss Minus Accounted for Loss = Net Loss	2.19
Operational Loss For this Batch	1.13%

Finally, it is my practice to simply withdraw the remnant bottles and pay tax on them rather than accumulate them in storage. One does this on line 13 of the storage report. I don't keep open cases on my premises for any reason and having one open case with 4, 6, 8, … bottles in it waiting for the next run means that I have to keep track of proof gallons in bottles as well as cases in my monthly reporting. If I also withdraw a case to use as samples or promotional purposes, I pay the tax on it as well. The net proof gallons then go into Part II of the processing account on line 29 as being "Received" into the finished products section:

Remnant Bottles Withdrawn Tax Determined	4.00		
Proof Gallons in Withdrawn Bottles	0.6340	$8.56	Tax on Withdrawn Bottles
Cases Withdwawn Tax Determined	1.00		
Total Cased Proof Gallons Withdrawn	1.90208	$25.68	Tax on Withdrawn Cases
Line 13 Total of Bottle and Case P.G. Withdawn	2.54		
Total Part I Tax on Withdrawn Proof Gallons	$34.24		Part I Tax on Withdrawn
Cases Remaining in Processing Part I to transfer	98.00		Cases from this batch
Proof Gallons in Cases Remaining in Processing	186.40		

Transfer these Proof Gallons to Part II Line 29 Received.

Bottling Log

Bottle Run Audit

After you do 4 or 5 runs in a month it can get a little hectic keeping track of all the proof gallons running through the shop. I've created a bottling run audit that helps to organize the data for the monthly processing report:

| | Run 1 | Run 2 | Run 3 | Run 4 | Run 5 | | |
	80 Proof 750	80 Proof 750	100 Proof 750	80 Proof 375	100 Proof 750	Total	
Alcohol 1	145.12	143.98	183.63	150.24	227.45	850.42	
Alcohol 2	26.06	26.44	33.72	27.59	33.73	147.54	
Total Proof Gal. Deposited	171.18	170.42	217.35	177.83	261.18	997.96	Total PG into Processing
Cases Produced	87.00	88.00	89.00	179.00	105.00	548.00	Cases
Proof Gallons Cased	165.48	167.38	211.60	170.23	249.64	964.34	Proof Gal in Cases
Extra Bottles	5.00	8.00	2.00	5.00	1.00	21.00	
Proof Gal in bottles	0.79	1.27	0.40	0.40	0.20	3.05	Proof Gal in Bottles
Total PG from Run	166.27	168.65	212.00	170.63	249.84	967.39	Total PG
Proof Gallon Losses By Run	4.91	1.77	5.35	7.20	11.34	30.57	Processing Losses
% Loss by run	2.87%	1.04%	2.46%	4.05%	4.34%	3.06%	% Loss of Total PG

Bottle Run Audit

Excise Tax, Storage & Processing

I try to bottle one type of whiskey at a time. It takes time to reset the bottling machine to 375-ml bottles from 750s and, when I bottle my Old Oak Recipe brown whiskey, I have to thoroughly rinse the machine and all hoses and pipes before I can go back to bottling clear moonshine. Also, I only keep one type of label at the distillery at any given time. I had a situation where I wasn't paying attention and sent someone to grab another roll of labels and he picked out the wrong ones. I was busy at my labeling machine and didn't check the other for an hour, at which point 10 cases had been bottled with the wrong label. The cases were already stacked and intermixed with the ones I'd been labeling. We had to go down layer by layer and find the bad cases until we were sure we were below the starting point. All the bottles had to be soaked in water for a day to get the labels off and then we had to start over. I say "we" but it was "me." So, lesson learned.

Diagram follows on the next page.

The following worksheet follows one month's production of 750-ml bottles of 100-proof alcohol through six runs, which is a big month at my shop. We will walk through the various parts of the form and recap how the reporting is handled:

2nd quarter Excise Tax on Production and Withdrawal				March Production		$13.50	Excise Tax Rate	
	Code #	Cases	Price	Gross	P.G. Per Case	P.G.	Tax	
Starting In Bond	1900	0.0	$67.00	$0.00	0.951031	0.00	$0.00	
	1901	31.0	$100.00	$3,100.00	1.902062	58.96	$796.01	
	1809	14.0	$110.00	$1,540.00	2.377530	33.29	$449.35	
	42	28.0	$94.00	$2,632.00	1.902062	53.26	$718.98	
	Total Cases	73.0		$7,272.00		145.51	$1,964.35	
				Rev in Bond		PG in Bond	Tax in Bond	
	Code #	Cases	Price	Gross	P.G. Per Case	P.G.	Tax	Wine Gal.
Produced this Period	1900	0.0	$67.0	$0.00	0.951031	0.00	$0.00	0.00
	1901	0.0	$100.0	$0.00	1.902062	0.00	$0.00	0.00
	1809	476.0	$110.0	$52,360.00	2.377530	1,131.70	$15,278.01	1,131.70
	42	0.0	$94.0	$0.00	1.902062	0.00	$0.00	0.00
				$52,360.00		1,131.70	$15,278.01	1,131.70
				Rev in Bond		PG in Bond	Tax in Bond	
Withdrawn This Period	Code #	Cases	Price	Gross		P.G.	Tax	
	1900	0.0	$67.0	$0.00	0.951031	0.00	$0.00	
	1901	0.0	$100.0	$0.00	1.902062	0.00	$0.00	
	1809	70.0	$110.0	$7,700.00	2.377530	166.43	$2,246.77	
	42	0.0	$94.0	$0.00	1.902062	0.000	$0.00	
		70.0		$7,700.00		166.427	$2,246.77	
				"rev" Qtr		PG Out Qtr	Tax This Qtr.	
	Code #	Cases	Price	Gross	P.G.	Total P.G.	Tax Paid	
Remaining In Bond	1900	0.0	$67.0	$0.00	0.951031	0.00	$0.00	
	1901	31.0	$100.0	$3,100.00	1.902062	58.96	$796.01	
	1809	420.0	$110.0	$46,200.00	2.377530	998.56	$13,480.60	
	42	28.0	$94.0	$2,632.00	1.902062	53.26	$718.98	
	Total Cases	479.0		$51,932.00		1,110.78	$14,995.59	
				Rev in Bond		PG in Bond	Tax in Bond	

Actual	Run 1	Run 2	Run 3	Run 4	Run 5	Run 6		
	100 Prf 750	100 Prf 750	100 Prf 750	100 Prf 750	100 Prf 750	100 Prf 750	Total	
Alcohol 1	34.24	28.86	36.44	36.40	36.40	36.40	208.74	
Alcohol 2	179.74	66.24	177.44	177.60	177.60	177.60	956.22	
Total Proof Gal.	213.98	95.10	213.88	214.00	214.00	214.00	1,164.96	Total PG Processing
								PG Per Case
Cases Produced	84.00	40.00	89.00	87.00	92.00	84.00	476.00	2.377530
						Loss %	2.9%	1,131.70	Tot PG

Excise Tax

We see that I bottled 476 cases that would sell for $52,360.00 but that the tax on that whiskey was $15,278.00. This is what I live with all the time. Roughly one third of my cash flow is excise tax, 29.2% in this case. I also shipped out 70 cases for net of 420 left in the shop at the end of the quarter. The average loss rate for the month was 2.9%, which is pretty good.

Looking at the storage report for the month shows that I received a shipment of 908 proof gallons of GNS (Alcohol 2) and then used even more than I received. Still, at the end of the month I still had 309.93 proof gallons in the plant and was liable for $4,184.06 in tax on it when it left the plant or if it was contaminated or lost through someone leaving a valve open or a barrel leaking:

Storage Report	Alcohol 1	Alcohol 2	Totals	Report	Alc 1 Tax	Alc 2 Tax	Total Tax
Beginning Proof Gallons	320.91	245.98	566.89	Line 1	$4,332.29	$3,320.73	$7,653.02
Received during month	0.00	908.00	908.00	Line 2			
Sub total	320.91	1,153.98	1,474.89	Line 6			
Transferred to Processing	208.74	956.22	1,164.96	Line 17	Tax on Unfinished Alcohol End of Month		
Ending	112.17	197.76	309.93	Line 23	$4,184.06		
Total	320.91	1,153.98	1,474.89	Line 24			

Following this month through to the processing report shows that I like to begin and end the month with a zero balance on the Bulk Ingredients side and that I transferred 1,131.70 proof gallons from the bulk to the finished side:

Bulk Ingredients Part I	From Last Month's Ending Balance				
On Hand First of Month	0.00		Part II Finished Products Processing Report		
Received	1,164.96	On Hand First of Month	145.52	From Last Month's Ending Balance	
Total	1,164.960	Received Line 29	1,131.70	Received as Bottled from Line 9	
Line 9 Bottled or Packaged	1,131.70		1,277.22	Total	
Line 17 Transfer	0.00		166.43	Withdrawn Tax Determined	
Losses Line 24	33.256		1,110.80	Calculated Remaining in Processing	
On Hand End of Month Line 25	0.00		1,110.80	Actual inventory result	
Difference	-33.2557	Report as gain or loss	Difference	0.0028	Report as gain or loss
ABS of Difference	33.2557			0.0028	ABS of Difference
On Hand End of Month	0.00	2.9%	Loss %	1,110.80	On Hand End of Month

Excise Tax

I will also take a quick snapshot of what my financial situation is by calculating my inventory with vendors and combining it with my current inventory:

		Cases	Price	Value	P.G.	Total P.G.	Tax Paid
Remaining With Vendors	1900	32	$63.00	$2,016.00	0.951031	30.433	$410.85
	1901	52	$95.00	$4,940.00	1.902062	98.907	$1,335.25
	1809	50	$105.00	$5,250.00	2.377530	118.877	$1,604.83
	42	10	$94.00	$940.00	1.902062	19.02	$256.78
Total Cases		144.00		$13,146.00		267.24	$3,607.70
				Rev from Vendors		P.G. ABC	Tax Paid

Ending in Bond	$51,932.00
Vendor Held Inventory	$13,146.00
Total Inventory Value	$65,078.00

Excise Tax

Then I combine my inventory with my cash on hand and, using my best estimates of how much it costs to produce my products and replace my inventory, get an idea of where I stand financially:

In Bank	$8,410.85
ABCA check	$984.62
Total Inventory value	$65,078.00
Total Inv & Cash Value	$74,473.47
Excise Tax on P.G in Bond	$14,995.59
Bills & Expenses	$4,000.00
Net Value	$55,477.88
Cost of New Inventory	$37,972.80
1 YR Expenses of Shop	$3,000.00
Surplus/Deficit	$14,505.08

Inventory Costs	Cost Per Bot.	Bot.-Case	Per Case	Cases	Total	
375 ml 80 Pr.	$4.25	12	$51.00	180.00	$9,180.00	
750 ml 80 Pf.	$6.40	12	$76.80	180.00	$13,824.00	
750 ml 100 Pr.	$6.93	12	$83.16	180.00	$14,968.80	
750 ml 80 Pr.	$6.40	12	$76.80	50.00	$3,840.00	
				540.00	$37,972.80	Cost to Make

Excise Tax

This month I'm doing OK. If I could produce $50k worth of whiskey a month (and sell it—that's the hard part), I could gross $500k a year.

Based on the last year's sales I can also do a little production planning if I want to make approximately a year's supply of each of my products:

Production Planning									16.0%	WV Spirits % of P.G.	
Item	12 Mo. Sales	On Hand	% of Yr.	Produce	Total	% of Yr. Sales	Proof Gal.	WV P.G.	GNS P.G.	Revenue
1900	325	0.0	0%	325	325	100%	309.1	49.5	259.6	$21,775.00
1901	460	31.0	7%	430	461	100%	817.9	130.9	687.0	$43,000.00
1809	610	420.0	69%	180	600	98%	428.0	68.5	359.5	$19,800.00
42	220	28.0	13%	190	218	99%	361.4	57.8	303.6	$17,860.00
	1,615.00	Tot. per Yr.				Total P.G.	1,916.3	306.6	1,609.7	$102,435.00

	Corn Whiskey Needed		GNS Needed				
On Hand	306.61		1,609.71			48	Empty 750 Ml Cases on Hand
WV. Corn on Hand	138.00		340.60	GNS on Hand		800	Planned 750 Cases
Excess or Deficit of Spirits	-168.61		-1,269.11	Excess or Deficit of GNS		-752	Excess or Deficit of Cases

Excise Tax

What this tells me is that I need to make 168 proof gallons of finished corn whiskey. I'll need 1,269.11 proof gallons of GNS and I'd better plan on acquiring 752 cases of 750-ml bottles in addition to the 325 cases of 375-ml bottles that I need.

Costs per Bottle

The following worksheet identifies the costs associated with making my products:

Cost per Bottle	100 Proof	Difference	80 Proof	375 80 Proof
Federal Tax	$2.67	0.53	$2.14	$1.07
City Privilege Tax	$0.01	0.00	$0.01	$0.01
Bottle	$0.75	0.00	$0.75	$0.67
Cap	$0.04	0.00	$0.04	$0.04
Bottle Labels	$0.19	0.00	$0.19	$0.19
Case Label Cost Per Bottle	$0.01	0.00	$0.01	$0.01
Grain Neutral Spirits	$0.79	0.16	$0.63	$0.32
Corn for WV Spirits	$0.11	0.00	$0.11	$0.06
WV Corn Spirits Energy Cost	$0.33	0.00	$0.33	$0.16
Shipping to ABC	$0.08	0.00	$0.08	$0.08
Bailment Plan Receiving	$0.07	0.00	$0.07	$0.07
Labor	$0.28	0.00	$0.28	$0.28
Overhead	$1.50	0.00	$1.50	$1.50
Sub Total - Bottle Costs	$6.83	0.69	$6.14	$4.460
Number of Bottles Per Case	12.00	Dif.	12.00	12.00
Sub Total Case Cost	$81.96	$8.28	$73.68	$53.52
	Current		Current	Current
Case Sales Price to ABC	$110.00		$100.00	$67.00
Working Case Profit Margin	$28.04		$26.32	$13.48
Working Profit Per Bottle	$2.34		$2.19	$1.12
Wholesale Markup Charges	Present ABCA Price			
Present Retail Price	$143.10		$130.30	$88.06
Present Bottle Price Calculated	$11.93		$10.86	$7.34
Present Catalog Bottle Price	$11.82		$10.75	$7.34
ABC Wholesale Gross Markup $	$33.10		$30.30	$21.06
ABC Markup Percentage	30.1%		30.3%	31.4%
Retailers' Cost Restated	$11.82		$10.75	$7.34
Retail Mark-up %	35.0%		35.0%	35.0%
$ Amount of Retail Markup	$4.14		$3.76	$2.57
Bottle Cost Sitting on Shelf	$15.96		$14.51	$9.91
Sales Tax %	11%		11%	11%
Sales Tax Amount	$1.76	Difference	$1.60	$1.09
Consumer Bottle Price In Hand	$17.71	$1.60	$16.11	$11.00
Case Price In Hand	$212.55		$193.31	$131.99

Costs per Bottle

What this worksheet shows is that by the time my product reaches my customers it costs twice as much as what I sell it for. It also demonstrates that I should probably quit making 375-ml bottles because my profit is so much less for each bottle sold ($1.12 v. ~ $2.20 per bottle) and the customer is paying 2/3 as much for half the product.

One also needs to check the actual in-store prices in order to calculate the actual markup retailers are using.

We can also use the worksheet prospectively in order to predict what effect an increase in price will have on the bottle cost in a consumer's hand:

	100 Proof	80 Proof	375 80 Proof
Proposed Case Cost	$122.25	$111.15	$74.50
Working Profit per Case	$40.29	$37.47	$20.98
Working Profit per Bottle	$3.36	$3.12	$1.75
Increase in Profit per Bottle	$1.02	$0.93	$0.63
Percentage Markup	10.02%	10.03%	10.07%
ABCA Wholesale Markup	30.1%	30.3%	31.4%
Anticipated Case Cost	$159.04	$144.83	$97.92
Anticipated Bottle Cost	$13.25	$12.07	$8.16
Retailers' Cost Restated	$13.25	$12.07	$8.16
Retail Mark-up %	35.00%	35.00%	35.00%
$ Amount of Retail Markup	$4.64	$4.22	$2.86
Bottle Cost Sitting on Shelf	$17.89	$16.29	$11.02
Consumer Shelf Cost	$17.89	$16.29	$11.02
Sales Tax %	11%	11%	11%
Sales Tax Amount	$1.97	$1.79	$1.21
Consumer Bottle Price In Hand	$19.86	$18.09	$12.23
Case Price In Hand	$238.32	$217.03	$146.73
Starting Shelf Price	$15.96	$14.51	$9.91
Ending Shelf Price	$17.89	$16.29	$11.02
Total Consumer Price Change	$1.93	$1.78	$1.11
Ratio of Costs	1.1212	1.1227	1.1117
Total Consumer Price Change %	12.12%	12.27%	11.17%
	Difference		
Starting In Hand Price	$17.71 $1.60	$16.11	$11.00
Ending Consumer Price In Hand	$19.86	$18.09	$12.23
Total Consumer Price Change	$2.15	$1.98	$1.23
Ratio of Costs	1.1212	1.1227	1.1117
Total Consumer Price Change %	12.12%	12.27%	11.17%

Costs per Bottle

What this worksheet shows is that a 10% increase in price translates into a 12% increase in consumer cost. I want to keep my products affordable and within reach of an average consumer who wants to try something new. If the price gets too high, I'm driving customers away and there is a big difference between a $16.00 shelf price and an $18.00 shelf price for my 100 proof. I always agonize over how much to increase my costs on an annual basis. With respect to spot checking the price of your products as they actually appear in liquor stores: it used to be that the retail markup was only 15% at some outlets but over the years they have increased the markup to where it is now averaging 35%. I have seen it go as high as 40%, which I resent.

Looking at how some of the component costs are calculated shows the tax payable on a per bottle basis:

Tax cost per bottle	PG per Case	Tax	Per Case	Tax Per Bot.
Excise Tax 80 proof Case	1.9020384	$13.50	$25.68	$2.14
Excise Tax 100 Proof Case	2.377548	$13.50	$32.10	$2.67
			375 Bottle Tax	$1.07

The following shows how bottle costs are calculated:

750 Bottles	2013	2012
Number of Cases	448.0	224.0
Cost per Case	$7.65	$7.65
Bottles per Case	12	12
Pallet Cost	$60.00	$30.00
Total Case & Pallet Costs	$3,487.20	$1,743.60
Cost	$0.65	$0.65
Shipping Costs	$525.00	$292.43
Shipping per Bottle	$0.10	$0.11
Total Cost at Facility	$4,012.20	$2,036.03
Tot. Cost per Bottle	$0.75	$0.76
Cost per Case	$8.96	$9.09

They used to give me the pallets for free, but then started charging extra for them.

As to label costs:

Label Costs	750 ml	750 ml	750 ml	750 ml	375 ml	375 ml	375 ml
Date	6/3/2013	2/1/2012	1/10/2011	2/25/2009	8/13/2012	9/24/2009	1/23/2008
Labels Purchased	6,500	4,500	4,500	2,400	4,800	4,800	5,000
Cost For Labels	$509.86	$284.99	$261.99	$295.99	$262.08	$264.00	$274.00
Freight	$18.20	$12.21	$21.60	$11.34	$8.28	$10.19	$8.00
Total Cost	$528.06	$297.20	$283.59	$307.33	$270.36	$274.19	$282.00
Cost Per Label Purchased	$0.08	$0.07	$0.06	$0.13	$0.06	$0.06	$0.06
Labels Per Bottle	2	2	2	2	2	2	2
Cost Per Bottle (Purchased)	$0.16	$0.13	$0.13	$0.26	$0.11	$0.11	$0.11
Cases Labeled	230	165	165	88	180	180	180
Bottles Per Case	12	12	12	12	12	12	12
Total Bottles	2,760	1,980	1,980	1,056	2,160	2,160	2,160
Total Labels Used	5,520	3,960	3,960	2,112	4,320	4,320	4,320
Labels Left Over or Spoiled	980	540	540	288	480	480	680
Cost Per Label Used	$0.10	$0.08	$0.07	$0.15	$0.06	$0.06	$0.07
Cost Per Bottle (Employed)	$0.19	$0.15	$0.14	$0.29	$0.13	$0.13	$0.13
Percentage Waste	17.8%	13.6%	13.6%	13.6%	11.1%	11.1%	15.7%

The thing about labels is that they don't keep for more than a year or so. The adhesive evaporates and they won't stick to bottles. I keep my labels in clear plastic bags with the tops tied shut to try and slow down the process and use them in the next bottling run. One always has to order more labels than are needed because some labels go on crooked and must be removed; some labels may be smudged, etc. Also, changing a roll of labels out in the machine takes time and you want to start with a full roll to save time during the bottling run. Trying to use the last of the roll from the last run costs more time to swap out for a new roll—half an hour into the run, when one has five people in the shop waiting around for the label machine to get a new roll put in it.

Corn cost and energy cost to make corn whiskey:

Corn Costs		Energy Cost	
Total Lb. Used	6,000.00		
Lb. Per Bag	50.00	$12.95	Energy Cost Per Proof Gal.
Bags Used	120.00	1.902	80 Proof Case Multiplier
Cost of 50 Lb. Bag	$8.00	$24.63	Energy Cost Per Case
Total Corn Cost	$960.00	$2.05	Cost Per Bottle 100% WV
Proof Gallons Produced	222.60	16.0%	Percentage Corn Whiskey
Corn Cost Per Proof Gallon	$4.31	$0.33	Cost Per Bottle
80 Proof Case Multiplier	1.902		
Cost Per Case	$8.20		
Cost Per Bottle @ 100% WV	$0.68		
Percentage Corn Whiskey	16.0%		
Cost Per Bottle	$0.11		

It is only fair to say that the energy cost is a huge fudge factor but it is a big number. In my distilling operations manual you will find ways to get close to an actual value. Between the cost of the oil, natural gas, electricity as well as cooking, fermenting, single run, double run and blending and to some extent the labor, you need to factor in all that it costs to make a proof gallon of alcohol.

Following is a rough and ready way to calculate the proof gallons one will receive in a shipment and the approximate cost per proof gallon:

GNS Costs	2013	2012	2011	2010
Per Dum $	$445.00	$365.00	$353.16	$315.00
Number of Drums	7.00	7.00	7.00	7.00
Total Drum Cost	$3,115.00	$2,555.00	$2,472.12	$2,205.00
Freight & Delivery	$315.00	$350.00	$330.00	$330.00
Freight Per Drum	$45.00	$50.00	$47.14	$47.14
Proof Gallons Per Drum	103.86	103.55	102.60	102.60
Total Proof Gallons	727.02	724.85	718.20	718.20
Total Cost	$3,430.00	$2,905.00	$2,802.12	$2,535.00
Cost per Proof Gallon	$4.72	$4.01	$3.90	$3.53

GNS Proof Gallons Per Bottle & Cost

Proof Gallons Per Case 80 Proof	Bottles Case	PG. Per Bot.	Cost Per PG	P.G. %	% of $ PG	Cost per
1.902061691	12	0.158505141	$4.72	84.00%	$3.96	$0.63

GNS Proof Gallons Per Bottle & Cost 100 Proof

Proof Gallons Per Case 100 Proof	Bottles Case	PG. Per Bot.	Cost Per PG	P.G. %	% of $ PG	Cost per
2.37753	12	0.1981275	$4.72	84.00%	$3.96	$0.79

Corn Whiskey Cost Proof Gallons Per Bottle & Cost -80 Proof

Proof Gallons Per Case 80 Proof	Bottles	PG. Per Bot.	Cost Per PG	P.G. %	% of $ PG	Cost per
1.902061691	12	0.158505141	$10.00	16.00%	$1.60	$0.25

Corn Whiskey Cost Proof Gallons Per Bottle & Cost 100 Proof

Proof Gallons Per Case 100 Proof	Bottles	PG. Per Bot.	Cost Per PG	P.G. %	% of $ PG	Cost per
2.37753	12	0.1981275	$10.00	16.00%	$1.60	$0.32

Costs per Bottle

Proof Gallon Ordering

GNS Proof Gallons Per Bottle & Cost				
Number of Cases of 80 Proof	250.00		500.00	Number of Cases of 100 Proof
Proof Gallons Per Case 80 Proof	1.902061691		2.37753	Proof Gallons Per Case 100 Proof
Total Proof Gallons	475.52		1,188.77	Total Proof Gallons
Percentage of GNS	84.0%		84.0%	Percentage of GNS
Proof Gallons Required	399.43		998.56	Proof Gallons Required
	Total P.G. Required	1,398.00		
Calculate Proof Gallons per Drum				
Gallons in Drum	54.50			
Proof of spirits	190.00			
Proof gallons per Fluod Gallon	1.900			
Proof Gallons per Drum	103.55			
Number of Drums	14.00			
Total fluid gallons	763.00			
Total proof Gallons	1,449.70			

Proof Gallon Ordering

Big shipments are better than small ones because each one really takes up the better part of a day of your time to do it. Trucks are late or don't arrive and you've wasted a day. On the other hand, ordering too much GNS can get you in trouble if you don't use it in a reasonable period of time. Six to nine months of storage is no problem but I've had spirits stored for more than a year that started to acquire an off taste from the plastic barrels or metal/Teflon-lined barrels. It therefore became less useable and in need of immediate use or a great deal of waste would have ensued.

Case and Label Planning

Nothing comes in the right quantity. Cases come in pallets containing different numbers of cases per row depending on how many can fit on a pallet. Labels come in spools that have varying amounts of labels on them and rarely do the spool sizes match the amount needed for any particular run. Also, there is waste of both items that needs to be accounted for. The following worksheet provides some means of accounting for some of this uncertainty:

Empty Case Plan and Label Plan for Production

Planned Production	Cases	Per Case	Lbl. Per. Bot.	Labels Req.	Waste %	Waste Labels	Tot.Lab les	Lbl per Spool	Spools	Per Spool Round up	Tot. Labels
80 Proof 750 #1901 UPC 002	150	12	2	3,600	7%	252.00	3,852	1,000	3.85	4.0	4,000
100 Proof 750 #1809 UPC 004	275	12	2	6,600	7%	462.00	7,062	1,000	7.06	8.0	8,000
Old Oak Recipe #42 UPC 005	84	12	2	2,016	7%	141.12	2,157	1,000	2.16	3.0	3,000
Total Planned Cases	509					Total Labels	13,071		Total Labels Ordered		15,000
Empty Cases on hand	30					Over Order %	12.86%				
Net Cases Needed	479										
Cases per Pallet	84										
Total Pallets Needed	5.70										
Roundup to Pallets Needed	6.0										
Total Cases to Arrive	504										
Surplus Cases	25										

Case Planning

It is unlikely that your printer will print spools with the number of labels you actually need but you can play with the waste label percentages to reduce the "Over Order" percentage to perhaps avoid over-ordering a spool you don't need.

You will need surplus cases. Some pallets come knocked over sideways on the truck and you can lose tens of cases to breakage that way. You also then have to inspect each bottle and throw out any case that has even one broken bottle in it, either that or rinse each bottle in that case to ensure no broken glass is in it.

We can also get some idea of how many proof gallons we will need and how our planned production relates to the requirements based on the prior year's sales:

% Proof Gal of Other Alc	15%						
Code #	Produce	12 Mo. Sales	% of Yr.	PG per Case	Proof Gal.	Other P.G.	GNS P.G.
1900	84	140	60%	0.951031	79.9	12.0	67.9
1901	150	130	115%	1.902062	285.3	42.8	242.5
1809	200	206	97%	2.377530	475.5	71.3	404.2
42	125	114	110%	1.902062	237.8	35.7	202.1
Planned	475	590		Total PG.	1,078.5	Alc. 1	GNS
				PG Needed		161.8	916.7
				Proof Gal. of Alc 1 On Hand		138.0	340.6	GNS on Hand
				Excess or Deficit of Alc 1		-23.8	-576.1	Net PG GNS

				Production Planning			
	Code #	On Hand	Price	P.G.	Mo. Sales	% of Yr.	Revenue
Remaining In Bond	1900	90.0	$67.0	0.951031	140	64%	$5,628.0
	1901	49.0	$99.0	1.902062	130	38%	$14,850.0
	1809	121.0	$109.0	2.377530	206	59%	$21,800.0
	42	7.0	$94.0	1.902062	114	6%	$11,750.0
					590		$54,028.0

Notes

Obscuration of Proof

Obscuration of proof is an arcane and somewhat boring topic. It is necessary to include it here because this is the first time we are starting to change density with something other than water.

We will be doing a fair amount of density changing with substances and temperature, particularly calibration temperatures, later on and this is a good introduction as to how those changes operate.

For those more interested in blending operations, they should skip this section and the Flavors section but look over the Alcohol Consumption section because it is kind of fun and interesting.

Any flavor or solids that are added to alcohol and water mixtures will affect the density of the alcohol they are added to. Flavor additions, like sugar, are denser than the alcohol they are added to; the additions increase the density of the mixture and consequently a hydrometer reads this as a lower proof. This change needs to be accounted for in order to preserve the tax payable on the alcohol itself and to ensure the customer is receiving what the proof or volume percentage on the bottle says is being sold to them. When flavorings are added they need to be contributed in such a way that the density changes but the alcoholic content stays the same. The proof is hidden but still there: it is only obscured by the flavor.

Obscuration by Weight

The following worksheet reproduces example 3 from §186.64 - §30.64 where the known obscuration of proof is added back to the gauged present "true" proof to increase the proof to the actual known alcohol content in the mixture:

Using Weight & Amount of Obscuration to Find Proof Gallons and Wine Gallons

Temp of Alcohol	75.00	0.000	Thermometer Cor.
Apparent Proof	92.00	0.000	Hydrometer Cor.
Weight of Alcohol (lbs.)	5,350.00	$13.50	Excise Tax Rate
Present Gauged "True" Proof	85.50	0.0015	Tolerance
Known Obscuration of Proof Added In	0.50	0.300	Tolerance ° of Proof
Corrected True Proof	86.00	85.5	Present Gauged True Proof

Find Proof Gallons	Corrected	Gauged Proof	
Proof Gal. per Lb. at Corrected True Proof	0.10905	0.10837	PG per Lb. at Present Gauged True Proof
Wt. * PG per Lb. = Proof Gallons	583.44	579.80	Wt. * PG per Lb. = Proof Gal @ Gauged Proof
Tax on P.G.	$7,876.43	-3.64	Dif. in Proof Gal. if Obscuration Ignored
		-$49.19	Tax not paid if Present Gauged True Proof is used

	Gauged Proof	Corrected Proof	
Calc. 60° Vol. From Present Gauged Proof	85.50	86.00	Vol. from Corrected True Proof
WG. per Lb. at Present Gauged True Proof	0.12675	0.12681	Wine Gal. Per lb at Corrected True proof
60° Vol. at Present Gauged True Proof	678.13	678.43	Weight * WG per Lb. = Vol @ Corrected proof
		0.30	Vol. Dif. If Corrected is Used

		Table 7 Vol. Cor.		
Find Present Volume From Apparent Proof	Apparent		678.13	Gauged Proof Vol.
WG per Lb. at Apparent Proof	0.12751	Gal. Dif.	1.0060	Factor to Present Volume.
Wt.* WG per Lb at Apparent = Present Vol.	682.19	0.01	682.20	60° Vol. * Factor = Present Gal.
Gal. Vol Dif. From 60° Vol.	4.06		4.07	

Back out Obscuration to Solids Added

Obscuration	0.50	0.30	Gal. Vol Dif.	0.30	Gal. Vol Dif.
Grams per liter per Degree of Proof Obscure	2.50	1.13	Liter Dif.	0.12007	WG per Lb H2O
Obscuration * Grams per Liter Value	1.25	2.50	Density of Solids	2.50	Inferred H20 Lbs.
60° F. Vol * Constant = Liters	2,567.00	1.25	Ratio per Liter		
(Liters*Grams per Liter)/1,000 = Kg. of Solids	3.209	3.125	Ratio * Density	0.30	Gal. Vol Dif.
Total Pounds Used to Obscure Alc.	7.07	3.54	Liters * Result	0.12675	WG per Lb Alc
		7.81	Inferred Lb. Mass	2.36	Inferred Alc. Lbs.

Obscured Alcohol Tracking Based on Weight

First, we gauge the alcohol and run a true proof correction on it to find what I've labeled the Present Gauged "True" Proof.[33]

Then we add the known amount of obscuration to it to arrive at what I've labeled as the "Corrected True Proof" As you can see in the example the proof gallon difference between the obscured and non-obscured spirits is 3.64 proof gallons, which is significant enough to require keeping track of. Obscuration is added to proof to retain full taxation of alcohol. The slight discrepancy between the proof gallon result of this spreadsheet and the regulation example is caused by the program using Table 4 values to calculate proof gallons and wine gallons, rather than the mathematical method of multiplying the Wine Gallons at 60°F by (Proof/100) to obtain 0.86. While the regulation procedure is a clever method of calculating proof gallons and apparently an approved method, the ABS program uses Table 4 for consistency.[34]

One other important point to note is that, with respect to volume, the regulations prescribe that the 60°F volume is calculated from the Gauged True Proof/Obscured Proof, not the Corrected True Proof (Taxable Proof). This seems counter intuitive.

The obscuration has increased the density and, while it has also increased the volume by the addition of some form of solids (probably sugar), because these solids had a density greater than alcohol or water the volume change was not as great compared to adding water or alcohol. But, these solids were present in the weight of the alcohol that was placed on the scale and in order to begin the process and so to "balance the equation" we have to get back to that mass in the end. Also, we have inferred that 7.07 lbs. of solids were added and this may have been up to several liters of a solid/powder like sugar.

In this case these concerns do not have that large an effect on the volume and we can still use the procedure of using the apparent proof and the temperature to calculate volume and when this result is compared to a Table 7 volume correction the results are essentially in agreement.

We will cover the *Grams per Liter per Degree of Proof* obscuration more fully when we predict the effect of adding solids to alcohol, but I have included here a means of backing out the solids in the alcohol to find their mass as a component of the alcohol using three different approaches. First, we just take the obscuration and the regulation obscuration value of 2.5 grams per liter per degree of proof and use it in a formula to build up from the 60°F gallons to liters and then mass in kilograms and pounds. Secondly, I've included a column that has a variable for the density of the solids in the event that they are known. We will use something like this when we use sugar as a flavoring and add it to alcohol. Lastly, I've included a way to use the wine gallons per pound of both the water and alcohol to infer the mass of the water or alcohol represented by the solids.

One might note that the value retrieved from the ABS tables is 0.12675 rather than the 0.12676 recited in the regulations. There are several reasons for this which we will discuss: part of the discrepancy is due to the differences in the definitions of a liter and the inch (which affect cubic inches), which have been accounted for in the ABS tables. Also, the ABS tables are ultimately derived from the International Alcohol Tables and the General Formula published in conjunction with those tables in order to avoid the duplicate wine-gallons-per-pound values contained in the TTB Table 4 along with other issues.

Obscuration by Volume

Keeping track of obscuration by volume requires that we transfer the present volume of 682.2 gallons at 75°F used in the prior worksheet (using the Table 7 volume of the gauged proof rather than the corrected proof) over to the *Obscuration by Volume* worksheet, where we apply a Table 7 volume correction to find the 60°F volume of 678.1. Then we find mass using Wine Gallons per Pound at the Gauged True Proof. The result of 5,349.83 pounds is only 0.17 lb. less than the 5,350 in the prior example.

---

[33] §186.64 "*Example*. 5,350 pounds of blended whisky containing added solids Temperature °F 75.0° Hydrometer reading 92.0° Apparent proof 85.5° Obscuration 0.5° True proof 86.0° 5,350.0 lbs.×0.12676 (W.G. per pound factor for apparent proof of 85.5°)=678.2 wine gallons 678.2 W.G.×0.86=583.3 proof gallons."

[34] Also, the regulation is somewhat confusing in that it refers to 85.5 as the Apparent Proof rather than the Presently Gauged True Proof.

Using Volume & Amount of Obscuration to Find Proof Gallons and Wine Gallons

Temp of Alcohol	75.00	0.000	Thermometer Cor.			
Apparent Proof	92.00	0.000	Hydrometer Cor.			
Present Gal. Volume of Alcohol	682.20	2,582.41	Liters	$13.50	Excise Tax Rate	
Table 7 Vol. Cor. Factor to be applied	0.9940			0.0015	Tolerance	
Present Gal. * Factor = 60° F. Gallons	678.11	2,566.91	Liters	0.300	Tolerance ° of Proof	
Gal. Dif. between Present Vol. & 60° Vol.	4.09	15.49	Liters	1.02	Gallon Tolerance	
Present Gauged "True" Proof	85.50			8.02	Lbs. of Tolerance	
WG. per Lb. at Pesent Gauged True Proof	0.12675					
60° F. Vol./WG per Lb. = Lbs. of Alcohol	5,349.83					
Known Obscuration of Proof Added In	0.50					
Corrected True Proof	86.00					

Find Proof Gallons	Corrected		Gauged		
Proof Gal. per Lb. at Corrected True Proof	0.10905		0.10837	PG per Lb. at Present Gauged True Proof	
Wt. * PG per Lb. = Proof Gallons	583.42		579.78	Wt. * PG per Lb. = Proof Gal @ Gauged Proof	
Tax on P.G.	$7,876.17		-3.64	Dif. in Proof Gal. if Obscuration Ignored	
			-$49.18	Tax not paid if Present Gauged True Proof is used	

Back out Obscuration to Solids Added			Corrected			
Obscuration	0.50		0.12681	WG PP @ Corrected		
Grams per liter per Degree of Proof Obscured	2.50		5,347.47	60° Vol * WGPP = Lb.s @ Corrected		
Obscuration * Grams per Liter Value	1.25		2.36	Lbs. Dif.		
60° F. Vol * Constant = Liters	2,566.91		0.300	Lb*WGPP=Gal. Dif	0.30	Gal. Vol Dif.
(Liters*Grams per Liter)/1,000 = Kg. of Solids	3.209		1.13	Liter Dif.	0.12007	WG per Lb H2O
Total Pounds Used to Obscure Alc.	7.07		2.50	Density of Solids	2.50	Inferred Lbs.
			1.25	Ratio per Liter		
			3.125	Ratio * Density	0.30	Gal. Vol Dif.
			3.54	Liters * Result = Kg	0.12675	WG per Lb Alc
			7.81	Inferred Lb. Mass	2.36	Inferred Lbs.

Obscured Alcohol Tracking Based on Volume

The reason I point this out is that, if we use the wine gallons per pound at the Corrected Proof (Taxable Proof), the mass of alcohol comes out to 5,347.46 lbs. which is 2.38 lbs. less than the actual pounds we began with. Were we to use this weight value to calculate proof gallons, that value would not match the starting value either.

Applying the same methods as in the prior example to back out the solids as best we can will give us some idea of what kind of mass we are dealing with apart from the alcohol and water.

Obscure Proof from Volume

Now that we've seen the effect of obscuration we would like to be able to control it and predict it as we take unobscured alcohol and add solids to it. Usually we start our examples using the weight of alcohol but in in this case it makes more sense to start by obscuring proof from volume.

The regulations describing the rules for tracking and also predicting obscuration of alcohol are particularly difficult to unravel but the first point to note is that whenever the solids content of alcohol is unknown, one must use one of the distillation/evaporation methods identified in §30.31[35] and §30.32.[36] Also, whenever the density

---

[35]§ 30.31 "Determination of proof. (a) General. The proof of spirits shall be determined to the nearest tenth degree which shall be the proof used in determining the proof gallons. (b) Solids content not more than 600 milligrams. Except as otherwise authorized by the appropriate TTB officer, the proof of spirits containing not more than 600 milligrams of solids per 100 milliliters of spirits shall be determined by the use of a hydrometer and thermometer in accordance with the provisions of §30.23 except that if such spirits contain solids in excess of 400 milligrams but not in excess of 600 milligrams per 100 milliliters at gauge proof, there shall be added to the proof so determined the obscuration determined as prescribed in §30.32. (c) Solids content over 600 milligrams. If such spirits contain solids in excess of 600 milligrams per 100 milliliters at gauge proof, the proof shall be determined on the basis of true proof determined as follows: (1) By the use of a hydrometer and a thermometer after the spirits have been distilled in a small laboratory still and restored to the original volume and temperature by the addition of pure water to the distillate; or (2) By a recognized laboratory method which is equal or superior in accuracy to the distillation method."

of solids is greater than 600 milligrams per 100 milliliters, a distillation method must be used. Between the values of 0 and 600 milligrams per 100 milliliters these two sections also define how obscuration is to be accounted for.

Quoting from §30.32 we can extract our first general rule of predicting obscuration. "Experience has shown that 0.1 gram (100 milligrams) of solids per 100 milliliters of spirits in the range of 80-100 degrees proof will obscure the true proof by 0.4 of one degree of proof." Scaling this example up to one liter shows that 1 gram per liter of alcohol will obscure proof by 0.4 degrees of proof. Accordingly, to obscure by a whole degree of proof requires 2.5 grams per liter:

Milliliters	100.0	1.00	Liter
Milligrams	100.0	1.00	Gram
Obscuration in Degrees of Proof	0.40	0.40	Obscuration
The value of 1	1.00		
.4 / 1 =	2.5		

Milliliters	100.0	1.00	Liter
Milligrams	250.0	2.50	Gram
Obscuration in Degrees of Proof	1.0	1.0	Proof Obscuration

Add Solids to Vol. of Alcohol & Find Obscuration

Now that we know the constants that we will be using, let's set up an example that will obscure 264.172 gallons, which is equal to 1,000 liters, by exactly 0.4 proof. To do so we will set the mass of solids added until it exactly produces 0.4 proof of obscuration, which conveniently turns out to be exactly 1 kilogram of solids.

---

[36] § 30.32 "Determination of proof obscuration. (a) General. Proof obscuration of spirits containing more than 400 but not more than 600 milligrams of solids per 100 milliliters shall be determined by one of the following methods. The evaporation method may be used only for spirits in the range of 80-100 degrees at gauge proof. (b) Evaporation method. Evaporate the water and alcohol from a carefully measured 25 milliliter sample of spirits, dry the residue at 100 degrees centigrade for 30 minutes and then weigh the residue precisely. Multiply the weight of the residue by 4 to determine the weight of solids in 100 milliliters. The resulting weight per 100 milliliters multiplied by 4 will give the obscuration. Experience has shown that 0.1 gram (100 milligrams) of solids per 100 milliliters of spirits in the range of 80-100 degrees proof will obscure the true proof by 0.4 of one degree of proof. For example, if the weight of solids remaining after evaporation of 25 milliliters 0.125 gram, the amount of solids present in 100 milliliters of the spirits is 0.50 gram (4 times 0.125). The obscuration is 4 times 0.50, which is **two degrees of proof**. This value added to the temperature corrected hydrometer reading will give the true proof. (c) Distillation method. Determine the apparent proof and temperature of the sample of spirits and then distill a carefully measured sample in a small laboratory still, and collect a quantity of the distillate, 1 or 2 milliliters less than the original sample. The distillate is adjusted to the original temperature and restored to the original volume by addition of distilled water. The proof of the restored distillate is then determined by use of a precision hydrometer and thermometer in accordance with the provisions of §13.23 to the nearest 0.1 degree of proof. The difference between the proof so determined and the apparent proof of the undistilled sample is the obscuration; or (d) Pycnometer method. Determine the specific gravity of the undistilled sample, distill and restore the samples as provided in paragraph (c) of this section and determine the specific gravity of the restored distillate by means of a pycnometer. The specific gravities so obtained will be converted to degrees of proof by interpolation of Table 6 to the nearest 0.1 degree of proof. The difference in proof so obtained is the obscuration."

Predict Obscuration of Proof by addition of solids to Volume of Alcohol

Temperature of Alcohol	60.00	0.00	Thermometer Cor.
Apparent Proof of Alcohol	100.00	0.00	Hydrometer Cor.
True Proof Result	100.00		
Present Gallon Volume of Alcohol	264.172		$13.50 Excise Tax Rate
Table 7 Vol. Cor. Factor to find 60° F. Vol.	1.0000		$3,566.27 Tax on Proof Gal.
Volume of that Alcohol corrected to 60° F.	264.172	1,000.00	Liters
Wine Gallons Per Lb. of Alcohol @ True Proof.	0.12853		
60° Vol. /W.G. per Lb. @ 60° F. True Pr. = Wt.	2,055.30	932.27	Kilograms
Wine Gallons Per Pound @ True proof	0.12853	0.12853	Proof Gal. per Lb.
Weight * W.G. per Lb. = 60° F. Wine Gal.	264.172	264.17	Proof Gal.
Pounds of Solids Added to Batch	2.20462	35.27	Ounces Avdp
Grams of Solids	1,000.00	1.00	Kilograms
Liters of Volume	1,000.00		
Grams Per liter	1.000	100.00	Milligrams per 100 Ml.
Grams per liter per 1° of Proof Obscuration	2.50		Ratio per 1.0° of Proof
Grams per Liter / 2.5 = Proof Obscuration	0.40		

Add Solids to Vol. of Alcohol & Find Obscuration

Setting the batch to 264.172 gallons produces 1,000 liters and the conversion factor from kilograms to pounds, 2.2046226, results in one kilogram of solids and 100 milligrams per 100 ml. Note that the mass in kilograms, 932.27, is the density in kilograms per meter cubed of 100-proof alcohol at 15.556°C under atmospheric conditions.

As we scale up the mass of solids added, we must remember that we are applying an essentially linear adjustment of grams per liter of obscuration. This is not quite true because the liters vary with proof (higher proof = more liters per unit of mass resulting in less obscuration for a given mass added). Perhaps for this reason among others the regulations state that this procedure is limited to the range of 80 to 100 proof for its accuracy. Within that range if we change the proof to 80 and look at the results we see that the density of 80-proof alcohol is 949.77 kilograms per meter cubed but the obscuration is still the same value at 0.4 degrees of proof:

Predict Obscuration of Proof by addition of solids to Volume of Alcohol

Temperature of Alcohol	60.00	0.00	Thermometer Cor.
Apparent Proof of Alcohol	80.00	0.00	Hydrometer Cor.
True Proof Result	80.00		
Present Gallon Volume of Alcohol	264.172		$13.50 Excise Tax Rate
Table 7 Vol. Cor. Factor to find 60° F. Vol.	1.0000		$2,853.02 Tax on Proof Gal.
Volume of that Alcohol corrected to 60° F.	264.172	1,000.00	Liters
Wine Gallons Per Lb. of Alcohol @ True Proof.	0.12616		
60° Vol. /W.G. per Lb. @ 60° F. True Pr. = Wt.	2,093.89	949.77	Kilograms
Wine Gallons Per Pound @ True proof	0.12616	0.10093	Proof Gal. per Lb.
Weight * W.G. per Lb. = 60° F. Wine Gal.	264.172	211.33	Proof Gal.
Pounds of Solids Added to Batch	2.20462	35.27	Ounces Avdp
Grams of Solids	1,000.00	1.00	Kilograms
Liters of Volume	1,000.00		
Grams Per liter	1.000	100.00	Milligrams per 100 Ml.
Grams per liter per 1° of Proof Obscuration	2.50		Ratio per 1.0° of Proof
Grams per Liter / 2.5 = Proof Obscuration	0.40		

Add Solids to Vol. of Alcohol & Find Obscuration

Having this tool available we want to see how far we can push it before we are outside the range of its approved use. Quoting from §30.31 we see that we can calculate obscured proof with a hydrometer up to 600 milligrams per 100 ml or six kilograms per 1,000 liters.

The proof of spirits containing not more than 600 milligrams of solids per 100 milliliters of spirits shall be determined by the use of a hydrometer and thermometer in accordance with the provisions of §30.23 except that if

such spirits contain solids in excess of 400 milligrams but not in excess of 600 milligrams per 100 milliliters at gauge proof, there shall be added to the proof so determined the obscuration determined as prescribed in §30.32.

I'm not sure the regulation says what it means. It appears to say that you can ignore proof obscuration up to 400 mg of solids but that degree of obscuration would amount to 1.6 degrees of proof and we already know that the blending tolerance for alcohol of less than 600 mg per liter is 0.015 proof; above that value the tolerance is relaxed to 0.25% or 0.50 proof.[37] Adding 4 kilograms of solids to our 80-proof alcohol does produce the desired obscuration of 1.6 proof:

Predict Obscuration of Proof by addition of solids to Volume of Alcohol			
Temperature of Alcohol	60.00	0.00	Thermometer Cor.
Apparent Proof of Alcohol	80.00	0.00	Hydrometer Cor.
True Proof Result	80.00		
Present Gallon Volume of Alcohol	264.172	$13.50	Excise Tax Rate
Table 7 Vol. Cor. Factor to find 60° F. Vol.	1.0000	$2,853.02	Tax on Proof Gal.
Volume of that Alcohol corrected to 60° F.	264.172	1,000.00	Liters
Wine Gallons Per Lb. of Alcohol @ True Proof.	0.12616		
60° Vol. /W.G. per Lb. @ 60° F. True Pr. = Wt.	2,093.89	949.77	Kilograms
Wine Gallons Per Pound @ True proof	0.12616	0.10093	Proof Gal. per Lb.
Weight * W.G. per Lb. = 60° F. Wine Gal.	264.172	211.33	Proof Gal.
Pounds of Solids Added to Batch	8.81849	141.10	Ounces Avdp
Grams of Solids	4,000.00	4.00	Kilograms
Liters of Volume	1,000.00		
Grams Per liter	4.000	400.00	Milligrams per 100 Ml.
Grams per liter per 1° of Proof Obscuration	2.50	Ratio per 1.0° of Proof	
Grams per Liter / 2.5 = Proof Obscuration	1.60		

Add Solids to Vol. of Alcohol & Find Obscuration

The regulations do speak of obscurations up to 2.0 proof so we know they contemplate at least that much obscuration, amounting to 5 kilograms per 1,000 liters. We will top out our examples at 6 kilograms per 1,000 liters (or 2.4 proof) equaling 600 milligrams per 100 ml using 13.22774 lbs. of solids.

[37] Title 27 Section 5.37 "(b) Tolerances. The following tolerances shall be allowed (without affecting the labeled statement of alcohol content) for losses of alcohol content occurring during bottling: (1) Not to exceed 0.25 percent alcohol by volume for spirits containing solids in excess of 600 mg per 100 ml; or (2) Not to exceed 0.25 percent alcohol by volume for any spirits product bottled in 50 or 100 ml size bottles; or (3) Not to exceed 0.15 percent alcohol by volume for all other spirits."

Predict Obscuration of Proof by addition of solids to Volume of Alcohol

Temperature of Alcohol	60.00	0.00	Thermometer Cor.
Apparent Proof of Alcohol	80.00	0.00	Hydrometer Cor.
True Proof Result	80.00		
Present Gallon Volume of Alcohol	264.172		$13.50 Excise Tax Rate
Table 7 Vol. Cor. Factor to find 60° F. Vol.	1.0000		$2,853.02 Tax on Proof Gal.
Volume of that Alcohol corrected to 60° F.	264.172	1,000.00	Liters
Wine Gallons Per Lb. of Alcohol @ True Proof.	0.12616		
60° Vol. /W.G. per Lb. @ 60° F. True Pr. = Wt.	2,093.89	949.77	Kilograms
Wine Gallons Per Pound @ True proof	0.12616	0.10093	Proof Gal. per Lb.
Weight * W.G. per Lb. = 60° F. Wine Gal.	264.172	211.33	Proof Gal.
Pounds of Solids Added to Batch	13.22774	211.64	Ounces Avdp
Grams of Solids	6,000.00	6.00	Kilograms
Liters of Volume	1,000.00		
Grams Per liter	6.000	600.00	Milligrams per 100 Ml.
Grams per liter per 1° of Proof Obscuration	2.50		Ratio per 1.0° of Proof
Grams per Liter / 2.5 = Proof Obscuration	2.40		

Find Proof Gallons Obscured

True Proof minus obscuration	77.600
Proof Gallons per Lb. at True P. - Obscuration	0.09772
Wt. * P. G./ Lb. @ Obs. Proof = PG @ Obscured	204.60
Starting "Unobscured" Proof Gallons Restated	211.33
Total Proof Gallons Obscured	-6.73
Tax Difference	-$90.85

Add Solids to Vol. of Alcohol & Find Obscuration

Summarizing these sections, we find that if the solids content is known because distillers (rectifiers) are adding them to their own batches, then at dissolved solids levels between 0 mg and 400 mg per 100 ml they can calculate the proof obscured by the addition of solids using the method outlined. Above 400 milligrams but less than 600 milligrams it is an optional method to use but it appears that the evaporation or distillation methods are also advised. If this was not the case one can imagine that rectifiers would use solids to obscure proof and thereby avoid taxation on spirits that were gauged by any revenue authority—in this case by 6.73 proof gallons and $90.85 in tax.

What this procedure does allow us to accomplish, if we begin with unobscured alcohol, is to accurately calculate the quantity of solids to be added before they are added to the batch. If the total is kept below 600 mg per 100 ml, then we can avoid doing a distillation to determine the proof obscuration when subsequently gauged.

For completeness we need to mention §30.71 and §30.72, which also indicate how obscured spirits are to be accounted for. Section §30.71 outlines an optional procedure. I think it is best to let the regulation speak for itself.

§30.71 "Optional method for determination of proof for spirits containing solids of 400 milligrams or less per 100 milliliters. The proof of spirits shall be determined to the nearest tenth degree which shall be the proof used in determining the proof gallons and all fractional parts thereof to the nearest tenth proof gallon. The proof of spirits containing solids of 400 milligrams or less per 100 milliliters shall be determined by the use of a hydrometer and a thermometer in accordance with the provisions of §30.23. However, notwithstanding the provisions of §30.31, the proprietor may, at his option, add to the proof so determined the obscuration determined as prescribed in §30.32."

This appears to make it permissible to ignore proof obscuration up to 400 mg per 100 ml. However, that idea appears to be contradicted by §30.72. Again, letting the words speak for themselves.

§30.72 "Recording obscuration by proprietors using the optional method for determination of proof. Any proprietor using the optional method for determination of proof for spirits containing solids of 400 milligrams or less per 100 milligrams as provided in §30.71 shall record the obscuration so determined on the record of gauge required by 27 CFR part 19."

I really can't make heads or tails out of these procedures. My practice is to simply accurately report the proof gallons employed in a batch and pay the tax. One last note: in the Gauging Manual (Publication #455, alternately as ATF-P5110.6 and published in 1978 as part of Title 27, Chapter I, Subchapter M, Part 186 of the Code of Federal Regulations) there is an addition, typed at the bottom of page 6 below §186.31- §30.31, which states the following:

"On May 5, 1976 the U.S. Customs Service issued Circular TEC-2-0:T:O H to allow the duties (and taxes, where applicable) to be calculated on the basis of apparent proof if the nonvolatile matter is equal to or less than 0.4 g/100 ml."

Now, 0.4 g/100 ml. = 4 grams per liter or 400 milligrams. This addendum appears to obviate the need to track proof obscuration below that amount. I would not rely too heavily on this since the regulations may have changed, but it might prove useful to someone to know about it.

Obscure Proof from Weight

It is a simple matter to arrange the volume-based method so that it will predict obscuration from the weight of solids added. Taking the weight calculated by our volume-based sheet of 2,093.89 lbs. and using that mass as the input weight prior to obscuration, we can then add the same ~13.22 lbs. of solids, resulting in the same obscuration value of 2.4 degrees of proof as in the *Add Solids to Vol. ...* worksheet:

Predict Obscuration of Proof by addition of solids to Weight of Alcohol			
Temperature of Alcohol	60.00	0.0000	Thermometer Cor.
Apparent Proof of Alcohol	80.00	0.0000	Hydrometer Cor.
True Proof Result	80.00		
Lbs. Weight of Alcohol to Be obscured	2,093.89	949.77	Kilograms
Wine Gallons Per Pound @ True proof	0.12616	$13.50	Excise Tax Rate
Weight * W.G. per Lb. = 60° F. Wine Gal.	264.17	$2,853.01	Tax on Proof Gal.
Pounds of Solids Added to Batch	13.228	6.000	Kilograms
Total Lbs.	2,107.12	955.77	Total Kilograms
60° F. Wine Gallons expressed as Liters	1,000.00		
Grams/Liters = Grams Per Liter	6.000	600.00	Milligrams per 100 Ml.
Grams per liter per 1° of Proof Obscuration	2.50	Ratio per 1.0 ° of Proof Obscuration	
Grams per Liter / Ratio = Proof Obscuration	2.40		
Find Proof Gallons Obscured			
True Proof minus obscuration	77.60		
Proof Gallons per Lb. at True P. - Obscuration	0.09772		
Wt. * P. G. per lb. @ Obs. Proof = PG @ Obscured	204.60		
Starting "Unobscured" Proof Gallons	211.33		
Total Proof Gallons Obscured	6.73		
Tax Paid if Wrong Proof Gallons Used	$2,762.16	Don't use this value	
Tax Difference	$90.85		

Add Solids to Wt. of Alcohol & Find Obscuration

Other Obscuration and Gauging Issues

§186.62-§30.62 describes a method of using Table 2 to determine the proof gallons and wine gallons of alcohol when obscuration is present. We will address it while we are discussing obscuration of proof in general. Table 2 is only marked in one-half-pound increments and I find Table 4 to be more accurate. So, following is the procedure to arrange the example so that it uses Table 4 methods. The example in the gauging manual itself has been superseded but as published it recites: 334 Lbs., 105 Proof, 0.8 Obscuration, 45.8 Proof Gal., 43.2 Wine Gal.

Using Weight & Amount of Obscuration to Find Proof Gallons and Wine Gallons

Temp of Alcohol	60.00	0.000	Thermometer Cor.
Apparent Proof	105.00	0.000	Hydrometer Cor.
Weight of Alcohol (lbs.)	334.00	$13.50	Excise Tax Rate
Present Gauged "True" Proof	105.00	0.0015	Tolerance
Known Obscuration of Proof Added In	0.80	0.300	Tolerance ° of Proof
Corrected True Proof	105.80	105.0	Present Gauged True Proof
Find Proof Gallons	Corrected	Gauged Proof	
Proof Gal. per Lb. at Corrected True Proof	0.13683	0.13568	PG per Lb. at Present Gauged True Proof
Wt. * PG per Lb. = Proof Gallons	45.70	45.32	Wt. * PG per Lb. = Proof Gal @ Gauged Proof
Tax on P.G.	$616.96	-0.39	Dif. in Proof Gal. if Obscuration Ignored
		-$5.20	Tax not paid if Present Gauged True Proof is used
	Gauged Proof	Corrected Proof	
Calc. 60° Vol. From Present Gauged Proof	105.00	105.80	Vol. from Corrected True Proof
WG. per Lb. at Present Gauged True Proof	0.12922	0.12933	Wine Gal. Per lb at Corrected True proof
60° Vol. at Present Gauged True Proof	43.16	43.20	Weight * WG per Lb. = Vol @ Corrected proof
		0.04	Vol. Dif. If Corrected is Used

Obscured Alcohol Tracking Based on Weight

The proof gallons come out slightly different from the example (45.7 rather than 45.8) because I have not rounded the obscuration up by 0.2 proof to 1.0 as indicated in the math of the Gauging Manual example. If one knows the obscuration is 0.8 there is no need to round it to more than 1/10 proof, which is the granularity of Table 4. Following is the new example from the regulations presently published online where the apparent proof is set to 96° F:[38]

Using Weight & Amount of Obscuration to Find Proof Gallons and Wine Gallons

Temp of Alcohol	60.00	0.000	Thermometer Cor.
Apparent Proof	96.00	0.000	Hydrometer Cor.
Weight of Alcohol (lbs.)	334.00	$13.50	Excise Tax Rate
Present Gauged "True" Proof	96.00	0.0015	Tolerance
Known Obscuration of Proof Added In	0.80	0.300	Tolerance ° of Proof
Corrected True Proof	96.80	96.0	Present Gauged True Proof
Find Proof Gallons	Corrected	Gauged Proof	
Proof Gal. per Lb. at Corrected True Proof	0.12401	0.12289	PG per Lb. at Present Gauged True Proof
Wt. * PG per Lb. = Proof Gallons	41.42	41.04	Wt. * PG per Lb. = Proof Gal @ Gauged Proof
Tax on P.G.	$559.16	-0.37	Dif. in Proof Gal. if Obscuration Ignored
		-$5.06	Tax not paid if Present Gauged True Proof is used
	Gauged Proof	Corrected Proof	
Calc. 60° Vol. From Present Gauged Proof	96.00	96.80	Vol. from Corrected True Proof
WG. per Lb. at Present Gauged True Proof	0.12801	0.12811	Wine Gal. Per lb at Corrected True proof
60° Vol. at Present Gauged True Proof	42.76	42.79	Weight * WG per Lb. = Vol @ Corrected proof
		0.03	Vol. Dif. If Corrected is Used

Obscured Alcohol Tracking Based on Weight

The §186.62-§30.62 examples simply restate that, in the context of obscuration, one can use the gauged true proof, weight and wine gallons per pound to determine 60°F volume. As to the apparent proof to volume correction, because the alcohol is obscured, so is the apparent proof; this method would become less reliable the more obscuration that is present. Note that there is the same arrangement where the proof gallons per pound at the Corrected Proof (Taxable Proof) are used to calculate the proof gallons and the wine gallons per pound at the Gauged True Proof are used to calculate the volume.

---

[38] §30.62 §186.62 "Table 2, showing wine gallons and proof gallons by weight. *Example*. 334 lbs. of distilled spirits. Apparent proof 96.0°. Obscuration + 0.8°. True Proof 96.0°+0.8°=96.8°. 334 lbs. at 96.0° apparent proof=42.8 wine gallons. 42.8 wine gallons×96.8°=41.4 proof gallons."

Turning to §186.63-§30.63 (which explains Table 3 and is used for determining the number of proof gallons from the weight and proof of spirituous liquor), this section encourages you to use a reductive method outlined in conjunction with Table 3. Using Table 4 allows for more consistent calculation of identical values as produced by Table 3. Following are the two gauging manual examples for this procedure:[39]

§30.63-§186.63

Using Weight & Amount of Obscuration to Find Proof Gallons	
Temp of Alcohol	60.00
Apparent Proof	190.00
Weight of Alcohol	60,378.00
Present Gauged "True" Proof	190.00
Known Obscuration of Proof Added In	0.00
Corrected True Proof	190.00
Find Proof Gallons	Corrected
Proof Gal. per Lb. at Corrected True Proof	0.27964
Wt. * PG per Lb. = Proof Gallons	16,884.10
Tax on P.G.	$227,935.40

	Gauged Proof
Calc. 60° Vol. From Present Gauged Proof	190.00
WG. per Lb. at Pesent Gauged True Proof	0.14718
60° Vol. at Present Gauged True Proof	8,886.43

§30.63-§186.63 Example 2

Using Weight & Amount of Obscuration to Find Proof Gallons a	
Temp of Alcohol	60.00
Apparent Proof	86.00
Weight of Alcohol	321.50
Present Gauged "True" Proof	86.00
Known Obscuration of Proof Added In	0.00
Corrected True Proof	86.00
Find Proof Gallons	Corrected
Proof Gal. per Lb. at Corrected True Proof	0.10906
Wt. * PG per Lb. = Proof Gallons	35.06
Tax on P.G.	$473.35

	Gauged Proof
Calc. 60° Vol. From Present Gauged Proof	86.00
WG. per Lb. at Pesent Gauged True Proof	0.12681
60° Vol. at Present Gauged True Proof	40.77

Obscured Alcohol Tracking Based on Weight

We can also use the *Obscured Alcohol Tracking Based on Weight* worksheet to check the § 30.64-186.64 *Example* 3 which combines all the elements together in one example.[40]

Diagram follows on next page.

---

[39] § 30.63 § 186.63 "Table 3, for determining the number of proof gallons from the weight and proof of spirituous liquor. When the weight or proof of a quantity of distilled spirits is not found in Table 2, the proof gallons may be ascertained from Table 3. The wine gallons (at 60 degrees Fahrenheit) may be ascertained by dividing the proof gallons by the proof. *Example.* **A tank car of spirits of 190 degrees of proof weighed 60,378 pounds net. We find Proof gallons 60,000 pounds equal to 16,778.4 - 300 pounds equal to 83.9 - 70 pounds equal to 19.6 - 8 pounds equal to 2.2 – Total 16,884.1 That is, the total weight of 60,378 pounds of spirits at 190 proof is equal to 16,884.1 proof gallons.** The equivalent gallonage for 70 pounds is found from the column 700 pounds by moving the decimal point one place to the left; that for 8 pounds from the column 800 pounds by moving the decimal point two places to the left.
[40] § 30.64 § 186.64 "*Example.* 5,350 pounds of blended whisky containing added solids: Temperature °F 75.0° Hydrometer reading 92.0° Apparent proof 85.5° Obscuration 0.5° True proof 86.0°. 5,350.0 lbs.*0.12676 (W.G. per pound factor for apparent proof of 85.5°) =678.2 wine gallons. 678.2 W.G.*0.86=583.3 proof gallons."

Using Weight & Amount of Obscuration to Find Proof Gallons and Wine Gallons

Temp of Alcohol	75.00	0.000	Thermometer Cor.
Apparent Proof	92.00	0.000	Hydrometer Cor.
Weight of Alcohol (lbs.)	5,350.00	$13.50	Excise Tax Rate
Present Gauged "True" Proof	85.50	0.0015	Tolerance
Known Obscuration of Proof Added In	0.50	0.300	Tolerance ° of Proof
Corrected True Proof	86.00	85.5	Present Gauged True Proof

Find Proof Gallons	Corrected	Gauged Proof	
Proof Gal. per Lb. at Corrected True Proof	0.10905	0.10837	PG per Lb. at Present Gauged True Proof
Wt. * PG per Lb. = Proof Gallons	583.44	579.80	Wt. * PG per Lb. = Proof Gal @ Gauged Proof
Tax on P.G.	$7,876.43	-3.64	Dif. in Proof Gal. if Obscuration Ignored
		-$49.19	Tax not paid if Present Gauged True Proof is used

	Gauged Proof	Corrected Proof	
Calc. 60° Vol. From Present Gauged Proof	85.50	86.00	Vol. from Corrected True Proof
WG. per Lb. at Present Gauged True Proof	0.12675	0.12681	Wine Gal. Per lb at Corrected True proof
60° Vol. at Present Gauged True Proof	678.13	678.43	Weight * WG per Lb. = Vol @ Corrected proof
		0.30	Vol. Dif. If Corrected is Used

			Table 7 Vol. Cor.	
Find Present Volume From Apparent Proof	Apparent		678.13	Gauged Proof Vol.
WG per Lb. at Apparent Proof	0.12751	Gal. Dif.	1.0060	Factor to Present Volume.
Wt.* WG per Lb at Apparent = Present Vol.	682.19	0.01	682.20	60° Vol. * Factor = Present Gal.
Gal. Vol Dif. From 60° Vol.	4.06		4.07	

Obscured Alcohol Tracking Based on Weight

Because we use Table 4 values to calculate the proof gallons, the 583.44 proof-gallon value is slightly different than the wine-gallon-times-proof/100 value used by the regulation. Also, I have not included rounding for the final 60°F gallon volume. Lastly, we see that the apparent proof and Table 7 volume correction methods are well within any conceivable blending tolerance one might need.

Blend Obscured Alcohol by Weight

Having sorted out the obscuration procedures we can turn our attention to blending obscured alcohol with water to a target proof. To do this we will set up our standard batch example to indicate that we know the alcohol we are going to use has been obscured by 1.0 proof. To do this we will take our standard *One Alcohol Blend to 80 Proof* batch of 102 cases and an alcohol mass of 872.15 pounds and obscure it with solids to the extent of 1.0 proof by adding 2.53 pounds of solids. We obtain the required ~250.0 mg per 100 ml and a total mass of 874.68 pounds:

Predict Obscuration of Proof by addition of solids to Weight of Alcohol

Temperature of Alcohol	60.00	0.0000	Thermometer Cor.
Apparent Proof of Alcohol	160.00	0.0000	Hydrometer Cor.
True Proof Result	160.00		
Lbs. Weight of Alcohol to Be obscured	872.15	395.60	Kilograms
Wine Gallons Per Pound @ True proof	0.13903	$13.50	Excise Tax Rate
Weight * W.G. per Lb. = 60° F. Wine Gal.	121.26	$2,598.65	Tax on Proof Gal.
Pounds of Solids Added to Batch	2.5300	1.148	Kilograms
Total Lbs.	874.68	396.75	Total Kilograms
60° F. Wine Gallons expressed as Liters	459.01		
Grams/Liters = Grams Per Liter	2.500	250.02	Milligrams per 100 Ml.
Grams per liter per 1° of Proof Obscuration	2.50	Ratio per 1.0 ° of Proof Obscuration	
Grams per Liter / Ratio = Proof Obscuration	1.00		

Add Solids to Wt. of Alcohol & Find Obscuration

Having obscured our 160-proof alcohol by 1.0 degree of proof, we are going to assume that the §186.23-§30.23 procedure will provide a Present Gauged True Proof of 159 when it is subsequently gauged. Adding in the 1 degree of proof obscuration we know is there brings us back up to 160 proof and we can proceed. The problem is that we are not able to use what we know to be the present weight of the alcohol, after we have added solids to it, in order to try and match the standard blend to target proof. Apparently, we have to keep track of the pounds of solids and subtract them from the alcohol weight we are using. This is impossible to do with a large tank obscured by solids. Using the original weight of 872.15 lbs. and adding in 1.0 degree of proof obscuration to achieve 160 proof does result in the same total batch values as our standard blend:

Blend One Obscured Alcohol - Reduce Proof by Weight using Table 4

Lbs. Weight of Alcohol	872.15	395.60	Kilograms
Temp of Alcohol	60.00	0.000	Thermometer Cor.
Apparent Proof of Alcohol	159.00	0.000	Hydrometer Cor.
Present Gauged True Proof Result	159.00	0.0015	Tolerance
Amount of Obscuration In ° of Proof	1.00	13.500	Tax per P.G.
Gauged Proof + Obscuration = Corrected Proof	160.00	$2,619.11	Tax
Results	Corrected	Gauged	
Proof Gallons per Lb. @ Corrected Proof	0.22245	0.22071	
Wt. * P.G. per Lb. @ Corrected Proof = P.G.	194.01	192.49	
	P.G. Obscured by Solids	1.52	
Blending Results			
Target Proof	80.0		
Gallons of Water to Add	126.09	477.29	Liters of H2O
Gross Volume of Batch	247.34	Other Results	
Volume Reduction due to Blending	-4.83	Net Liters	-4.83	Volume Difference in Gallons
Actual "net" Volume available for Bottling	242.51	918.01	-18.28	Volume Dif. In Liters
Weight of Water to Add	1,050.07	-24.37	# of 750 ML bottles "Short"
Weight of Alcohol Restated	872.15	-1.95%	Vol Difference/Gross Gallons
Total Fluid Weight of Batch w/o Solids	1,922.22	1.99%	(Gross Gal/Net Gal)-1

Blend One Obscured Alcohol by Weight

If we take the 874.68 lbs. that the alcohol now weighs (after the addition of 2.53 pounds of solids) and use it to blend with, we will end up including an extra 0.5 proof gallons and adding almost a 0.75 gallon more water. The total batch weight will increase from 1,922.22 lbs. to 1,927.80 lbs:

Blend One Obscured Alcohol - Reduce Proof by Weight using Table 4

Lbs. Weight of Alcohol	874.68	396.75	Kilograms
Temp of Alcohol	60.00	0.000	Thermometer Cor.
Apparent Proof of Alcohol	159.00	0.000	Hydrometer Cor.
Present Gauged True Proof Result	159.00	0.0015	Tolerance
Amount of Obscuration In ° of Proof	1.00	13.500	Tax per P.G.
Gauged Proof + Obscuration = Corrected Proof	160.00	$2,626.71	Tax
Results	Corrected	Gauged	
Proof Gallons per Lb. @ Corrected Proof	0.22245	0.22071	
Wt. * P.G. per Lb. @ Corrected Proof = P.G.	194.57	193.05	
	P.G. Obscured by Solids	1.52	
Blending Results			
Target Proof	80.0		
Gallons of Water to Add	126.45	478.67	Liters of H2O
Gross Volume of Batch	248.06	Other Results	
Volume Reduction due to Blending	-4.84	Net Liters	-4.84	Volume Difference in Gallons
Actual "net" Volume available for Bottling	243.22	920.68	-18.33	Volume Dif. In Liters
Weight of Water to Add	1,053.12	-24.45	# of 750 ML bottles "Short"
Weight of Alcohol Restated	874.68	-1.95%	Vol Difference/Gross Gallons
Total Fluid Weight of Batch w/o Solids	1,927.80	1.99%	(Gross Gal/Net Gal)-1

What this means is that one cannot obscure a large batch of alcohol and then remove a mass of that alcohol without keeping track of the mass per unit of volume in the obscured alcohol.

By including the weight of the solids for purposes of calculating proof gallons and also adding the degrees of proof obscuration, one is double correcting. More weight = more tax; higher proof = more tax. This is truly a no-win situation.

As noted above it is also possible to reverse the obscuration and find the quantity of solids that were originally added to the alcohol to obscure it by using the original 60°F volume and the amount of obscuration per degree of proof:

Find Mass of Solids used to Obscure Alc.	Corrected
Wine Gal. per Lb. at Corrected Proof	0.13903
Lbs.* WGPP = 60° Gal. @ Corrected Proof	121.26
Source Alcohol Original 60° F. Liters	459.00
Grams per liter per Degree of Proof Obsc.	2.50
Obscuration * Grams per Liter Value	2.50
(Liters*Grams per Liter)/1,000 = Kg.	1.148
Kg. * Kg to Lbs. = Tot. Lbs. to Obscure Alc.	2.530

Blend One Obscured Alcohol by Weight

Finally, we must note that the target-proof alcohol will still be obscured when gauged:

Find Proof Obscuration After Blending	
Origional Grams per Liter of Obscuration	2.50
60° F. Gal at Corrected Proof	121.26
Source Alcohol Original 60° F. Liters	459.00
Ending Net Blended Liters @ 60° F.	917.99
Original Liters/ Net Liters = Ratio	0.50
Ratio * Grams per Liter = Blended g/Liter	1.25
Proof ° Obscured by 1 Gram per Liter	0.40
g/L * Proof Obscured per Gram = Obscuration °	0.50
Target Proof - ° Obscured= Predicted Proof	79.50
Grams per Liter *100 = mg. per 100 ml	125.00

Predict & Blend by Weight

It seems a shame to know all the mechanisms that affect proof obscuration and not assemble them into one worksheet that predicts obscuration both at the source alcohol level and the target proof level so one can adjust the mass of solids added and see the results before they are added to the batch. Here is one way to set up such a procedure.

Predict Obscuration Before and After Blending to Target Proof Using Weight of Alcohol			
Lbs. Weight of Alcohol	872.15	395.60	Kilograms
Temp of Alcohol	60.00	0.000	Thermometer Cor.
Apparent Proof of Alcohol	160.00	0.000	Hydrometer Cor.
True Proof Result	160.00	0.0015	Tolerance
Pounds per Proof Gallon	4.4954	13.500	Tax per P.G.
Lbs. * PGPP = Proof Gallons	194.01	$2,619.11	Tax

Known Lbs. of Solids to be Added to Alc.	2.53		7.19259	Lbs. per WG at Gauged Proof
Kilograms of Solids	1.15		121.257	Wt. /PPWG = 60° Gal.Vol. of Alc.
(kg/Liters)*1,000 = Grams Per Liter	2.500		459.01	Liters of Alcohol
Grams per liter per 1° of Proof Obscuration	2.50		874.68	Total Lbs of Alc. & Solids
2.5/Grams per Liter= ° of Proof Obscured	1.00		0.29%	Solids % of Total Weight
True Proof - Proof Obscured by Solids =	159.00	Obscured Proof of Source Alcohol if Gauged		
Grams per Liter *100 = mg. per 100 ml	250.02	Source Alc. gm/100 ml Obscuration		

			600.0	Tolerance Limit mg. per 100 ml.
Blending Results			Find Proof Obscuration After Blending	
Target Proof	80.0	Liters of H2O	2.50	Origional Grams per Liter of Obscuration
Gallons of Water to Add	126.09	477.29	459.01	Original Liters
Gross Volume of Batch	247.34		918.01	Ending Net Blended Liters @ 60° F.
Volume Reduction due to Blending	-4.83	Net Liters	0.50	Original Liters / Net Liters = Ratio
Actual "net" Volume available for Bottling	242.513	918.01	1.25	Ratio * Grams per Liter = Blended g/Liter
Weight of Water to Add	1,050.07		0.40	Proof ° Obscured by 1 Gram per Liter
Weight of Alcohol Restated	872.15	Kilograms	0.50	g/L * Proof Obscured per Gram = Obscuration °
Total Fluid Weight of Batch	1,922.22	871.91	79.50	Target Proof - ° Obscured= Predicted Proof
Total Fluid & Solids Weight of Batch	1,924.75	873.05	125.01	Blend Grams per Liter *100 = mg. per 100 ml

Predict Obscuration & Blend 1 Alc. Weight

We see that while the source alcohol is obscured by 250 milligrams per 100 ml, the blended alcohol only contains 125.01 milligrams per 100 ml, which is well below our blending tolerance of 600 milligrams per 100 ml.

Stretching the procedure to the limit of 600 milligrams per 100 ml we see that it is possible to add up to 12.14 pounds of material and, even though the source alcohol is obscured by 1,199.7 milligrams per 100 ml, the blended batch is still beneath the threshold of 600 milligrams per 100 ml at 599.84 total. The reason we can't use the 13.22 lbs. of solids is that in this case we must use the net blended volume to gauge the limit of obscuration, not the gross volume contributed.

Predict Obscuration Before and After Blending to Target Proof Using Weight of Alcohol

Lbs. Weight of Alcohol	872.15	395.60	Kilograms
Temp of Alcohol	60.00	0.000	Thermometer Cor.
Apparent Proof of Alcohol	160.00	0.000	Hydrometer Cor.
True Proof Result	160.00	0.0015	Tolerance
Pounds per Proof Gallon	4.4954	13.500	Tax per P.G.
Lbs. * PGPP = Proof Gallons	194.01	$2,619.11	Tax

Known Lbs. of Solids to be Added to Alc.	12.14		7.19259	Lbs. per WG at Gauged Proof
Kilograms of Solids	5.51		121.257	Wt. /PPWG = 60° Gal.Vol. of Alc.
(kg/Liters)*1,000 = Grams Per Liter	11.997		459.01	Liters of Alcohol
Grams per liter per 1° of Proof Obscuration	2.50		884.29	Total Lbs of Alc. & Solids
2.5/Grams per Liter= ° of Proof Obscured	4.80		1.37%	Solids % of Total Weight
True Proof - Proof Obscured by Solids =	155.20		Obscured Proof of Source Alcohol if Gauged	
Grams per Liter *100 = mg. per 100 ml	1,199.68		Source Alc. gm/100 ml Obscuration	

			600.0	Tolerance Limit mg. per 100 ml.
Blending Results			Find Proof Obscuration After Blending	
Target Proof	80.0	Liters of H2O	12.00	Origional Grams per Liter of Obscuration
Gallons of Water to Add	126.09	477.29	459.01	Original Liters
Gross Volume of Batch	247.34		918.01	Ending Net Blended Liters @ 60° F.
Volume Reduction due to Blending	-4.83	Net Liters	0.50	Original Liters / Net Liters = Ratio
Actual "net" Volume available for Bottling	242.513	918.01	6.00	Ratio * Grams per Liter = Blended g/Liter
Weight of Water to Add	1,050.07		0.40	Proof ° Obscured by 1 Gram per Liter
Weight of Alcohol Restated	872.15	Kilograms	2.40	g/L * Proof Obscured per Gram = Obscuration °
Total Fluid Weight of Batch	1,922.22	871.91	77.60	Target Proof - ° Obscured= Predicted Proof
Total Fluid & Solids Weight of Batch	1,934.36	877.41	599.84	Blend Grams per Liter *100 = mg. per 100 ml
			OK	Logical Test for Tolerance
Volume Difference in Gallons	-4.83	-18.28	Liters	
Vol Difference/Gross Gallons	-1.95%			

Predict Obscuration & Blend 1 Alc. Weight

The interesting point to note is that the volume difference in gallons remains the same as does the gallons of water because those variables are tied to the target proof. We will address that issue shortly.

Blend Obscured Alcohol by Volume

The question then becomes whether it is possible to blend obscured alcohol by volume rather than weight in order to get around the problems associated with measuring additional pounds per unit of mass that adding solids generates and which flows down into the blending worksheets as an increase in proof gallons and total batch mass.

Returning to a simple blend using a known degree of proof obscuration, let's see how well we can translate the prior weight example into volume blending. Starting with the same standard blend we measure out the 121.257 gallons of alcohol called for by our prior blends:

Blend One Obscured Alcohol - Using Volume of Alcohol				
Temp of Alcohol	60.00	0.000	Thermometer Cor.	
Apparent Proof of Alcohol	159.00	0.000	Hydrometer Cor.	
Present Gauged True Proof Result	159.00	0.00150	Tolerance	
What is the Target Proof of the Batch	80.00	$13.50	Tax Rate	
Amount of Obscuration	1.00	$2,619.11	Tax	
True Proof @ 60 + Obscuration	160.00			
Present Volume of Alcohol to be used	121.257			
Table 7 Volume Correction Factor applied	1.0000			
Volume of Alcohol Corrected to 60° F.	121.26			
60° F. Volume Difference from Present Vol.	0.00			
Proof Gallons of Alcohol	194.01		192.49	P.G. at Unobscured proof
Weight of Alcohol	872.15		1.52	P.G. Obscured by Solids
Blending Results				
60° F. Gallons of Water to Add	126.09	477.29	Liters of H2O	
Gross Volume of Batch	247.34		Other Results	
Volume Reduction due to Blending	-4.83	Net Liters	-4.83	Volume Difference in Gallons
Actual "net" Volume available for Bottling	242.51	918.01	-18.28	Liters
Weight of Water to Add	1,050.07		-24.37	# of 750 ML bottles "Short"
Lbs. Weight of Alcohol	872.15		-1.95%	Vol Difference/Gross Gallons
Total Weight of Batch	1,922.22		1.99%	(Gross Gal/Net Gal)-1

Blend One Obscured Alcohol by Volume

The weight of alcohol is not the actual weight because we know that it has been obscured by the addition of 2.53 lbs. of solids. By ignoring that fact we can blend to correct batch volume, ensuring that the correct amount of alcohol and water is in the batch for the consumer by not overstating the mass of alcohol (actually alcohol including solids) and then adding too much water.

In this case volume has come to the rescue and made things simpler than using weight and one can obscure a large batch of alcohol and then withdraw amounts by volume for subsequent blending to target proofs. What we haven't addressed so far is how to keep track of the volume of solid material that is added to the batch and whether and how we should reduce the volume of water required; but, at the concentrations we have been dealing with, that aspect is apparently ignored by the regulations.

Predict & Blend by Volume

Concluding, we see that we can change the temperature, apparent proof and present volume to values so that a true proof correction matches the 160-proof alcohol we are using as our source alcohol and that the batch itself blends with the same results as in prior examples:

Predict Obscuration Before and After Blending to Target Proof Using Volume of Alcohol

Temp of Alcohol	75.00	0.000	Thermometer Cor.
Apparent Proof of Alcohol	164.90	0.000	Hydrometer Cor.
Apparent Proof Made True @ 60°	160.00	0.00150	Tolerance
Present Volume of Alcohol to be used	122.29	$13.50	Tax Rate
Table 7 Volume Correction Factor applied	0.9915	$2,618.97	Tax
Volume of Alcohol Corrected to 60° F.	121.25	458.98	Liters
60° F. Volume Difference from Present Vol.	-1.04		
Proof Gallons of Alcohol	194.00		
Weight of Alcohol	872.11		

Known Lbs. of Solids to be Added to Alc.	2.53	7.19259	Lbs. per WG at Gauged Proof
Kilograms of Solids	1.15	121.25	Wt. /PPWG = 60° Gal.Vol. of Alc.
(kg/Liters)*1,000 = Grams Per Liter	2.500	458.98	60° F.Liters of Alcohol
Grams per liter per 1° of Proof Obscuration	2.50	874.64	Total Lbs of Alc. & Solids
2.5/Grams per Liter= ° of Proof Obscured	1.00	0.29%	Solids % of Total Weight
True Proof - Proof Obscured by Solids =	159.00		Obscured Proof of Source Alcohol if Gauged
Grams per Liter *100 = mg. per 100 ml	250.03		Source Alcohol Obscuration

Blending Results		Liters of H2O	Find Proof Obscuration After Blending	
			600.0	Tolerance Limit mg. per 100 ml.
Target Proof	80.00		2.50	Origional Grams per Liter of Obscuration
60° F. Gallons of Water to Add	126.08	477.26	458.98	Original Liters
Gross Volume of Batch	247.33		917.97	Ending Net Blended Liters @ 60° F.
Volume Reduction due to Blending	-4.83	Net Liters	0.500	Original Liters / Net Liters = Ratio
Actual "net" Volume available for Bottling	242.50	917.97	1.25	Ratio * Grams per Liter = Blended g/Liter
Weight of Water to Add	1,050.02		0.40	Proof ° Obscured by 1 Gram per Liter
Lbs. Weight of Alcohol	872.11		0.50	g/L * Proof Obscured per Gram = Obscuration °
Total Fluid Weight of Batch	1,922.12	Kg.	79.50	Target Proof - ° Obscured= Predicted Proof
Total Fluid & Solids Weight of Batch	1,924.65	873.01	125.01	Blend Grams per Liter *100 = mg. per 100 ml
			OK	Logical Test for Tolerance

Volume Difference in Gallons	-4.83	-18.28	Liters
Vol Difference/Gross Gallons	-1.95%		

Predict Obscuration & Blend 1 Alc. Volume

The volume correction operates to correct to the 60°F volume. From that we can acquire the proof gallons and mass of the alcohol, allowing the blend to proceed normally since the pounds of water are calculated based upon the target proof.

Analysis of Obscuration

To complete our discussion, I have put together an Analysis of Obscuration worksheet. In this case we will compare the assumed density of 2.5 grams per cubic centimeter to a known density of the material actually added. Let's assume we are using 8 pounds of calcium carbonate (in essence powdered limestone like what one finds in hard water). This compound has a density of 2.711 grams per cubic centimeter and is also equal to 2.711 kilograms per liter.

Label	Value	Kg. Solids	Value	Label
Lbs. Weight of Alcohol	872.15		395.60	Kilograms
Temp of Alcohol	60.00		0.000	Thermometer Correction
Apparent Proof of Alcohol	160.00		0.000	Hydrometer Correction
True Proof Result	160.00		0.0015	Tolerance
Pounds per Proof Gallon	4.4954		13.500	Tax per P.G.
Lbs. * PGPP = Proof Gallons	194.01		$2,619.13	Tax
Known Lbs. of Solids to be Added to Alc.	8.00	3.63	7.19269	Lbs. per WG at True Proof
Kilograms of Solids	3.63		121.26	Wt. /PPWG = 60° Gal.Vol. of Alc.
(kg/Liters)*1,000 = Grams Per Liter	7.906		459.00	Liters of Alcohol
Grams per liter per 1° of Proof Obscuration	2.50		880.15	Total Lbs of Alc. & Solids
2.5/Grams per Liter= ° of Proof Obscured	3.16		0.91%	Solids % of Total Weight
True Proof - Proof Obscured = Obscured Prf.	156.84		7.22909	Lb. per WG at Obscured Proof
g/l *100 = mg./100 ml = Obscuration	790.57		120.645	Wt./ Lb. per WG = Vol. @ 60° F.
			0.611	Gal. Dif.
Density of Material Added (grams per CC)	2.711		2.31	Liters Dif.
Kg. Mass/Density = Liter Volume	1.339		4.39	Lbs. per WG * Gal. = Mass
Gal. Vol.	0.35		1.99	Kilograms
			861.87	(Mass/Vol)*1,000 =Kg. m3
Source Alcohol Find Kg. per Liter and Meter Cubed				
Pounds per Wine Gallon	7.1927		0.86364	SG Air of True Proof
Pounds Avdp to Kilograms	0.45359		0.9979462	Water Den @ 60° F. (Air)
Lbs Per WG * Factor = Kg. per Gallon	3.26		861.870	(SG * Water Den)*1,000 =Kg. m3
Gallons per Liter	0.2642			
Kg. per Gal. * Gal. per Liter = Kg. per Liter	0.86187		0.86380	SG Vac of True Proof
Kg. per Liter * 1,000 = Kg. per m3 @ 60° F. (Air)	861.87		0.999020	Water Density @ 60° F. (Vac)
Kg. per Liter of Air at 87.5 % Std. Atm.	1.070		0.862953	(SG * Water Den)*1,000 =Kg. m3
Add 7/8 Air Value = Kg. m3 Vac	862.94		862.95	Kg. per m3
			1.080	Air |Vac Dif. As Kg. m3

Analysis of Obscuration

We see that our 8 pounds of material occupy 1.339 liters of volume before adding it to the batch. It may or may not actually take up this much space in the blend. Either way, it is a nominal amount for a batch of the size we are blending. In comparison, the volume difference that is obscured by changing the measurable gauged proof accomplished by the material is far greater at 2.31 liters and 1.99 kilograms of alcohol.

Because the density is far less, the volume is greater despite the mass being less. The mass/volume ratio equals the density of the alcohol and amounts to 861.87 kg per meter cubed (kg/m^3). Looking up the specific gravity in air of the True Proof of the Alcohol yields 0.86364, which when divided by the density of water in air at 60°F also results in 861.87 kg/m^3.

Applying the SG Vac value from Table 6 to the density of water in vacuum yields 862.95 kg/m^3 and the difference between them is 1.080 kg/m^3. This represents the air buoyancy between air and vacuum densities; this is not quite the same as the total mass of air displaced but is approximately 7/8 of that density because of the volume occupied by the balance weights used to make these calculations. We will have more to say about why 7/8 (87.5%) of the air pressure is a good all-around number to use instead of full atmospheric pressure of 0.001223 kilograms per liter or 1.223 kg per meter cubed a little later on.

This is just a first run through how to navigate changing values from one form to another.

To confirm these values, we can follow the pounds per wine gallon down through a conversion to metric units and derive the same value of 861.87 kg/m^3 in air derived from the Specific Gravity in air. Using a rule-of-thumb standard value of 1.070 kg/m^3 for the mass of air displaced by a cubic meter of material yields 862.94 kg/m^3 in a vacuum, which is within 10 grams per meter cubed of the value obtained by using the specific gravity in vacuum of 862.95.

One last value we can extract from Table 4 is to determine the mass percent of the source alcohol. To do this we find that the ratio of Wine Gallons to Proof Gallons is indeed 1.6 for our 160-proof alcohol and then divide that number by two in order to reduce the value to how much absolute alcohol is contained in each wine gallon of 160 proof. That value is 0.8 wine gallons of absolute alcohol or 200-proof alcohol:

Source Alcohol Find Mass %

Lbs. per Wine Gal. @ True	7.1927	3.26255	Kg.
Lbs. per Proof Gal. at True Proof	4.4954		
Lbs. per WG/Lbs. per PG = Proof Gal. per WG.	1.6000		
Divide by 2 = Absolute Alcohol Gal. per WG	0.8000		AA Lbs. per PG
200 Proof Alc. Lbs. per WG = Absolute Alc.	6.60982	Kg.	3.30491
AA Lb/WG * AA Gal per WG = Lbs. AA per WG	5.2879	2.39855	2.6440
Lbs. AA per WG/Lbs. per WG = Fraction AA	0.7352		0.3676
Result as a Percentage Alcohol % Mass (Air)	73.52%	Kg.	73.52%	Times 2
Lbs. per WG True Proof - Lbs AA = Lbs. H2O	1.905	0.86400	

Find Mass of Alcohol Obscured by Solids Added

Lbs. of Solids * Mass % Alc = Lbs. Alc. Mass.	5.88	2.67	Kg. Alc
Lbs. per Wine Gal. - Alc. Mass = H2O Mass	2.12	0.961	Kg. of H2O
Tot. Lbs. Used to Obscure Alcohol	8.00	3.629	Total Kg.

Analysis of Obscuration

Table 4 will also provide the value for pounds per wine gallon of 200-proof alcohol; we can use that value to multiply our 0.8 absolute alcohol gallons per wine gallon value to find the actual pounds of absolute alcohol contained in one wine gallon of our source alcohol. The resulting value of 5.2879 pounds absolute alcohol is then divided by Pounds per Wine Gallon to find the fraction of absolute alcohol contained in each wine gallon. Then that number is expressed as a percentage of the mass of each gallon. Finally, we can determine just how much water is contained as well. These values can also be arrived at by using the pounds per proof gallon of absolute alcohol but one must then multiply by two to find the mass percentage. Note that the mass percentage is in air and not vacuum. There is a way to change one to the other that will be covered a little later on.

We can also find the pounds of alcohol actually obscured by using the mass percentage we have just found and relating it to the total pounds of solids that were added to the batch.

To conclude our discussion, we will follow the same analysis after our blend and see the changes engendered by adding water to our batch in order to bring it down to target proof:

Blending Results				Find Proof Obscuration After Blending	
Target Proof	80.0	Liters of H2O		7.91	Origional Grams per Liter of Obscuration
Gallons of Water to Add	126.09	477.29		459.00	Original Liters
Gross Volume of Batch	247.34			917.99	Ending Net Blended Liters @ 60° F.
Volume Reduction due to Blending	-4.83	Net Liters		0.50	Original Liters / Net Liters = Ratio
Actual "net" Volume available for Bottling	242.51	917.99		3.95	Ratio * Grams per Liter = Blended g/Liter
Weight of Water to Add	1,050.07			0.40	Proof ° Obscured by 1 Gram per Liter
Weight of Alcohol Restated	872.15	Kilograms		1.58	g/L * Proof Obscured per Gram = Obscuration °
Total Fluid Weight of Batch	1,922.22	871.90		78.42	Target Proof - ° Obscured= Predicted Proof
Total Fluid & Solids Weight of Batch	1,930.22	875.53		395.29	Blend Grams per Liter *100 = mg. per 100 ml
				OK	Logical Test for Tolerance

Blended Alcohol Find Kg. per Liter and Meter Cubed					
Pounds per Wine Gallon	7.9264			7.93651	Lb. per WG
Pounds Avdp to Kilograms	0.45359			242.200	Wt./ Lb. per WG
Lbs Per WG * Factor = Kg. per Gallon	3.60			0.308	Gal. Dif.
Gallons per Liter	0.2642			1.16	Liters Dif.
Kg. per Gal. * Gal. per Liter = Kg. per Liter	0.94980				
Kg. per Liter * 1,000 = Kg. per m3 @ 60° F. (Air)	949.80				
Kg. per Liter of Air at 87.5 % Std. Atm.	1.070				
Add 7/8 Air Value = Kg. m3 Vac	950.87				

Blend Alcohol Find Mass %					
Lbs. per Wine Gal. @ Target	7.9264	3.5954	Kg.		
Lbs. per Proof Gal. at Target	9.9079				
Lbs. per WG/Lbs. per PG = Proof Gal. per WG.	0.8000				
Divide by 2 = Absolute Alcohol Gal. per WG	0.4000		AA Lbs. per PG		
200 Proof Alc. Lbs. per WG = Absolute Alc.	6.60982	Kg.	3.30491		
AA Lb/WG * AA Gal per WG = Lbs. AA per WG	2.6440	1.19929	1.3220		
Lbs. AA per WG/Lbs. per WG = Fraction AA	0.3336		0.1668		
Result as a Percentage Alcohol % Mass (Air)	33.36%	Kg.	33.36%	Times 2	
Lbs. per WG True Proof - Lbs AA = Lbs. H2O	5.2825	2.3961			

Find Mass of Alcohol Obscured by Solids Added				
Lbs. of Solids * Mass % Alc = Lbs. Alc. Mass.	2.67	1.21	Kg. Alc	
Lbs. per Wine Gal. - Alc. Mass = H2O Mass	5.33	2.418	Kg. of H2O	
Tot. Lbs. Used to Obscure Alcohol	8.00	3.629	Total Kg.	

Analysis of Obscuration

The grams of obscuration are below the limit of 400 at 395.29. The mass percentage has decreased to 33.36% and the pounds of alcohol displaced is also roughly half of the original value as well.

Dissolved Solids in Water

Another variable we will address is the water weight constant we use to determine the mass of water to use in our blends. Unless water is distilled, it contains some dissolved solids content. Most distillers do not desire dissolved solids in their blending water and to the extent possible, short of distillation, try to filter them out. However, any dissolved solids remaining can affect the blend weight. The regulations require you to use the weight of 60°F pure water in your calculations so the closer you can get to that weight by filtering out dissolved solids, the better.

My water at the distillery comes into the shop with about 200–250 parts per million of calcium carbonate and other dissolved solids in it. I process the water through a reverse osmosis system that removes most of that—as a method to obtain good blending water it works well. For our example let's assume that one has water containing 250 parts per million of dissolved solids that have not been filtered out and also assume these solids have the density of calcium carbonate at 2.711 grams per cubic centimeter.

Using our standard batch value of 126.09 US gallons for the quantity of water required we find that at 250 ppm our water contains 0.713 lbs. of solids and this amounts to 45.28% of the water weight blending tolerance:

60° F. Gallons of Water Needed for Batch	126.09		0.0015	Tolerance
Gallons to Liters Converter	3.7854118		0.189	Gallon Tolerance
Liters of Water Required	477.29		0.72	Liters of Tolerance
Water Lbs. per Gal. @ 60° (§ 5.47a)	8.32823		1.58	Lbs. of Tolerance
Lbs. Mass of Pure Water	1,050.07			
Find PPM Relationship to Cubic Centimeters				
Parts Per Million of Dissolved Solids in Water	250.00			Or
Parts per Million * Liters = Total Parts	119,321.75		0.000250	Parts / 1,000,000
Total Parts / 1,000,000 = Decimeters Cubed	0.119	Liters	0.119	Result * Tot. Liters = Solids Vol. in Liters
Dm3 * 1,000 = Cubic Centimeters	119.32			
Find Weight of Solids in Water from PPM				Find Obscuration Caused by Water
Density of Solids in Grams per Cubic Centimeter	2.711		0.323	Kilograms of Solids
CC's * Grams per Cm Cubed = Total Grams	323.48		0.678	(kg/Liters)*1,000 = Grams Per Liter
Divide by 1,000 to Find Kilograms of Solids	0.323		2.50	Grams per liter per 1° of Proof Obscuration
Pounds of Solids in Water	0.713		0.27	2.5/Grams per Liter= ° of Proof Obscured
Percent of Wt. Tolerance used by Solids in H2O	45.28%			

Dissolved Solids in Water

Almost more significantly, the obscuration of proof caused by the water is equal to 0.27 proof and since our blending tolerance is 0.3 proof this is almost at the limit. In our standard batch from 160 to 80 proof the total blended volume will be roughly twice the volume of water and so the obscuration will be halved to ~0.14 proof. This is just about the detectable limit for hydrometer use but it always causes the alcohol to read a lower proof than true. If not accounted for, this will result in the distiller consistently adding more proof gallons to bring the proof up when that is not necessary if they understand the obscuration caused by their water.

Next, we will take the calculated liters of water and convert them directly to gallons and fluid ounces to find the volume of water required to make up the molecular amount of water for the batch on a volume basis. This value turns out to be only 4.03 fluid ounces. We can also note that the volume itself is only 16.67% of the tolerance and that the ratio of tolerance closely matches the density of material we assumed for the solids added:

Find Actual Gallons per Lb. & Batch Mass H2O				Use Volume Displaced Find % of Vol. Used
Number of Gallons	126.09		0.1193	Net increase in Liters restated
Lbs. Excess caused by Solids	0.713		0.032	Gallon Volume of Water
Lbs. Excess/Gal. = Lbs. per Gal. Increase	0.0057		4.03	Fluid Ounces of Water
Water Lbs. per Gal. @ 60° (§ 5.47a)	8.32823		16.67%	Percent of Gallon Tolerance
Wt. per Gal. + Increase = Actual Lbs. per Gal.	8.33389			
Gal. * Actual Lbs. of Water = Mass of H20	1,050.79			
Lbs. Extra Mass	0.713			
Kg. Extra Mass	0.323			

Dissolved Solids in Water

What is useful about this is that we can predictably change the mass of water per gallon to include the parts of solids contained in it. The total mass calculated from the increased value matches the values calculated above for the total batch mass with the solids expressed as parts per million.

This suggests that one could change the Wine Gallons per Pound of water in a blending sheet to more accurately determine the mass of water required if the parts per million of the water employed were measured or simply a consistent value over the regulation standard.

I hate to drag this out but there is one more approach that can be taken and that is to regard the mass as the primary controlling factor and try to determine the mass of water required to provide the molecular amount of water called for by a blend to target proof. In this case we should probably use the relative density of solids to water but I worked it out from density:

Using Ratio of Density find Mass of Water and Volume	
Kilograms of Solids Restated	0.323
CC's as "Liters" of Volume Restated	0.1193
Ratio of Solids to Volume	2.71
Displaced Vol * Ratio = Liters of Vol. at 1.0 SG	0.32336
Water Density at 60° F. (Air)	0.99795
Vol * Density = Kg. Mass of Water	0.3227
Vol. in "Liters" * Constant = Gal. of Water	0.0854
Fluid Ounces of Water	10.93
Gal. of fluid * Lbs. per Gal. of Water = Lbs. H2O	0.711

Dissolved Solids in Water

We take the kilograms in the water and divide that value to find the ratio of volume to mass—same as when we started since it is in the definition of what density means. Then we multiply the volume times the ratio to find the volume as if the other material had an SG of 1 relative to the solids. Then we take the actual density of water and multiply it by the volume: we have just found the mass in kilograms. The volume in liters can be converted directly to a gallon value and with a tiny shortcut we can covert that gallon volume to a mass in pounds by multiplying it by the mass of water in pounds per gallon.

On a volume basis the open question is whether one should "make up" the deficiency in volume caused by the solids. I think the answer is no. On a volume basis we should keep track of the obscuration caused by solids in water rather than trying to get the molecular amount of water correct. If we add water on top of the obscuration, it will dilute the mixture further and we will be that much farther off of our target proof.

I think the answer is the same as when blending obscured alcohol. When blending by mass one discounts the mass of solids added in order that the mass of water required will come out correctly. When blending obscured alcohol by volume, only volume is used, and the additional mass per unit volume is ignored.

On a mass basis it might make more sense to make up the mass difference. This is a little convoluted but let's start with it and refine it later in the flavor additions section coming up. First, we start with a volume but simply generate the mass required if it was pure water. Then using the actual pounds of water, we find the difference yet again. Now we net out the difference because the solids do not need to be counted to the extent that they have the mass of water itself and supply part of the mass necessary to the blend:

Gallons of Water	126.09
Water Lbs. per Gal. @ 60° (§ 5.47a)	8.32823
Lbs. of Water Total	1,050.07
Actual Lbs. per Gal. of Water	8.3339
Lbs. on Scale * Actual Lbs. = Actual Gal.	126.00
Gallon Deficiency Based on Weight	0.086
Lbs. of H20 Deficiency	0.713
Liter deficiency based on weight	0.324
Liters taken up by solids	0.119
Net Liter Volume not accounted for	0.205
Net Gal. Volume Deficiency	0.054
Fluid Ounces of make up water	6.92
Net Lbs. Deficiency	0.4505
Net Kg. Deficiency	0.204
Actual Lbs. to Load on Scale.	1,050.52
Lbs. per Gal. to use for blend	8.3318

Dissolved Solids in Water

Flavors

Batch Flavor Additions

This section is going to get a little boring because there are a lot of technical requirements surrounding the addition of flavors to alcohol. This section also attempts to address some of the volume and mass issues that we sidestepped in the last section regarding how to account for the volume of the solids that are added.

The requirements regarding the addition of components other than alcohol and water to make beverage alcohol products are partially provided in §5.23 of part 5 of 27 CFR.[41] The TTB also provides the definitions of flavoring materials footnoted below and are cited for completeness.[42]

In addition to flavors and solids the use of sugar to sweeten alcohol is required for many beverage blending formulas and is becoming more popular. The regulations even contemplate adding sugar to vodka and we will use that as our first example.

Title 27 Chapter I Part 24 Subpart F Sec. 24.181 quantifies the amount of sugar necessary to displace one US gallon of water as 13.5 pounds of sugar. There is also a definition of sugar itself contained in 27 CFR Subpart B: Definitions, §4.10 which states: "Sugar. Pure cane, beet, or dextrose sugar in dry form containing, respectively, not less than 95% of actual sugar calculated on a dry basis." That regulation also contains the following cryptic definition of Brix: "Total solids. The degrees of Brix of the dealcoholized wine restored to its original volume."

Relating the mass of sugar per gallon volume to metric units allows us to calculate the density in kilograms per meter cubed, in air:

Find Density of Sugar			
Sugar Lbs. per Gal.	13.50	13.50	Lbs. per Gal. of Volume Increase
Kilograms to Pounds Avdp	0.4535924	16.00	Wt. Oz per Lb.
Kg. per Gallon	6.123	216.00	Total Oz. per Gal of Vol. Increase
m3 to Gal.	264.17205	28.3495	Ounces To Grams Conversion Factor
Kg m3 Air	1,617.66	6,123.49	Total Grams per Gal.
Liters per m3	1,000.00	6.123	Kg. per Gal.
Kg. per Liter	1.618	2.204623	Kg. to Lbs.
Grams per Liter	1,617.66	13.50	Lbs. per Gal.

---

[41] 27 CFR Part 5 §5.23 "Alteration of class and type. (a) *Additions.* (1) The addition of any coloring, flavoring, or blending materials to any class and type of distilled spirits, except as otherwise provided in this section, alters the class and type thereof and the product shall be appropriately redesignated. (2) There may be added to any class or type of distilled spirits, without changing the class or type thereof, (i) such harmless coloring, flavoring, or blending materials as are an essential component part of the particular class or type of distilled spirits to which added, and (ii) harmless coloring, flavoring, or blending materials such as caramel, straight malt or straight rye malt whiskies, fruit juices, sugar, infusion of oak chips when approved by the Administrator, or wine, which are not an essential component part of the particular distilled spirits to which added, but which are customarily employed therein in accordance with established trade usage, if such coloring, flavoring, or blending materials do not total more than 2 ½ percent by volume of the finished product. (3) "Harmless coloring, flavoring, and blending materials" shall not include (i) any material which would render the product to which it is added an imitation, or (ii) any material, other than caramel, infusion of oak chips, and sugar, in the case of Cognac brandy; or (iii) any material whatsoever in the case of neutral spirits or straight whiskey, **except that vodka may be treated with sugar in an amount not to exceed 2 grams per liter and a trace amount of citric acid.** (b) *Extractions.* The removal from any distilled spirits of any constituents to such an extent that the product does not possess the taste, aroma, and characteristics generally attributed to that class or type of distilled spirits alters the class and type thereof, and the product shall be appropriately redesignated. In addition, in the case of straight whisky the removal of more than 15 percent of the fixed acids, or volatile acids, or esters, or soluble solids, or higher alcohols, or more than 25 percent of the soluble color, shall be deemed to alter the class or type thereof. (c) *Exceptions.* (1) This section shall not be construed as in any manner modifying the standards of identity for cordials and liqueurs, flavored brandy, flavored gin, flavored rum, flavored vodka, and flavored whisky or as authorizing any product which is defined in §5.22(j), Class 10, as an imitation to be otherwise designated."

[42] The TTB provides the following definition of flavoring materials: "Alcoholic flavoring materials: Distilled Spirits: Any nonbeverage product on which drawback has been or will be claimed under 26 U.S.C. 5131-5134 or flavors imported free of tax which are unfit for beverage purposes. The term includes eligible flavors but does not include flavorings or flavoring extracts manufactured on the bonded premises of a distilled spirits plant (DSP) as an intermediate product." Also the following: "Alcoholic beverage: This includes any beverage in liquid form which contains not less than one-half of one percent (0.5%) alcohol by volume and is intended for human consumption." Also the following: "Drawback: A return or rebate, in whole or in part, of excise taxes previously paid. A drawback is granted when the claimant has complied with certain statutory requirements. For TTB purposes, drawback is not a 'refund.'"

The result of 1,617.66 k/m^3 is consistent with other measurements obtained from technical literature.[43]

As we add sugar and flavors to alcohol we will be using the same standard batch values for consistency. The procedure requires that you know the net blended volume of your batch so you will need to calculate these values in a blending worksheet and then enter them into this one. I've set the parameters to the maximum amount of sugar that we could add to a batch and still have the resulting blend qualify for the standard of identity of vodka by keeping the sugar added below the defined limit of 2 grams of sugar per liter:

Batch Flavor Additions to Volume That Reduce the Volume of Water Required in a Batch

60° F. Volume of Alcohol in Batch	121.26			0.0015	Tolerance
60° F. Gallons of Water to be used in Batch	126.09			0.364	Blended Gal. Tolerance
Total 60° F. Vol. to be Contributed to batch	247.35			1.38	Liters of Tolerance
Target Proof of the Batch	80.00			2.88	Lbs. of Tolerance
Net Blended Volume	242.507	917.99	Liters	$13.50	Tax Rate
				$2,619.04	Tax on Proof Gal.
Pounds of Sugar to be used in batch	4.000	1.81	Kg.	Brix Change	from Sugar additions
Sugar Pounds Per Gallon of Volume	13.50			64.00	Lbs. of Sugar * 16 Oz. per Lb. = Tot. Oz.
Sugar Vol. in Fluid Gal. (Wine Gal.)	0.30	1.12	Liters	0.26	Ounces / Blended Gal. = Oz. per Gal.
Gallons of Water used to make Sugar Syrup	0.50			0.75	% Oz. of Sugar per 1% (°) Brix per Gal.
Additions to Vol. Eligible Flavors or Colors	0.00			0.35	Oz. per Gal. / Sugar per 1% of Brix = Brix
Total Fluid Added	0.50				
Gross Gallons needed - Fluid Vol. = Net Gal.	125.59			2.50%	% by Vol. Flavors & Solids Vol. Limit
Total Additions to Vol. including Sugar	0.80			0.328%	% By Volume Additions
H2O Gal. - Vol. Added = Net Gal.of Water	**125.29**			OK	Logical Test for Limit
Find Obscuration					
Grams of Solids	1,814.37				
Liters of Volume	917.99				
Grams/Liters = Grams Per liter	**1.976**				
mg. per 100 Ml.	197.65				
Grams per liter per 1° of Proof Obscuration	2.50			600.0	Limit for mg. per 100 ml.
Grams per Liter / 2.5 = ° of Proof Obscuration	0.79			197.65	Blended mg. per 100 ml
True Proof - Proof Obscured = Gauged Proof	79.21			OK	Test for Tolerance

Flavor Additions That Reduce Vol. of H$_2$O Needed

With four pounds of sugar added to the batch, the volume will increase by 0.30 gallons. We used 0.5 gallons of water in order to make sugar syrup out of our dry sugar so that amount will also have to be subtracted. Then a net volume can be determined and converted into the weight of water remaining to be added. Using the grams and liters of volume we find that the grams per liter are just under the 2.0 maximum at 1.976 and we can maintain the standard of identity of vodka assuming our source alcohol began as Grain Neutral Spirits.

Looking at the right-hand column we see that the Brix has only changed by 0.35 using the value of 0.75 ounces of change in sugar per degree of Brix. We will cover this a little more fully in our second flavor worksheet. We also have to be aware that the total limit for adding material is 2.5% of the volume; I take this to mean the blended volume rather than the gross, so that is the comparison made. In this case it is de-minimus at 0.328% of the blended volume. The net gallons of water are the 60°F gallons that can be added to the batch if one is blending by volume. Calculating the weight of water is complicated by the fact that the weight of sugar is now accounted for as an increase in volume rather than mass.

Examining this problem, we see that the gallons of water used to make sugar syrup amounted to 4.16 pounds. We have not used any other flavors in this mixture but, if we had, I have made an initial assumption that those flavors would have the density of water. This value can be changed if one knows the density of the actual flavors added or their combined density.

---

[43] A. E. Flood and S. Puagsa, "Refractive Index, Viscosity, and Solubility at 30°C, and Density at 25°C for the System Fructose + Glucose + Ethanol + Water," *J. Chem. Eng. Data*, 2000, 45 (5), pp 902–907. Harvey Wilson was able to determine a value of 1,615 kg/m^3 for the density of sugar used in their work.

Find Weight Reduction In Water Required				
Water Constant Lbs. per Gal. @ 60° F.	8.32823			
Original Lbs. of Water Required for Batch	1,050.11			
Gal. H2O for Syrup * H2O Wt. = Lbs H2O	4.16			
Net Lbs. H20	1,045.95			
Density of other flavors in Lbs. per Gal.	8.3282			
Vol. of Flavors or Colors * Density = Lbs.	0.000		Net Lbs./ Lbs. per Gal. = Net Gal.	
Net Lbs. of H2O to Add.	1,045.95		125.59	Net Gal. by Wt.
Lbs. of Sugar	4.00		0.30	Gal. Dif. = Sugar Vol.
Net Lbs. of H2O + Sugar = Tot. Sugar & H2O	1,049.95			

Flavor Additions That Reduce Vol. of H₂O Needed

After accounting for the water used to make syrup, we subtract that value from the water required to arrive at a net weight of water to use; this works out to 1,045.95 pounds. For confirmation we take the net pounds and divide by the pounds per gallon; we see that the net gallons, calculated by weight, are different from those calculated by volume by the same amount of sugar that has been added and which resulted in an increase in the volume of the batch. Adding in pounds of sugar gives us the total mass of sugar and water to add to the batch.

Now we will increase the amount of sugar we could add to the batch and still stay under to 2.5% limit of volume additions that are permissible under the regulations and see just how much sugar could be added:

Batch Flavor Additions to Volume That Reduce the Volume of Water Required in a Batch					
60° F. Volume of Alcohol in Batch	121.26			0.0015	Tolerance
60° F. Gallons of Water to be used in Batch	126.09			0.364	Blended Gal. Tolerance
Total 60° F. Vol. to be Contributed to batch	247.35			1.38	Liters of Tolerance
Target Proof of the Batch	80.00			2.88	Lbs. of Tolerance
Net Blended Volume	242.507	917.99	Liters	$13.50	Tax Rate
				$2,619.04	Tax on Proof Gal.
Pounds of Sugar to be used in batch	81.500	36.97	Kg.	Brix Change from Sugar additions	
Sugar Pounds Per Gallon of Volume	13.50			1,304.00	Lbs. of Sugar * 16 Oz. per Lb. = Tot. Oz.
Sugar Vol. in Fluid Gal. (Wine Gal.)	6.04	22.85	Liters	5.38	Ounces / Blended Gal. = Oz. per Gal.
Gallons of Water used to make Sugar Syrup	0.00			0.75	% Oz. of Sugar per 1% (°) Brix per Gal.
Additions to Vol. Eligible Flavors or Colors	0.00			7.17	Oz. per Gal. / Sugar per 1% of Brix = Brix
Total Fluid Added	0.00				
Gross Gallons needed - Fluid Vol. = Net Gal.	126.09			2.50%	% by Vol. Flavors & Solids Vol. Limit
Total Additions to Vol. including Sugar	6.04			2.489%	% By Volume Additions
H2O Gal. - Vol. Added = Net Gal.of Water	**120.05**			OK	Logical Test for Limit
Find Obscuration					
Grams of Solids	36,967.78				
Liters of Volume	917.99				
Grams/Liters = Grams Per liter	**40.270**				
mg. per 100 Ml.	4,027.03				
Grams per liter per 1° of Proof Obscuration	2.50			600.0	Limit for mg. per 100 ml.
Grams per Liter / 2.5 = ° of Proof Obscuration	16.11			4,027.03	Blended mg. per 100 ml
True Proof - Proof Obscured = Gauged Proof	63.89			Not OK	Test for Tolerance

Flavor Additions That Reduce Vol. of H₂O Needed

It turns out that we can add 81.5 pounds of sugar to our batch and still stay below the 2.5% limit for volume additions at 2.489. To get this much sugar into the batch on a blended basis, I've had to assume that no water was used to make the sugar syrup. This problem can be solved by simply using some of the net gallons of water required after calculating what that value is and keeping track of it. I've also had to assume that no other flavors or colors were added to the batch. The interesting point to note is that we have exceeded the limit of 600 mg per 100 ml that requires evaporation methods to proof by a considerable margin while still staying within the volume tolerance limit of 2.5%. Because of this, the degrees of proof that are obscured by the sugar are probably not accurate and the distiller should gauge the blend and determine for themselves what the actual obscuration is based upon the "known" proof we are blending to. Note that we have only raised the Brix of the mixture to 7.17.

Looking at one final example that contemplates using water to make syrup and adding two gallons of other flavorings while still staying below the limit of 600 mg per 100 ml and 2.5% by volume uses 12.1 pounds of sugar and amounts to a change of 1 degree of Brix (which can barely be tasted):

Batch Flavor Additions to Volume That Reduce the Volume of Water Required in a Batch					
60° F. Volume of Alcohol in Batch	121.26			0.0015	Tolerance
60° F. Gallons of Water to be used in Batch	126.09			0.364	Blended Gal. Tolerance
Total 60° F. Vol. to be Contributed to batch	247.35			1.38	Liters of Tolerance
Target Proof of the Batch	80.00			2.88	Lbs. of Tolerance
Net Blended Volume	242.507	917.99	Liters	$13.50	Tax Rate
				$2,619.04	Tax on Proof Gal.
Pounds of Sugar to be used in batch	12.100	5.49	Kg.	Brix Change from Sugar additions	
Sugar Pounds Per Gallon of Volume	13.50			193.60	Lbs. of Sugar * 16 Oz. per Lb. = Tot. Oz.
Sugar Vol. in Fluid Gal. (Wine Gal.)	0.90	3.39	Liters	0.80	Ounces / Blended Gal. = Oz. per Gal.
Gallons of Water used to make Sugar Syrup	1.00			0.75	% Oz. of Sugar per 1% (°) Brix per Gal.
Additions to Vol. Eligible Flavors or Colors	2.00			1.06	Oz. per Gal. / Sugar per 1% of Brix = Brix
Total Fluid Added	3.00				
Gross Gallons needed - Fluid Vol. = Net Gal.	123.09			2.50%	% by Vol. Flavors & Solids Vol. Limit
Total Additions to Vol. including Sugar	3.90			1.607%	% By Volume Additions
H2O Gal. - Vol. Added = Net Gal.of Water	**122.19**			OK	Logical Test for Limit
Find Obscuration					
Grams of Solids	5,488.47				
Liters of Volume	917.99				
Grams/Liters = Grams Per liter	**5.979**				
mg. per 100 Ml.	597.88				
Grams per liter per 1° of Proof Obscuration	2.50			600.0	Limit for mg. per 100 ml.
Grams per Liter / 2.5 = ° of Proof Obscuration	2.39			597.88	Blended mg. per 100 ml
True Proof - Proof Obscured = Gauged Proof	77.61			OK	Test for Tolerance

Flavor Additions That Reduce Vol. of H₂O Needed

Finding the weight of water to add to the blend results in 1,025 pounds, assuming that the density of the flavors is the same as water:

Find Weight Reduction In Water Required				
Water Constant Lbs. per Gal. @ 60° F.	8.32823			
Original Lbs. of Water Required for Batch	1,050.11			
Gal. H2O for Syrup * H2O Wt. = Lbs H2O	8.33			
Net Lbs. H20	1,041.79			
Density of other flavors in Lbs. per Gal.	8.3282			
Vol. of Flavors or Colors * Density = Lbs.	16.656		Net Lbs./ Lbs. per Gal. = Net Gal.	
Net Lbs. of H20 to Add.	1,025.13	123.09	Net Gal. by Wt.	
Lbs. of Sugar	12.10	0.90	Gal. Dif. – Sugar Vol.	
Net Lbs. of H20 + Sugar = Tot. Sugar & H2O	1,037.23			

Again, the volume of water calculated from the net pounds required differs only by the mass of sugar that has been converted to volume.

Turning our attention to simply changing Brix of a mixture, I have put together a procedure that accomplishes only that function:

Change Brix of Alcohol and track volume change by addition of Sugar and Other Solids			
Gallon Volume	242.51	917.99	Liters
Present ° of Brix of Batch	0.00	° of Brix	
% Ounce of Sugar per 1% change in Brix per Gallon	0.750		
Target Brix as a Degree	7.170		
Brix Differential Start to End	7.17		
Brix Dif. * % Ounce of Sugar per Gal = Oz. per Gal.	5.38		
Gallons * Ounces per Gallon = Total Ounces	1,304.08		
Ounces per Pound	16.00		
Ounces/ Ounces per Lb. = Pounds of Sugar	81.51	36.97	Kilograms
Pounds of Sugar Per Gallon Volume increase.	13.50		
Pounds / Sugar Per Gallon Volume increase = Gallons	6.04		
Gallons of Water used to make Sugar Syrup	0.00		
Additions to Volume - Liquids - Color, Flavor	0.00		
Additions to Volume - Wine or Alcohol	0.00		
Other Additions	0.00		
Total Additions to Volume	6.04	22.85	Liters
Total Blended Wine Gallons	248.54	940.84	Liters

% by Vol. Flavors & Solids Vol. Limit	2.50%	
Volume Additions as % of Starting Volume	2.49%	
Logical Test for Limit	OK	

Flavor Add 2 Change Brix

Maxing out the solids to the 2.5% maximum of volume additions will permit the Brix to move by 7.17° using 81.5 pounds of sugar as in the prior worksheet. This requires that no water is used to make syrup or that water is taken from blend water already designated for the batch so that no *additional* water is used.

Surprisingly little water is needed to make sugar into sugar syrup.[44] When heated gently, sugar dissolves easily into a syrup. You only really need a quart or two of water per 50 pounds of sugar. The amount of water that is actually used to make sugar syrup must also be accounted for and subtracted from the total water volume of the batch.

---

[44] Making sugar syrup and how to do so is covered in various candy making articles. We will have more to say on how and when to blend sugar with alcohol and water mixtures in Volume 2. My initial advice is to add the sugar syrup to the blending water first and make sure the water is still absolutely clear before blending with the alcohol. However, remember I never add alcohol to water—I always add water to alcohol. So, one will have to use a vat large enough to contain most if not all of the water one is to use when adding the sugar syrup to the water and then pump that water into the blending tank wherein all the alcohol to be used is already located.

Adding Wine to a Batch

If one is using wine or other eligible flavors in their batches, the *Tax on Adding Flavors & Wine* spreadsheet will assist in determining the tax payable for using these components in blending. We first set up our worksheet to reproduce the example given in §19.34 of Title 27 subpart C – Taxes of the Code of Federal Regulations:[45]

Taxes on Adding Wine and Eligible Flavors			
Proof Gallons of Distilled spirits	2,249.10	2.50%	Limit of Eligible Flavors
Proof Gallons of Eligible flavors	100.90	$13.50	Proof Gal. Excise Tax Rate
		$1.07	14% Alcohol or Less
Wine Gallons of Wine #1	2,265.00	$1.57	Over 14 to 21%
Percent alcohol by volume of Wine 1	14%	$3.15	Over 21 to 24%
Tax Rate for Wine 1	$1.07	$3.40	Naturally Sparkling
Wine Gallons of Wine #2	1,020.00	$3.30	Artificially Carbonated
Percent alcohol by volume of Wine 2	19.0%	$0.23	Hard Cider
Tax Rate for Wine 2	$1.57		

Calculate Tax Rate				
Numerator 1			Denominator 1	
Distilled Spirits Proof Gallons	2,249.10		2,249.10	Proof Gallons of Distilled Spirits
Tax Rate per Proof Gallon on Distilled Spirits	$13.50		100.90	Proof Gallons of Eligible flavors
Calculated Tax on Spirits	$30,362.85		2,265.00	Wine Gallons of Wine #1
Taxable Proof Gallons of Eligible Flavors	16.60		0.280	Proof Equivalent of Wine 1 as a Proportion
Tax Rate per Proof Gallon on Distilled Spirits	$13.50		634.20	Wine Gallons * Proportion =
Result	$224.10		1,020.00	Wine Gallons of Wine #2
Wine Gallons of Wine #1	2,265.00		0.380	Proof Equivalent of Wine 1 as a Proportion
Tax Rate per Wine Gallon of Wine 1	$1.07		387.60	Wine Gallons * Proportion =
Tax Payable Result	$2,425.82		3,371.80	Sum the 4 Proof Gallon Results
Wine Gallons of Wine #2	1,020.00		$10.27	Tax Sum / Proof Gal. Sum = Effective Tax Rate
Tax Rate per Wine Gallon of Wine 2	$1.57			
Tax Payable Result	$1,602.42			
Sum of 4 Tax Results	34,615.19			
			Calculate Tolerance for Eligible Flavors	
Tax Payable on Distilled Spirits	$30,362.85		3,371.80	Total Proof Gallons
Tax Payable on Eligible Flavor Proof Gallons	$224.10		2.5%	Limit on Proof Gallons in Eligible Flavors
Tax Payable on Wine 1	$2,425.82		84.30	Total Proof Gallons of Flavors Tolerance
Tax Payable on Wine 2	$1,602.42		100.90	Actual Proof Gallons of Eligible Flavors
Total Tax on Batch	$34,615.19		16.61	Proof Gal. by Which Flavor PG. Exceed Tol.
Effective Tax Rate Result	$10.27		2.99%	Percentage of Proof Gal. Actually Added

---

[45] §19.34 "Computation of effective tax rate. (a) The proprietor shall compute the effective tax rate for distilled spirits containing eligible wine or eligible flavors as the ratio of the numerator and denominator as follows: (1) The numerator will be the sum of: (i) The proof gallons of all distilled spirits used in the product (exclusive of distilled spirits derived from eligible flavors), multiplied by the tax rate prescribed by 26 U.S.C. 5001; (ii) The wine gallons of each eligible wine used in the product, multiplied by the tax rate prescribed by 26 U.S.C. 5041(b)(1), (2), or (3), which would be imposed on the wine but for its removal to bonded premises; and (iii) The proof gallons of all distilled spirits derived from eligible flavors used in the product, multiplied by the tax rate prescribed by 26 U.S.C. 5001, but only to the extent that such distilled spirits exceed 2 1/2% of the denominator prescribed in paragraph (a)(2) of this section. (2) The denominator will be the sum of: (i) The proof gallons of all distilled spirits used in the product, including distilled spirits derived from eligible flavors; and (ii) The wine gallons of each eligible wine used in the product, multiplied by twice the percentage of alcohol by volume of each, divided by 100. (b) In determining the effective tax rate, quantities of distilled spirits, eligible wine, and eligible flavors will be expressed to the nearest tenth of a proof gallon. The effective tax rate may be rounded to as many decimal places as the proprietor deems appropriate, provided that, such rate is expressed no less exactly than the rate rounded to the nearest whole cent, and the effective tax rates for all products will be consistently expressed to the same number of decimal places. In such case, if the number is less than five it will be dropped; if it is five or over, a unit will be added. (c) The following is an example of the use of the formula. BATCH RECORD Distilled spirits 2249.1 proof gallons. Eligible wine (14% alcohol by volume). 2265.0 wine gallons. Eligible wine (19% alcohol by volume). 1020.0 wine gallons. Eligible flavors 100.9 proof gallons. 2249.1($13.50) +2265.0 ($1.07)+1020 ($1.57)+16.6 1 ($13.50) = 2249.1+100.9+(2265.0×.28)+(1020×.38) $30,362.85+$2,423.55+$1,601.40+$224.10 = 2,350.0+634.2+387.6 $34,611.90 = $10.27, the effective tax rate. Proof gallons by which distilled spirits derived from eligible flavors exceed 2 1/2% of the total proof gallons in the batch = (100.9 – (2 1/2% × 3,371.8)) = 16.6."

Functionally the procedure works. I have used the 2.5% limit on flavors as also a proof gallon limit and in this context the blend has exceeded the tolerance limit (100.9 proof gallons added compared to a "limit" of 84.3 proof gallons). This approach may not the best way to evaluate the procedure but does give some indication as to how much of the final product has been added to the base stock relative to a fixed standard, albeit proof gallons rather than fluid gallons.

TTB Flavor Additions Public Domain Sheet

The Tax and Trade Bureau (TTB) authored a flavor addition sheet which tracks the limits for other types of additions:

Flavor Ingredient Data Sheet

Flavor Producer Information

TTB Co. Code: _____ Date: _____
Company Name: _____ Contact Person: _____
Address: _____ Phone: _____
_____ Fax: _____
_____

Check Appropriate Box:
Approved for Drawback (DrB)
Approved as No Action (N/A)

Flavor Name: _____ Fit for Beverage Purposes (Fit)
TTB Drawback Number: _____ Submitted for TTB Approval
Alcohol Range by Volume: _____ Not Yet Submitted for TTB Approval
Flavor Product Number: _____

Classification

Total Artificial Flavor Content: _____ 1000 ppm (Excluding Synthetic Vanillin, Ethyl Vanillin, Synthetic Maltol, and Ethyl Maltol)

Flavor Components

Additive	TTB Limitation in Finished Product	Amount of Additive or Agent Present in Flavor	Maximum Use Rate	Beverage Label Information based of functionality (check or list label ingredients that affect the beverage label)
1. Synthetic Vanillin	40.0 ppm	_____ ppm	_____ %	1. FD&C Yellow #5
2. Ethyl Vanillin	16.0 ppm	_____ ppm	- _____ %	2. FD&C Yellow #6
3. Synthetic Maltol	250.0 ppm	0 ppm	- _____ %	3. FD&C Blue #1
4. Ethyl Maltol	100.0 ppm	0 ppm	- _____ %	4. FD&C Blue #2
5. Ester Gum	100.0 ppm	0 ppm	- _____ %	5. FD&C Green #3
6. BVO	15.0 ppm	0 ppm	- _____ %	6. FD&C Red #40
7. Sodium Benzoate	1000.0 ppm	0 ppm	- _____ %	7. Grapeskin Extract
8. Gum Arabic/Acacia	200000.0 ppm	0 ppm	- _____ %	8. Caramel Color
9. Propylene Glycol	50000.0 ppm	0 ppm	- _____ %	9. Annatto
10. BHA	(<0.5% Essential Oil)	check if contained		10. Elderberry Extract
11. Acetic Acid	1500.0 ppm	_____ ppm	- _____ %	11. Beet Extract
12. _____	_____ ppm	_____ ppm	- _____ %	12. Oak Extract
13. _____	_____ ppm	_____ ppm	- _____ %	13. _____
14. _____	_____ ppm	_____ ppm	- _____ %	14. _____
15. Total Vanillin	40.0 ppm	- _____ ppm	- _____ %	
16. Total Maltol	250.0 ppm	- _____ ppm	- _____ %	

Flavor Additions TTB Public Sheet

In my distilled spirits plant, I do not go so far as to use any of these components in my blend, but this will be very useful for some distillers to know about.

Notes

Alcohol Consumption

Alcoholic Units Calculator

As distillers we have a duty to encourage the responsible consumption of our products. I post a blood alcohol chart on my mountainmoonshine.com website and of course the government-mandated warning is always on the labels of my products. It is also convenient to be able to calculate the amount of alcohol one is consuming and there are several easy ways to do so.

First, we need to define some terms. The value of what "One Alcoholic Unit" comprises has varied over time from 8 grams of absolute alcohol (200 proof) per drink to 14 grams per drink. For starters we will assume that one alcohol unit is measured as 8 grams of pure alcohol. By making the grams of alcohol per unit an assumption we can find the milliliters per alcoholic unit (25.25 ml) and the fluid ounces per unit as well (0.854 fl. oz.):

From Grams per Alc Unit to ml per Unit.	
Grams Alc per Alc. Unit	8.0
Percent Alcohol by Volume	40.00
Kg. per m3 (Air)	949.78
Mass % of Alc.	33.35%
Kg m3 * Mass % = Grams AA per Liter	316.8
Grams AA per ml	0.317
Grams per Unit / g per ml = Tot. ml	25.25
ml to Fluid Ounces	0.0338140
Fluid Ounces per Alcoholic Unit	0.854

Alcohol Units Calculator

Standard "drinks" do not conform to these calculated values. In the US, alcohol is still consumed by fluid ounces and a "drink" is generally regarded as comprising 1.5 fluid ounces (44.36 ml), 5 fluid ounces of table wine or 12 ounces of beer. Using those values, we can establish a standard for alcoholic units per drink of ~1.756 alcoholic units per standard drink:

Alcoholic Units Calculator			
Find Alcoholic Units per Drink	Liquor	Wine	Beer
Alcohol by Volume (ABV)	40.0%	12.0%	5.0%
Proof Equivalent	80.00	24.00	10.00
Fluid Ounce Size of Drink	1.50	5.0	12.0
Fluid Ounces to Milliliters	29.574	29.574	29.574
Total ml. Per Drink	44.36	147.87	354.88
Mass % of Alc.	33.36%	9.68%	4.00%
Density of Alc in Kg. per dm3 (liter)	0.94978	0.98227	0.99081
Kg dm3 * Mass % = Grams AA per ml	0.317	0.095	0.040
Grams of Pure Alcohol per Drink	14.05	14.06	14.05
Grams Absolute Alcohol. Per Alc. Unit	8.00	8.00	8.00
Alcoholic Units per Drink	1.7567	1.7569	1.7562
Food Cal. (Kilocalories) per Gram of Alc.	7.0	7.0	7.0
Total Alcohol Calories per Drink	98.38	98.39	98.35

Alcohol Units Calculator

This calculator can be used to find any standard drink size one desires and the alcoholic units it contains.

Having established the milliliters per total grams of alcohol per alcoholic unit, we can use a simple method employing the alcohol by volume to convert any number and size of drink(s) to alcoholic units:

Number of Drinks	3.0
Milliliters per drink	44.36
Fluid Oz per drink	1.500
Alcohol by Volume (ABV)	40.00
Total ml consumed	133.08
Total Fluid Oz.	4.500
% ABV * ml = Result	5,323
Result/1000 = Alcoholic Units	5.32

Grams of Alcohol per Alcoholic Unit	8.0
Food Cal. (Kilocalories) per Gram of Alc.	7.0
Total Food Calories	298.10
Total Food Calorie Intake per Day	2,500
Percentage of Daily Food Calorie Intake	11.92%

You can see that we have also introduced the idea that it is meaningful to look at the food calories in alcohol in terms of the percentage of total food calories consumed during the course of a day and I have assigned a value of 7.0 food calories (kilocalories) per gram of alcohol.

The Royal College of Physicians (RCP) advises no more than 21 units per week for men and 14 units per week for women. But also, you should allow 2–3 alcohol-free days a week to allow the liver time to recover after drinking anything but the smallest amount of alcohol. The National Health Service, UK, advises that men should not regularly consume more than three or four units a day, 21–28 per week, while women should not exceed two to three, 14 to 21 per week. From a study conducted by fans of Ian Fleming's books about James Bond, it appears that Bond consumed an average of 92 units per week (13.14 per day for a 7-day drinking week) and had just 12.5 alcohol-free days out of the 87.5[46] days he was able to drink and he frequently drove when over the limit.

Taking this advice and turning it into a procedure shows what three drinks on four days per week of 80-proof whiskey means in terms of overall alcoholic consumption:

Number of Drinks	3.00				
Fluid Ounce Size of Drink	1.50				
Fluid Ounces to milliliters converter	29.5735				
Milliliters per drink	44.36				
Alcohol by Volume (ABV)	40.00	80.00	Proof	125.0	Calories per. Unit
Grams of Alcohol per Alcoholic Unit	8.0			5.0	Ounces per Unit
Total ml consumed	133.08	4.50	Oz.	29.574	Ml per Oz.
% ABV * ml = Result	5,323			147.87	ml per unit
Result/1000 = Alc. Units	5.32			0.90	# of Oz. Portions
Food Cal. (Kilocalories) per Gram of Alc.	7.0			112.5	Total Calories
Total Food Calories	298.10			-185.6	Dif. In Calories
Total Food Calorie Intake per Day	2,500		% Dif.	2,500	Daily Calories
Percentage of Daily Food Calorie Intake	11.92%		-7.42%	4.50%	% Daily Calories
Days per Week Consumed	4.00			1,192.40	Calories per week
Alcoholic Units per Week	21.3			532.32	ml per week
Comparison to Recommended Values	Low		Middle	High	James Bond
Healthy Units per day	2.0		3.0	4.0	13.1
Units per Week	14.0		21.0	28.0	92.0
% of Healthy Units of Alcohol	152.1%		101.4%	76.0%	23.1%

---

[46] One day a week = 14.28% of a week.

We see that three drinks amount to 5.32 alcoholic units per drinking session and when scaled up to four days per week result in matching the middle recommended value for men at 101.4%, but constitutes only 23.1% of James Bond's drinking.

Looking at wine consumption, a standard glass of wine contains five (5) fluid ounces (147.87 ml); it also consists of approximately 125 food calories:

Drinks	1.000
Fluid Oz. per Glass	5.0
Ounces to ml.	29.57353
ml per Drink	147.87

Using a similar procedure, we can find the "glasses" of wine in half a bottle of wine (750 ml capacity) and this results in 2.54 glasses per half bottle or 375 ml:

Milliliters per Drink	375.00
ml to Fluid Oz.	0.03381402
Fluid Oz. US	12.68
Fluid Oz. per Glass	5.0
"Glasses" of Wine	2.54
Ounces to ml.	29.57353
ml. per "Glass"	147.87

At the same consumption frequency of four days of drinking per week, the total consumption is less than the middle value recommended for men at 85.7% of the recommended middle consumption value:

Number of Drinks	2.54				
Fluid Ounce Size of Drink	5.00				
Fluid Ounces to milliliters converter	29.5735				
Milliliters per drink	147.87				
Alcohol by Volume (ABV)	12.00	24.00	Proof	125.0	Calories per. Unit
Grams of Alcohol per Alcoholic Unit	8.0			5.0	Ounces per Unit
Total ml consumed	374.99	12.68	Oz.	29.574	Ml per Oz.
% ABV * ml = Result	4,500			147.87	ml per unit
Result/1000 = Alc. Units	4.50			2.54	# of Oz. Portions
Food Cal. (Kilocalories) per Gram of Alc.	7.0			317.0	Total Calories
Total Food Calories	251.99			65.0	Dif. In Calories
Total Food Calorie Intake per Day	2,500		% Dif.	2,500	Daily Calories
Percentage of Daily Food Calorie Intake	10.08%		2.60%	12.68%	% Daily Calories

Days per Week Consumed	4.00		1,007.98	Calories per week
Alcoholic Units per Week	18.0			
Comparison to Recommended Values	Low	Middle	High	James Bond
Healthy Units per day	2.0	3.0	4.0	13.1
Units per Week	14.0	21.0	28.0	92.0
% of Healthy Units of Alcohol	128.6%	85.7%	64.3%	19.6%

Alcohol Units Calculator

In the above example we are comparing the mass-based method using kilocalories per gram of alcohol to the assigned-calories-per-unit-by-volume method and the results are similar with only 2.6% difference between them.

Increasing the consumption to a bottle of wine a day *for seven days* blows the lid off of all the recommended consumption recommendations but is still only 68.5% of the "James Bond" value:

By the Bottle of Wine				
Portion or full of Bottle (s) Consumed	1.0			
ml size of bottle	750.00	25.36	Fluid Oz.	
Alcohol by Volume (ABV)	12.00	24.00	Proof	125.0	Calories per. Unit
Grams of Alcohol per Alcoholic Unit	8.0			5.0	Ounces per Unit
Total ml consumed	750.00	25.36	Oz.	29.574	Ml per Oz.
% ABV * ml = Result	9,000.0			147.87	ml per unit
Result/1000 = Alcoholic Units	9.00			5.07	# of Oz. Portions
Food Cal. (Kilocalories) per Gram of Alc.	7.0			634.0	Total Calories
Total Alcohol Calories	504.0			130.0	Dif. In Calories
Total Food Calorie Intake per Day	2,500		% Dif.	2,500	Daily Calories
Percentage of Daily Food Calorie Intake	20.16%		5.20%	25.36%	% Daily Calories

Days per Week Consumed	7.00			3,528.00	Calories per week
Alcoholic Units per Week	63.0			5,250.0	ml per week

Comparison to Recommended Values	Low	Middle	High	James Bond
Healthy Units per day	2.0	3.0	4.0	13.1
Units per Week	14.0	21.0	28.0	92.0
% of Healthy Units of Alcohol	450.0%	300.0%	225.0%	68.5%

Turning to beer we see that three 12-ounce bottles of beer matches the three drinks per day of 80 proof in terms of alcoholic units but entails far more calories than either wine or liquor. The calories are calculated based upon the alcohol consumed and are most accurate for whiskey. Both beer and wine have additional calories not associated with the purely alcohol content:

By Bottles of Beer				
Bottles	3.0		5.00	Alc. by Vol.
Grams of Alcohol per Alcoholic Unit	8.0		145.0	Calories per. Bot.
Total ml consumed	1,064.6	36.00 Oz.	12.0	Oz. per Bot.
% ABV * ml = Result	5,323.2		29.574	Oz. to ml
Result/1000 = Alcoholic Units	5.32		354.9	ml. per bottle
Food Cal. (Kilocalories) per Gram of Alc.	7.0		435.0	Total Calories
Total Alcohol Calories	298.1		136.9	Dif. In Calories
Total Food Calorie Intake per Day	2,500	% Dif.	2,500	Daily Calories
Percentage of Daily Food Calorie Intake	11.92%	5.48%	17.40%	% Daily Calories

Days per Week Consumed	4.00
Alcoholic Units per Week	21.3

Comparison to Recommended Values	Low	Middle	High	James Bond
Healthy Units per day	2.0	3.0	4.0	13.1
Units per Week	14.0	21.0	28.0	92.0
% of Healthy Units of Alcohol	152.1%	101.4%	76.0%	23.1%

Alcohol Units Calculator

Just so you don't have to wonder, I worked out how much James Bond was actually drinking:

James Bond's Drinking per Week					
Number of Alcoholic Units	92.0				
milliliters per Unit	25.00				
Total ml per Week	2,300.0				
Bottle Size	750.00		8.0	Grams of Alcohol per Alcoholic Unit	
Bottles	3.07		7.00	Food Cal. (Kilocalories) per Gram of Alc.	
Days	7.00		736.0	Total Grams of Pure Alcohol	
Bottles per Day	0.44		5,152	Total Calories	
milliliters per day	328.57		2,500	Total Food Calorie Intake per Day	
Alcoholic Units per Day	13.1		29.44%	Percentage of Daily Food Calorie Intake	

	Low	Middle	High	James Bond
Healthy Units per day	2.0	3.0	4.0	13.1
Units per Week	14.0	21.0	28.0	92.0
% of Healthy Units of Alcohol	657.1%	438.1%	328.6%	100.0%

It turns out that he was drinking three 750 ml bottles of 80-proof whiskey a week comprising almost a pint (328 ml) a day. Compared to Winston Churchill that's a pretty light load, but still it is a lot of booze. The telling figure is how large a percentage of his nutrition was coming from alcohol—almost 30%. That level of alcohol metabolism is what the liver cannot easily accommodate for long periods of time.

There are two other ways to look at alcohol consumption. The easiest way for me to keep track of it is by how long it takes me to go through a 750 bottle of whiskey. If I'm going to the liquor store every 10 days or so then I'm consuming about three alcoholic units a day:

By ml Consumed over Time					
ml consumed	750.00		7.0	Food Cal. (Kilocalories) per Gram of Alc.	
ABV	40.00		8.0	Grams of Alcohol per Alcoholic Unit	
Alcoholic Units	30.00		1,680	Total Calories	
Days	10.00		168.0	Calories per Day	
ml per day	75.00		2,500	Total Food Calorie Intake per Day	
Alcoholic Units per Day	3.00		6.72%	Percentage of Daily Food Calorie Intake	
Units per Week	21.00				

	Low	Middle	High	James Bond
Healthy Units per day	2.0	3.0	4.0	13.1
Units per Week	14.0	21.0	28.0	92.0
% of Healthy Units of Alcohol	150.0%	100.0%	75.0%	22.8%

Alcohol Units Calculator

The last way is to turn the procedure around and calculate from ml per day consumed through a period of time and it is apparent that it doesn't take very much volume of alcohol to supply 3 units a day at 2.54 fluid ounces per day:

By ml per day		Oz.
ml consumed per Day	75.00	2.54
ABV	40.00	
Alcoholic Units per Day	3.00	
Food Cal. (Kilocalories) per Gram of Alc.	7.0	
Grams of Alcohol per Alcoholic Unit	8.0	
Total Calories per Alcoholic Unit	56	
Calories per Day	168.0	
Total Food Calorie Intake per Day	2,500	
Percentage of Daily Food Calorie Intake	6.72%	
Days	10.00	
Total ml over time	750.00	
Calories over Time	1,680.0	

Units per Week as Input	21.0			
Comparison to Recommended Values	Low	Middle	High	James Bond
Healthy Units per day	2.0	3.0	4.0	13.1
Units per Week	14.0	21.0	28.0	92.0
% of Healthy Units of Alcohol	150.0%	100.0%	75.0%	22.8%

Air and Vacuum Introduction

We are going to fact check this procedure in order to introduce a few ideas that we will be using for the rest of this book. We need to confirm the volume of one alcoholic unit. To do this we need to first find the density of our alcohol and then find its mass percentage:

Proof of Alcohol	80.00	
Find Density Air & Vac	Air	Vac
SG Vac at Calib °	0.95174	0.95180
H2O Density Kg. per dm3 at 60° F.	0.997939	0.99902
SG * Density = Kg. dm3	0.94978	0.95086
Kg. m3 @ 60° F.	949.78	950.86

Find Kg. per Meter Cubed Air from 15.556 C. values	
Pounds per Wine Gallon	7.92626
Pounds Avdp to Kilograms	0.45359
Lbs Per WG * Factor = Kg. per Gallon	3.59529
Gallons per Liter	0.26417205
Kg. per Gal. * Gal. per Liter = Kg. per Liter	0.94978
Kg. per Liter * 1,000 = Kg. per m3	949.78

Density per decimeter cubed (dm^3 ~ 1 liter) is obtained by taking the specific gravity of 80-proof alcohol and multiplying it by the density of water at our calibration temperature of 60°F in both air and vacuum. We then convert that density to kilograms per meter cubed.

It is also possible to take the pounds per wine gallon from the 15.556°C (Air) tables and convert it to the same value obtained by the specific gravity method simply by using our English to metric converters to find kilograms per gallon, then kilograms per liter and then scaling up to kilograms per meter cubed under atmospheric conditions.

Having found the density, we now use it to find the mass percentage of our 80-proof alcohol. We do this by finding the ratio of density between our alcohol and absolute alcohol and then multiplying that ratio by the volume percentage:

	Air	Vac
Density of Alcohol	949.78	950.86
Density of Pure Alc. (AA)@ Calib Temp	791.93	793.03
Density of AA/ Den of Sample = ratio	0.83380	0.83402
Vol % Restated	40.00%	40.00%
Vol % * Ratio = Mass %	33.352%	33.361%

	TTB Air
Lb. per WG Air @ True Proof	7.9263
PG per Lb.	9.90786
Lb per WG /Lbs. per PG = PG per WG	0.800
Divided by 2 = AA Gal per WG	0.4000
AA Lb. per Gal.	6.6089
AA Gal per WG/AA Lb. per Gal = Mass of AA	2.64
Lb. Mass AA/ Lb. per WG = Mass %	33.352%

Alcohol Units Calculator

We note that the air mass percentage matches using both methods and we will pass over the difference between air and vacuum mass percentages for the moment. It is a real effect and needs to accounted for when calculating volume and proof gallons and we will spend considerable time on it. Working from grams of alcohol per alcoholic unit back up to volume we can confirm that 8 grams of pure absolute alcohol (AA) is in fact close to 25.26 ml:

From Grams per Alc Unit to ml per Unit.	Air	Vac
Grams Alc per Alc. Unit	8.0	8.0
Kg. per dm3	0.94978	0.95086
Mass % of Alc.	33.352%	33.361%
Kg dm3 * Mass % = Grams AA per Liter	316.8	317.2
Grams AA per ml	0.317	0.317
Grams per Unit * Grams per ml = Tot. ml	25.26	25.22

On a density basis vacuum volume values are in fact less than those calculated using air densities.

To a reasonable degree of accuracy eight grams of absolute alcohol is indeed 25.25 ml of 80-proof alcohol. We've gone to this trouble to confirm this value because some countries use 10 grams of alcohol per alcoholic unit and other countries use 7 grams and calculate the units of alcohol for any consumption regimen. For instance, if you were in a country where 7 grams of alcohol was the standard, one would have to set the milliliters per alcoholic unit to 22.1 rather than 25:

From Grams per Alc Unit to ml per Unit.	Air	Vac
Grams Alc per Alc. Unit	7.0	7.0
Kg. per dm3	0.94978	0.95086
Mass % of Alc.	33.352%	33.361%
Kg dm3 * Mass % = Grams AA per Liter	316.8	317.2
Grams AA per ml	0.317	0.317
Grams per Unit * Grams per ml = Tot. ml	22.10	22.07

In a country with 10 grams per alcoholic unit, one could have a good old time with 31.57 ml per unit:

From Grams per Alc Unit to ml per Unit.	Air	Vac
Grams Alc per Alc. Unit	10.0	10.0
Kg. per dm3	0.94978	0.95086
Mass % of Alc.	33.352%	33.361%
Kg dm3 * Mass % = Grams AA per Liter	316.8	317.2
Grams AA per ml	0.317	0.317
Grams per Unit * Grams per ml = Tot. ml	31.57	31.52

Even with all the strict regulation and disclosures of alcoholic content it seems like the entire alcohol business is designed to make it hard to determine just how much alcohol one is drinking.

Blood Alcohol Concentration

Blood alcohol concentration (BAC), also known as blood alcohol content, is used as a means of gauging alcohol intoxication. Blood alcohol content is expressed as the portion of alcohol in the blood.

While the values published for BAC are accurate, there is a lot of mis-identification of what units are referred to with respect to the concentration of alcohol in the blood. The easiest way to determine what quantities are being referred to and in what concentration is to refer to the statute books.

In general the United States and my state of West Virginia define alcohol concentration as "the number of grams of alcohol per one hundred cubic centimeters of blood"; also defined are the limits for impairment and intoxication: "(3) Evidence that there was, at that time, eight hundredths of one percent or more, by weight, of alcohol in his or her blood, shall be admitted as prima facie evidence that the person was under the influence of alcohol."[47] Eight hundredths of one percent is expressed numerically as 0.0008 and as a percentage as 0.08%.

Neither of these values correspond to the published values in alcohol consumption tables generally as 0.08 as the numerical value for intoxication, but the statute is controlling, fortunately.

Because the assumption is made that blood has a SG of 1.0, the same as for water,[48] it follows that a liter of blood has a mass of 1,000 grams as well as a volume of 1,000 cubic centimeters. This allows one to express blood alcohol content on a volume basis and a mass basis at the same time as units of mass per units of volume.

To calculate estimated blood alcohol concentration (EBAC), a version of the Widmark formula,[49] can be used. The formula is:

$$EBAC = \left(\frac{0.806 \times SD \times 1.2}{BW \times Wt} - MR \cdot DP \right) \times 10$$

where: 0.806 is a constant of body water contained in an individual's blood, SD is the number of standard drinks, 1.2 is a factor to convert the amount in grams to Swedish standards set by The Swedish National Institute of Public Health, BW is a body water constant (0.58 for men and 0.49 for women), Wt. is body weight (kilograms), MR is the metabolism constant (0.015 for men and 0.017 for women) and DP is the drinking period in hours. Multiplying by 10 converts the result to permillage of alcohol which is the common way to refer to BAC as in "He blew a 1.2 on the breathalyzer." Permillage is essentially parts out of 1,000.

In the Alcohol Units Calculator that we just discussed, each alcoholic unit comprised 8 grams of absolute alcohol per 25-ml-serving of 80-proof whiskey and we will use that standard drink size in our example.

Putting all this information together we see that, after consuming four (4) standard drinks, our standard drinker has placed a total of 0.82 grams of alcohol per 100 cubic centimeters of blood into his bloodstream and has a BAC of 0.082, just above the "under the influence" standard of the statute. During a one-hour drinking session he has metabolized a portion of that alcohol to have remaining 0.67 grams per 100 cc corresponding to a BAC

---

[47] West Virginia Code 17C-5-8 (2016 Edition) "(a) Upon trial for the offense of driving a motor vehicle in this state while under the influence of alcohol …, evidence of the amount of alcohol in the person's blood at the time of the arrest or of the acts alleged, as shown by a chemical analysis of his or her blood or breath, is admissible, if the sample or specimen was taken within the time period provided in subsection (g). (b) The evidence of the concentration of alcohol in the person's blood at the time of the arrest or the acts alleged gives rise to the following presumptions or has the following effect:
"(1) Evidence that there was, at that time, five hundredths of one percent or less, by weight, of alcohol in his or her blood, is prima facie evidence that the person was not under the influence of alcohol;
"(2) Evidence that there was, at that time, more than five hundredths of one percent and less than eight hundredths of one percent, by weight, of alcohol in the person's blood is relevant evidence, but it is not to be given prima facie effect in indicating whether the person was under the influence of alcohol;
"(3) Evidence that there was, at that time, eight hundredths of one percent or more, by weight, of alcohol in his or her blood, shall be admitted as prima facie evidence that the person was under the influence of alcohol.
"(c) A determination of the percent, by weight, of alcohol in the blood shall be based upon a formula of:
"(1) The number of grams of alcohol per one hundred cubic centimeters of blood;
"(2) The number of grams of alcohol per two hundred ten liters of breath; or
"(3) The number of grams of alcohol per eighty-six milliliters of serum."
[48] For whole blood it was found to be 1.0621 (95% confidence interval: 1.0652–1.0590) at 4°C and 1.0506 (95% confidence interval: 1.0537–1.0475) at 37°C. Plasma specific gravity was 1.0310 (95% confidence interval: 1.0324–1.0296) at 4°C and 1.0205 (95% confidence interval: 1.0216–1.0193) at 37°C. clinchem.aaccjnls.org/content/20/5/615
[49] https://en.wikipedia.org/wiki/Blood_alcohol_content#Widmark_formula

of 0.067 at the end of the drinking session. His intoxication level at that time is ~83.4% of the legal limit for driving but 133.4% of the impaired limit:

Blood Alcohol Content

					Alc %	Fl. Oz.	ml	
Alcohol by Vol. Consumed	40.0%				40.0%	1.50	44.36	Liquor
Convert actual Drink Volume to Std. Vol.					12.0%	5.0	147.87	Wine
Fluid Oz. of Alcohol per Drink Made	1.50	44.36	ml per Drink Made		5.0%	12.0	354.88	Beer
Actual Drinks Made	4.00	177.44	Total ml Consumed		1.75673	Alcoholic Units per Std. Drink.		

Find Alcoholic Units

Grams Alc per Alc. Unit	8.0				
Alc Kg. per dm3	0.94978				
Mass % of Alc.	33.36%		Second Method Find Units @ 8 Gram Alc. Units		
Kg dm3 * Mass % = Grams AA per ml	0.3168		177.44	Total ml consumed	
Total grams of Absolute Alcohol Consumed	56.22		40.0%	Alcohol by Volume (ABV)	
Grams per Unit * g per ml = ml per Unit	25.25		71.0	% ABV * ml = Result	
Tot ml consumed / ml per Alc Unit = Alc Units	7.03		7.10	Result/10 = Alc. Units	
Std. 8 Gram Alcoholic Units Per Drink	1.75673		0.07	Units Dif. In Methods	
Consumed Units / Calc. Units = Std.Drinks	4.00				

Find Blood Alcohol Concentration

					Second Method Using Blood Mass	
Lbs. Body Weight	180.00	81.65	Kilograms		81.65	Kilograms of Body Mass
Kg. of Body Weight as Input	81.65				8.43%	% of Blood Mass
"Standard" Drinks Consumed (Restated)	4.00				6.883	Kg. of Blood = (Liters at SG 1.0)
Body Water in Blood (% Constant)	80.60%				6,883	Grams of Blood
Body Water in Blood % * Std Drinks=Result 1	3.22				56.22	Tot. Grams AA Consumed
Constant to Convert to Swedish NIPH Value	1.20		Body Water Avg.%		0.0082	AA Fraction of Blood Mass
Result 1 * Constant = Result 2	3.87		0.58	Men	0.82%	% of Blood Mass
Body Water Avg. Percentage (Constant)	0.58		0.49	Women	8.2	Grams per Liter
Kg. Weight * Body Water % = Result 3	47.36				0.82	Grams per 100 cc (deciliter)
Result 2/Result 3 = Result 4 (Total BAC)	0.082		0.82	* 10= Permillage (parts per 1,000)		

Find Present Blood Alcohol Contant (Since Drinking Began)

Drinking Period (Hours)	1.0		Metabolism Constant (MR)		
Metabolism Constant (MR)	0.015		0.015	Men	
Metabolism Rate * Hours = Result 5	0.015		0.017	Women	
Find Grams of Alcohol per Liter of Blood	Pesent BAC		* 10 = permillage g/liter		
Result 4 - Result 5 = Present Blood Alc. (BAC)	0.067	g/dl	0.67	Grams per Liter	

Compare BAC to Driving Limit

Legal Driving Limit (BAC)	0.080		Impaired Driving Limit	
Present Compared to Limit	-0.013		0.05	
% of Limit	83.4%		133.4%	% of Limit

Hours Since Last Drink	1.0			
Metablolism Rate * Hrs.Post Drinking =	0.015			
Present Blood Alcohol Content	0.052	BAC Hours After End of Drinking		
Legal Driving Limit	0.080			
Present Compared to Limit	-0.028		0.05	Impaired Driving Limit
% of Limit	64.6%		103.4%	% of Limit

Blood Alcohol Concentration

After an hour since the last drink the individual has reduced his/her alcohol content to 64.6% of the legal limit but is still over the impaired limit at 103.4%.

The formula runs off of the number of "Standard Drinks" with a variable value for grams of absolute alcohol per alcoholic unit which is set to 8 grams per unit. This works for legal purposes since in the US if one is caught driving, that is the basis on which the authorities will judge the individual's intoxication. There is also a second method to calculate alcoholic units that is only valid at 8 grams per unit. If one decides to adjust the grams per unit value in the first column it will show the difference between 8-gram units and the chosen units.

Varible Grams per Unit	Variable Units	Std. Units
Grams Alc per Unit	10.0	8.0
% by Vol	40.0%	40.0%
Alc Kg. per dm3	0.94978	0.94978
Mass % of Alc.	33.36%	33.36%
Kg dm3 * Mass % = Grams AA per ml	0.317	0.317
Grams per Unit * g per ml = ml per Unit	31.56	25.25
Ounces per Drink	1.5	1.5
ml per Standard Drink	44.36	44.36
Units per Drink	1.41	1.76

One would adjust this grams per alcoholic unit value at their peril because the intoxication calculated will go down even though the consumption will be going up. However, if one is in a different country where the values are different than the US, one can calculate the alcoholic units per standard drink for that country and use that value in conjunction with the different grams per unit of a standard drink to get an accurate picture of one's intoxication level for a given intake of alcohol with respect to another country's regimen. When changing calibration temperatures as well one would require the density and mass percent values at that calibration temperature.

To be sure I got it right I also worked the problem from blood mass and volume related to grams of alcohol consumed. The value for the percentage of blood in the body has been tuned to 8.43% to better match the formula results but represents a value within bounds of the rule of thumb value of ~7%:

Second Method Using Blood Mass

81.65	Kilograms of Body Mass
8.43%	% of Blood Mass
6.883	Kg. of Blood = (Liters at SG 1.0)
6,883	Grams of Blood
56.22	Tot. Grams AA Consumed
0.0082	AA Fraction of Blood Mass
0.82%	% of Blood Mass
8.2	Grams per Liter
0.82	Grams per 100 cc (deciliter)

If one is dehydrated and drinking they will have a higher blood alcohol content than otherwise. Food in the stomach can increase the time it takes to reach its highest level and thus give the body more time to metabolize alcohol so the highest level will not be as great as when one is drinking on an empty stomach.

It is also important to note that for routine drinkers no amount of "built up tolerance" to alcohol will affect the BAC in their bloodstream even though subjectively they may not feel as inebriated as a novice drinker.

With respect to the metabolism rate, "… females demonstrated a higher average rate of elimination (mean, 0.017; range, 0.014–0.021 g/210 L) than males (mean, 0.015; range, 0.013–0.017 g/210 L). Women on average have a higher percentage of body fat (mean, 26.0; range, 16.7–36.8%) than males (mean, 18.0; range, 10.2–25.3%). Additionally, men are, on average, heavier than women but it is not strictly accurate to say that the water content of a person alone is responsible for the dissolution of alcohol within the body, because alcohol does dissolve in fatty tissues as well. When it does, a certain amount of alcohol is temporarily taken out of the blood and briefly stored in the fat. For this reason, most calculations of alcohol to body mass simply use the weight of the individual, and not specifically his/her water content."[50]

Following is a standard blood alcohol chart from West Virginia that is representative of the standard used in the United States for determining BAC. In these charts the number of standard drinks is compared to the chart to find the peak alcohol level and then the metabolism rate is used to find out how much sobering up has occurred:

---

[50] Wikipedia: https://en.wikipedia.org/wiki/Blood_alcohol_content#cite_note-PMC2724514-2, retrieved, 3/14/17.

Referencing the Blood-Alcohol Chart below and SUBTRACTING the % of alcohol "burned up" during the time elapsed since your first drink, you will be able to estimate your Blood Alcohol Content (BAC).

No. Hours Since 1st Drink	1	2	3	4	5	6
SUBTRACT	.015%	.030%	.045%	.060%	.075%	.090%

Example – 180 lb. man - 8 drinks in 4 hours .167% minus .060% = .107%

% OF BLOOD ALCOHOL	INTOXICATED?	IF YOU DRIVE A CAR.
.000 to .050	You Are Not	Take It Easy
.050 to .080	You Probably Are	Better Not
.080 & above	You Are	Illegal to Drive

FOR BEST RESULTS - DON'T DRINK AND DRIVE

FEMALE   MALE

Drinks	Body Weight in Pounds																
	90	100	120	140	160	180	200	220	240	100	120	140	160	180	200	220	240	
	.00	.00	.00	.00	.00	.00	.00	.00	.00	.00	.00	.00	.00	.00	.00	.00	.00	
1	.05	.05	.04	.03	.03	.03	.02	.02	.02	.04	.03	.02	.02	.02	.02	.02	.02	IMPAIRMENT BEGINS
2	.10	.09	.08	.07	.06	.05	.05	.04	.04	.08	.06	.05	.05	.04	.04	.03	.03	DRIVING SKILLS SIGNIFICANTLY AFFECTED
3	.15	.14	.11	.10	.09	.08	.07	.06	.06	.11	.09	.08	.07	.06	.06	.05	.05	
4	.20	.18	.15	.13	.11	.10	.09	.08	.08	.15	.12	.11	.09	.08	.08	.07	.06	
5	.25	.23	.19	.16	.14	.13	.11	.10	.09	.19	.16	.13	.12	.11	.09	.09	.08	POSSIBLE CRIMINAL PENALTIES
6	.30	.27	.23	.19	.17	.15	.14	.12	.11	.23	.19	.16	.14	.13	.11	.10	.09	
7	.35	.32	.27	.23	.20	.18	.16	.14	.13	.26	.22	.19	.16	.15	.13	.12	.11	—
8	.40	.36	.30	.26	.23	.20	.18	.17	.15	.30	.25	.21	.19	.17	.15	.14	.13	LEGALLY INTOXICATED
9	.45	.41	.34	.29	.26	.23	.20	.19	.17	.34	.28	.24	.21	.19	.17	.15	.14	CRIMINAL PENALTIES
10	.51	.45	.38	.32	.28	.25	.23	.21	.19	.38	.31	.27	.23	.21	.19	.17	.16	

The expression of BAC as "0.08" is essentially parts per 100 and in that respect can be expressed as a percentage since a BAC of 1.0 is 1 part out of 100 of the bloodstream on both a mass and volume basis.

There are also wrong solutions and expressions of BAC that one runs into. For instance, when retrieved on 5/31/17 Wikipedia's solution to the Widmark formula expresses BAC as the equivalent of 0.080 grams per deciliter (g/dl),[51] when in fact I have solved for 0.80 grams per deciliter in accordance with the West Virginia statute.

Another incorrect example is on display at the Austin Texas police department: "A BAC describes the amount of alcohol in a person's blood, expressed as weight of alcohol per unit of volume of blood. For example, 0.08 percent BAC indicates 80 mg of alcohol per 100 ml of blood"[52] when my solution calls for 800 mg per 100 ml.

A correct statement can is phrased as: "Blood Alcohol Concentration (BAC) levels represent the percent of your blood that is concentrated with alcohol. A BAC of 0.10 means that 0.1% (one tenth of one percent) of your bloodstream is composed of alcohol."[53]

Because of the equivalence of mass and volume in the definition of BAC this means that at a BAC of 1.0 your blood contains 1% by both mass and volume and an essentially lethal dose is achieved if one's BAC reaches 5% of blood content.

---

[51] Wikipedia Solution to Widmark formula g/dl

[52] https://www.austintexas.gov/faq/what-does-blood-alcohol-concentration-bac-measure

[53] http://awareawakealive.org/educate/blood-alcohol-content

Alcohol and Cancer

Alcohol is a cancer-causing agent and should be treated as such in all respects whether on the job or off it. The carcinogenic properties of alcohol are not so much due to the ethanol itself but its metabolite, acetaldehyde. Acetaldehyde is produced by the liver as it breaks down ethanol. The liver then normally eliminates 99% of the acetaldehyde. An average liver can process 7 grams of ethanol per hour. For example, it takes 12 hours to eliminate the ethanol in a bottle of wine, giving 12 hours or more of acetaldehyde exposure. A study of 818 heavy drinkers found that those who are exposed to more acetaldehyde than normal through a defect in the gene for alcohol dehydrogenase are at greater risk of developing cancers of the upper gastrointestinal tract and liver. There are many associations between alcohol drinking and different types of cancer. Data from 2009 indicated 3.5 percent of cancer deaths in the U.S. were due to consumption of alcohol.[54]

Color of Water

The color of water varies with the ambient conditions in which that water is present. While relatively small quantities of water appear to be colorless, pure water has a slight blue color that becomes a deeper blue as the thickness of the observed sample increases. The blue hue of water is an intrinsic property and is caused by selective absorption and scattering of white light. Impurities dissolved or suspended in water may give water different colored appearances.[55]

---

[54] https://en.wikipedia.org/wiki/Alcohol_and_cancer

[55] https://en.wikipedia.org/wiki/Color_of_water

Modules and Blending

Alcohol Modules

Let us review a few old methods and introduce a few new ones that are used to determine various characteristics of alcohol. First, we will look at our basic method of finding mass percent using the TTB table values (*Alcohol Modules Mass % and Vol %*). This can be done with values of pounds per gallon or gallon per pound. We will use 125.7-proof alcohol for this test since it is very close to 55% alcohol on a mass percent basis:

True Proof @ 60° F.	125.70					
Find Mass % Using TTB Tables						
Lbs. per Wine Gal. @ Proof	7.5528	3.42590	Kg.	0.132401	Wine Gal. per Lb.
Lbs. per Proof Gal. at True Proof	6.0087			0.16642	Proof Gal. per Lb.
Lbs. per WG/Lbs. per PG = Proof Gal. per WG.	1.2570			1.2570	WGPP/PGPP = Proof Gal. per WG
Divide by 2 = Absolute Alcohol Gal. per WG	0.6285		AA Lbs. per PG	0.6285	Divide by 2 = AA Gal per WG
200 Proof Alc. Lbs. per WG = Absolute Alc.	6.60982	Kg.	3.30491	0.30258	200 Proof Alc. PG per Lb. = AA.
AA Lb/WG * AA Gal per WG = Lbs. AA per WG	4.1542	1.88430	2.0771	0.4814	AA PG per lb. / AA Gal per WG
Lbs. AA per WG/Lbs. per WG = Fraction AA	0.55002		0.2750	0.2750	Wine Gal. per Lb/Ratio = Result
Result as a Percentage Alcohol % Mass (Air)	55.0017%	Kg.	55.0017%	Times 2	55.0017%	Result * 2 = Alcohol % Mass (Air)
Lbs. per WG True Proof - Lbs AA = Lbs. H2O	3.399	1.542	kg. H20		

From Air Density Find Mass % in Vac	Vac Values	Air Values		From Vac Density Find Mass % in Air		
				Vac Values	Air Values	
Density of Sample	906.12	905.03	Density	906.12	905.03	
Density of AA @ Calib Temp	793.03	791.92	AA Den.	793.03	791.92	
Vac Density/Air Density	100.121%			99.88%	Air Density/Vac Density
Vac Density AA/Air Density AA	100.140%			99.86%	Air Density AA/Vac Density AA
AA Density Ratio/Alc Den Ratio	100.019%			99.981%	AA Density Ratio/Alc Den Ratio
TTB Mass % Air Restated	55.0017%			55.0122%	Vac Mass % Restated
Vac Mass % * % Change = Vac mass %	55.0122%			55.0017%	Vac Mass % * % Change = Air mass %

Below is a procedure for finding mass percent in vacuum from mass percent in air. This transformation is reciprocal. To change mass percent on a density basis one simply looks up the closest density in air to the vacuum density. In this case 126.6 proof returns a vacuum density of 905.09 kg/m^3 which is close to 905.03 in air above. Accordingly, the air mass percent on a density basis is the vacuum mass percent at the air density, or 55.459. This topic will be examined further in the second volume when we have all-metric units available.

True Proof @ 60° F.	126.60			
Find Mass % Using TTB Tables				
Lbs. per Wine Gal. @ Proof	7.5442	3.42200	Kg.	
Lbs. per Proof Gal. at True Proof	5.9592			
Lbs. per WG/Lbs. per PG = Proof Gal. per WG.	1.2660			
Divide by 2 = Absolute Alcohol Gal. per WG	0.6330		AA Lbs. per PG	
200 Proof Alc. Lbs. per WG = Absolute Alc.	6.60982	Kg.	3.30491	
AA Lb/WG * AA Gal per WG = Lbs. AA per WG	4.1839	1.89780	2.0920	
Lbs. AA per WG/Lbs. per WG = Fraction AA	0.55459		0.2773	
Result as a Percentage Alcohol % Mass (Air)	55.459%	Kg.	55.459%	
Lbs. per WG True Proof - Lbs AA = Lbs. H2O	3.360	1.524	kg. H2O	

From Air Density Find Mass % in Vac	Vac Values	Air Values
Density of Sample	905.09	903.99

Generally mass percent is found with reference to metric units rather than pounds-per-gallon methods and the metric method is set forth below.

Find Mass % Traditional Method	Vac	Air
Density of Sample (Vac)	906.12	905.03
Density of AA @ Calib Temp	793.03	791.92
Density of AA/Den of Sample = Ratio	0.87520	0.87503
(Proof/2)/100 = Vol %	62.85%	62.85%
Vol % * Ratio = Mass %	55.006%	54.996%

Following are three methods of finding volume percent which are all effective to the same degree. Note that with respect to volume percent there is no difference between air and vacuum regimens:

Vol % Module 1 (IAT 3b Method)	Vac	Air
Sample Density	906.12	905.03
AA Density at Temp	793.03	791.92
Sample Den/AA Den = Ratio	1.14260	1.14282
Mass % Restated	55.012%	55.002%
Ratio * Mass % = Vol % @ Temp	62.8570%	62.8570%

Vol % Module 2		
Mass % of Sample	55.012%	55.002%
AA Density at Temp	793.03	791.92
(Mass % / Sample Den) * 100 = Result 1	0.06937	0.06945
Sample Density	906.12	905.03
100/Sample Density = Result 2	0.1104	0.1105
Result 1/Result 2 = Vol %	62.8570%	62.8570%

Vol % Module 3		
Mass % of AA	55.01%	55.00%
Sample Density	906.12	905.03
Mass * Density = Result	498.47	497.78
Density of Alcohol @ Temp	793.03	791.92
Result/Density of Alc @ Temp = Vol %	62.8570%	62.8570%

Alcohol Modules Mass % and Vol %

One can also use a known mass to find the mass of absolute alcohol and the volume of the entire sample.

Using the mass of absolute alcohol and the density of absolute alcohol, the liters of absolute alcohol can be found. This value is convenient for calculating the proof gallon content of a blend:

Find Volume % From Mass		
Find Kg. of Absolute Alcohol	Vac	Air
Kilograms of Alcohol	1,000.00	1,000.00
Mass % Restated	55.012%	55.002%
Kg. * % Ethanol = Kg. AA	550.12	550.02

Find Volume of Alcohol	Vac	Air
Sample Density	906.12	905.03
Kg Mass/Sample Density = m3 Volume	1.1036	1.1049
m3 Volume * 1000 = Liters of Blend	1,103.61	1,104.94

Calculate Vol % of Blend (Module 4)	Vac	Air
Density of Absolute Alc. At 15.556	0.79303	0.79192
Tot. kg. AA	550.12	550.02
Kg. AA / Den AA = Liters AA Vac	693.70	694.53
Volume of Alcohol	1,103.61	1,104.94
Liters of AA / Net Vol = Vol %	62.8570%	62.8570%

Find Volume of Absolute Alcohol		
Mass AA Restated	550.12	550.02
Density of AA at Temp	793.03	791.92
(Mass/Density)*100 = Liters of AA	693.70	694.53

Alcohol Modules Mass % and Vol %

One last look at a way to convert between volume percent and mass percent:

Convert Between Vol % & Mass %	Find Vol %	Find Mass % Vac	
Mass %	55.012%	793.03	Density of AA @ 15.556
Sample Density at 15.556 (Vac)	906.12	906.12	Sample Density
Mass % * Sample Den = Result 1	498.47	0.8752	AA Den/ Sample = ratio
Density of AA	793.03	62.8570%	Vol % Restated
Result 1/Den of AA = Vol %	62.8570%	55.012%	Vol % * Ratio = Mass %

Find Volume of Absolute Alcohol		Find Volume %	
Mass %	55.012%	906.12	Sample Density
Density of Absolute Alcohol at 15.556	793.03	0.01104	10/Sample Den.
Mass %/ AA Den = m3 Vol AA	0.6937	0.6937	m3 Vol. of AA
Liters of AA at 15.556	693.70	62.8570	m3 AA/ (10/Sample Den) = Vol %

Alcohol Modules Mass % and Vol %

Finally, one can find the grams of alcohol per 100 cc:

Find Grams per Liter	Vac	Air
Kg. per Liter	0.90612	0.90503
Alc Mass %	55.006%	54.996%
(Alc % * Kg. per liter)*100 = Grams per 100 CC	49.84	49.77
Grams per 100 CC*10 = Grams per liter (Kg m3)	498.42	497.72
Alc Kg. per Liter * 1000 =Vac Kg. per m3 of Alc.	906.116	905.025
Kg. per m3 Alc -Alc Kg = Kg/m3 H2O	407.70	407.30
Grams per 100 CC of H2O	40.77	40.73

It is also possible, to a fair degree of accuracy to use our 7/8 air mass procedure to convert vacuum and air densities back and forth:

7/8 Rule Find Air Density from Vac Density	Vac to Air	Air to Vac	
7/8 Air Kg/Ltr. Of Mass Displaced	0.00106982	0.00106982	
Vack Kg. per Liter - 7/8 Air = Air Kg per Liter.	0.90505	0.90609	Air Kg. + Air Mass = Vac Den
Rounded 5	0.90505	0.90609	Rounded 5
Kg. m3 Air, Rounded	905.046	906.095	Kg. m3 Air, Rounded
Table Value Air	905.025	906.116	Table Value Vac
Grams per m3 difference from Air Value	21.18	-21.18	Grams per m3 Dif. from Vac

Find Grams and kg per Gallon	Vac	Air
Kg. per m3 of Alcohol	906.116	905.025
Liters Per Gal.	3.7854118	3.7854118
Grams per Liter * Constant = Grams per Gal.	3,430.02	3,425.89
Kg. per Gal.	3.43002	3.42589

Lbs. per Gal. Air	7.5528
7/8 Air Lbs. per Gal.	0.00892809
Lb. WG Vac	7.56174
Vac WG lb.	0.13224

In this case the difference is about 20 g/m^3.

I have also derived a value for the pounds per gallon of air which one can use to approximate the pounds per Wine Gallon in vacuum. It is important to note that this procedure will not derive a value for pounds per proof gallon in a vacuum. Every time I've tried to do so I have failed. The unit simply will not tolerate a straight up air-to-vacuum adjustment. To get around this problem, I was able to derive a moles-per-proof-gallon value that can be used with the metric system to calculate the proof gallons of our sample.

Moles and Kilograms per Proof Gallon

As we transition to using the IAT tables and metric blending procedures using mass percent, we need to continue to keep track of alcohol on a proof-gallon basis for verification that our procedures will be consistent with those outlined previously.

Since the TTB tables are the source we must use those values. Beginning with the Table 4 value for the proof gallons per pound for 100 proof we use 1/0.12583 to find the pounds in air for one fluid gallon at 15.556°C.

Using that unrounded value as an input, we convert it to kilograms in air. By using the mass percent in vacuum and air respectively we can use the same mass value as the starting value to calculate the kilograms of absolute alcohol in each proof gallon. The equivalent value in grams is then divided by the molecular weight of alcohol to arrive at the moles per proof gallon:

Find Proof Gallon Values	TTB	
Table 4 Proof Gallons per Lb.	0.12853	
1/Proof Gal. per Lb = Lbs. per Proof Gal.	7.7802848	Unrounded
Pounds (Air) As Input (100 Proof Lbs. Gal.)	7.78028	
Table 4 Proof Gallons per Lb.	0.12853	
Lbs. * PG per Lb. = Proof Gallons	1.00000	
Pounds Avdp to Kilograms	0.45359237	
Kilograms of Alcohol (Air)	3.5291	

	Vac	Air	
Index Mass % Vac of Density at 15.556	42.4879	42.4780	Mass % Air
Mass * Mass % = Kg. of Absolute Alc. Per PG	1.49943	1.49908	Kg. AA Alc.

Find Moles of Alcohol	Vac	Air
Grams of Absolute Alc. Per Proof Gal.	1,499.43	1,499.08
Molecular Wt. per Gram	46.07	46.07
Grams/Molecular Wt.= Moles per Proof Gal.	32.54680	32.53923

Find Moles H2O and Molar %	Vac	Air
(Mass Input Kg. -Kg. AA)* 1000 = Grams H2O	2,029.65	2,030.00
Molecular Wt. of Water	18.01	18.01
Grams/Molecular Wt.=Moles of H2O	112.70	112.72
Total Moles of Alcohol & Water	145.243	145.255
Moles H20/Total Moles = Molar % Water	77.59%	77.60%
1- Molar % Molar % H20 = Molar % Alcohol	22.41%	22.40%

Find Kg. per Proof Gallon (Second Method)	Air	Air	
Wine Gal. per Lb. TTB for 200 Proof T4	0.151290	Table 5 Value	
Pounds Per Gallon T4 Quotient	6.60982	6.60970	
Pounds Avdp to Kilograms	0.45359237	0.45359237	
Kilograms per Gallon 200 proof	2.99816	2.99811	
Divide by 2 = Kg per Proof Gal.	1.499082	1.499055	
Kg. Dif.	0.00000	-0.00003	
Grams Dif	0.0000	-0.03	

Alcohol Modules Mass % and Vol %

Since we are examining the value of a proof gallon so closely, it is helpful to determine the amount of variance in the TTB published values in comparison to the ABS values.

Ultimately the ABS values were generated using the IAT General Formula that we will discuss in detail a little later. But that was after a long process of development in which I tried every way I could to use the TTB tables as the basis of the program. The interim ABS values were derived from TTB values for mass percent in air by using Table 4 as the operative terms. These mass percent values were then converted to vacuum values and run through the general formula to a density at 15.556°C. The results in meters-cubed-vacuum were then converted back to air densities and the pounds per unit calculated from the metric values for density.

The results allow the general formula to become the operative equation for changing density with temperature rather than the Table 37a procedure which must round any mass percent input to the nearest whole number before operating; the latter is not as accurate as the general formula, which permits fractional mass percent inputs to any degree of specificity desired. Much later I learned that it was possible to modify the general formula to accommodate air-based mass percent densities by modifying the term A_1 in the equation for the density of water from 998.20123 to its air density of 997.1237 and also uses an air density as an input. See the General Formula Air & Vac section, infra.

While we are on the topic I will mention that there was one final problem with the TTB mass percent values in that the TTB tables did not generate unique mass percent(s) for each degree of proof because of duplicate values for wine gallons per pound in some portions of Table 4 where density was changing very slowly. This resulted in adjacent proofs with essentially identical mass percent(s), unless one wanted to use 8 decimal places for mass percent, which was not reasonable.

In order to obtain a set of tables that were consistent and reciprocal between air and vacuum regimens and in conformity with the modern definition of the liter and the inch, and, by implication, the cubic inch, I had to build up tables from the general formula as the starting point, generate vacuum densities and then convert them to air values, calculate the air mass percents[56] from those values and also the air densities. This process is covered extensively in Volume 2. Because I was concerned about matching my results as closely as possible to the TTB tables, I wanted to determine the variability existing in the TTB tables themselves:

100 Proof Values			
Proof Gal. per Lb. TTB Table 4	0.128530		
Pounds Per Gal. (Wine & Proof) T4 Quotient	7.780285	3.52908	Kg.
Pounds Per Gal. (Wine & Proof) ABS	7.780258	3.52907	Kg.
Proof Gal. per Lb. TTB Table 5	7.780070	3.52898	Kg.
Kg. Dif. T4 and T5	0.09741	97.41	g m3 Dif.
m3 Air Den	932.273		
Tolerance Value	0.0015		
Grams of Tolerance	1,398.41		
Percent of Tolerance	6.97%		

Alcohol Modules Mass % and Vol %

Looking at the difference between the TTB published values and the ABS values for 100-proof alcohol we see that the ABS value is slightly less than the Table 4 value and greater than the Table 5 value. The total difference between Table 4 and 5 values are ~97 g/m^3 and 6.97% of the tolerance per meter cubed. Since the ABS value is between the two TTB values it is within tolerance as well.

The values in kilograms per proof gallon for air and the moles per proof gallon in air can be used at different calibration temperatures. If one finds the mass percent and calculates the mass of absolute alcohol or the moles at any calibration temperature, these two values will work to find proof gallons, at least within tolerance.

However, while the values in kilograms per gallon (in vacuum) and subsequent moles per proof gallon generated are both quite serviceable, as density decreases the air becomes a larger fraction of the mass displaced and to some degree it also affects the mass percent values generated in air and vacuum as well. It is an intractable problem and with so many moving parts including the molecular weight of the components, it is difficult to find a one-size-fits-all value that is valid at all proofs. The discrepancies are slight but the vacuum value generated tends to generate the largest discrepancies at higher proofs.

---

[56] It is also possible to obtain air mass percents on a density basis by using the K&L equation to boost the vacuum density to an even higher density by adding the air rather than subtracting it to obtain air density. It was found that by this "boosting" procedure one could obtain the air mass percent equivalent of the alcohol. See Volume 2 (*Rules of the Road* Section)

Notes

Blending Formulas 2

Now that we have introduced the procedures necessary to reliably obtain mass percent and volume percent, we can apply these tools to blend alcohol using those parameters.

With respect to blending by mass percent, the first example will be to conduct the simplest blend we can, which is a blend of known mass percent and quantity of alcohol with a known quantity of water to an unknown target mass percent. To give some anchor to the test we will take our standard batch of 160-proof alcohol being blended to 80 proof because we know quantities and the results of that blend. First, we will convert the pounds avoirdupois values to kilograms:

Standard Batch Values	
Lbs. of Alcohol	872.151
Converter to Kg.	0.45359237
Kg. Alc.	395.6010
Lbs. of H20	1,050.07
Kg. H20	476.3047

Blend to Unknown Target Mass Percent

Using our standard batch and starting with 160-proof concentrate we index the corresponding mass percent in both air and vacuum. Using mass and mass percent we calculate the mass of absolute alcohol and then, using our standard air value and the modified vacuum value for the kilograms of absolute alcohol per proof gallon, we verify that the proof gallons match in air to the 194.01 value generated by our standard batch blend:

Blend to Unknown Target Mass %	Concentrate		Dilute					
Proof of Alcohol 1	160.00		0.00		Proof of Alc. 2	871.91	Total Kg. Mass	
Mass In Kilograms	395.601		476.3047		Kg. Mass			
	Vac		Vac			Air		Air
Index Mass % Vac of Density at 15.556	73.517		0.000		Mass % Vac	73.5082	Mass %	0.0000
Mass * Mass % = Kg. of Absolute Alcohol	290.84		0.00		Kg. AA Alc. 2	290.80	Kg. AA	0.00
Kg. of Absolute Alcohol per PG.	1.4991	Vac Kg.	1.4991			1.499	Air Kg.	1.499
Kg. AA/Kg of AA per Proof Gal. = Proof Gal.	194.01		0.00			194.00		0.00

Blend to Unknown Target Mass %	Vac					Air		
Alc. 1 Kg. AA + Alc 2 Kg AA = Tot. Kg. AA	290.84					290.80		
Total Kg. Mass of Alcohol	871.91	1,922.22	Lbs.			871.91	1,922.22	Lbs.
Mass of AA/Total Mass = Mass % of Blend	33.356%	33.361	Confirm w/Index			33.352%	33.352	Index
From Mass % Index Vol. % at 15.556	40.00%	80.00	Proof at 15.556					
From Mass % Index Density % at 15.556	950.86					949.78	Target Density Air	

Find Net Volume	Concentrate		Dilute			Concentrate		Dilute
	Vac		Vac			Air		Air
Kg m3 Density of Alcohol at 15.556 (Vac)	862.96		999.02		Density	861.86	kg. m3 Air	997.94
(Mass/Den)*1000 = Vac Liter Vol. of Alc. 1	458.42		476.77		Vac Liters Alc. 2	459.01	Air Liters	477.29
Total Vac Liters	935.20		1.10		Air Liters Dif.	936.30	Total Air Liters	
Vac Gallons	247.05		0.29		Air Gal. Dif.	247.34	Air Gallons	

Find Blended Volume	Vac					Air		
Target Density Restated	950.86					949.78		
Batch kg. Mass	871.91	1,922.22	Lbs.			871.91	1,922.22	Lbs.
Mass/Density = Net Liter Vol. (Vac)	916.96	242.24	Net Gal.			918.01	242.51	Net Gal.
Total Liters Contributed	935.20	247.05	Gross Gal.			936.30	247.34	Gross Gal.
Liter Volume Change Due to Blending	-18.23	-4.82	Gal.Change			-18.29	-4.83	Gal. Dif.
Percentage of Volume Change	-1.95%					-1.95%		
Find Ideal Case Values								
Milliliters Bottle Size	1,000.00		Vac Bottles Dif.			1,000.00		
Ideal Bottles	916.96		-1.05			918.01		

Blend to Unknown Mass %

The blend to unknown target mass percent procedure simply sums the total kilograms of absolute alcohol (AA) and divides it by the total kilograms employed to find the mass percent of the blend.

This blend mass percent is then used to find the closest mass percent (vacuum) to that contained in the lookup table for each component and from those values one can obtain the equivalent density in air and vacuum.

Using the densities obtained one can find the total liters of the batch and by using the density of the resulting blend one can find the net blended volume.

We see that by this method we have found essentially the same output values in air as for our standard batch. Note that the vacuum liters are less than those calculated in air. This is because we have assumed true values (accurate in both air or vacuum modes) for the mass input on both an air and vacuum basis and this assumption allows us to focus on whether the density and mass percentages as well as the values in kilograms per proof gallon will work together and produce results consistent with the TTB-based results, but while using metric units and vacuum mass percents as the operative values. Comparing the results to our standard batch shows that we have achieved a good correspondence between the two different regimens:

Compare to Std. Batch	Standard	Calculated
Proof Gallons	194.01	194.00
Gross Volume	247.34	247.34
Volume Reduction	-4.83	-4.83
Actual Net Volume	242.51	242.51
Lbs Water to Add	1,050.07	Input Value
Weight of Alcohol	872.15	Input Value
Total Batch Wt.	1,922.22	1,922.22
Tax	2,619.13	2,618.94
Liters	917.99	918.01

Blend to Unknown Mass %

If one is just blending with water it is possible to use the following equation to achieve virtually identical results to the *Blend to Unknown Target Mass Percent* method (Travagli, Equation 1[57]):

$$\%_{w/w\,diluted} = \%_{w/w\,conc.} \times \frac{\rho_{conc} \times V}{\rho_{conc} \times V + \rho_{water} \times V}$$

Turning the equation into a procedure shows that it is functionally equivalent to the prior method, but it only works with water and alcohol. With the right water temperature, it will work at any calibration temperature:

Equation 1 Blend with Water to Unknown Target Mass %					
	Vac			Air	
Density of Concentrate Kg. per dm3 (Vac)	0.86296			0.86186	
liter Volume of Alcohol 1 (Concentrate)	458.42	Vac Liters		459.01	Air Liters
Density * Concentrate Vol. = Result 1	395.60			395.60	
Liter Volume of Alcohol 2 (Dilute)	476.77	Vac Liters		477.29	Air Liters
Density of H2O @ 15.556 (Vac) kg. dm3	0.999016			0.997939	
Density of Water * Dilute Vol. = Result 2	476.30			476.30	
Sum Results 1 and 2 = Result 3	871.91			871.91	
Result 1/ Result 3 = Result 4	0.4537			0.4537	
Result 4 * Mass % of Alc 1 = Mass % Vac	33.361%			33.356%	

Blend to Unknown Mass %

---

[57] V. Travagli, "The Alcohol Dilution," paper, self-published, no date, https://www.scribd.com/doc/70865498/Alcohol-Dilution. This is the first of seven equations cited in G. Spedding, A. Weygandt, and M. Linske, "Alcohol Dilution Practices for Distillers: New and Older Approaches," *Artisan Spirit Magazine*, Winter 2016, pp. 65–70.

With both these procedures one can find mass percent but that is all; one must use some form of lookup for volume percent and density in order to continue. The converse holds true if you do it by volume percent: one can find volume percent and then must look up mass percent.

There is also a second method to find the net volume after finding the volume percents of the components:

Second Method to find Net Volume
Use Found Volume to find Volume % of Source Alcohols

	Vac			Vac			Air			Air	
Density of Absolute Alc. At 15.556 (Vac)	0.79303			0.79303	Density of AA		0.79193			0.79193	Den. of AA
Kg. AA / Den AA = Liters AA Vac	366.74			0.00	Liters AA		367.21			0.00	Liters AA
Liters of AA Vac / Liter Vol of Alc 1 = Vol %	80.000%			0.000%	Vol. % Alc 2		80.000%			0.000%	Vol. % Alc 2

Using Volume % of Blend, Find Volume Change due to Blending

Vol of Concentrate Liters	458.42			459.01		
Volume % of Concentrate	80.00%			80.00%		
Volume % of Blend Result (Target Vol %)	40.00%			40.00%		
Concentrate Vol % / Dilute % = Ratio	2.00			2.00		
Ratio * Starting Vol. of Conc = Net Volume	916.85	242.21	Net Gal.	918.01	242.51	Net Gal.
Total Contributed Liter Volume	935.20	247.05	Tot. Gal.	936.30	247.34	Tot. Gal.
Liter Volume Change Due to Blending	-18.35	-4.85	Gal. Change	-18.28	-4.83	Gal. Change
Percentage of Volume Change	-1.96%			-1.95%		

Blend to Unknown Mass %

To find the volume change due to blending, we need to find the total volume of our alcohol and the volume percent. The volume percent must be indexed from the known or ("Found") destination mass percent. To find volume we will use the found mass percent to index the density at 15.556°C and then divide the mass of alcohol by the density to obtain the volume. Note that the air liters are greater by about 1 part in 1,000. Independently, we can use the known kilograms of absolute alcohol and its density to find the volume percents of the source concentrate and dilute components.

The equation that governs the volume change due to mixing can be stated as follows (Travagli, Equation 4):

$$ xV_{dilute} = \frac{\%_{v/v\,conc}}{\%_{v/v\,dilute}} \times V_{conc} $$

The net result is that using vacuum densities results in calculating less volume than the distiller will actually be dealing with. Using vacuum density values underestimates total volume when blending by mass.

The opposite effect occurs when blending by volume. Since distillers work exclusively in the atmosphere the volumes they combine are atmospheric volumes with atmospheric densities. If this air volume and air density convention is followed, the volume and mass relationships will work out to a very close degree of tolerance as we can see from our standard batch values when entered as volumes and converted to mass so that the mass percent equations can work:

Blend by Volume (Air)

Std. Batch Alc. Gallons	121.255	126.084	Water Gal.
Alc Liters	459.00	477.28	Water Liters
Alc Density Air	861.86	997.94	Air Den
Den * Vol = Alc Kg.	395.595	476.296	Water Kg.
Std. Batch Alc. Kg.	395.601	476.305	Std. H20 Kg.
Alc. Grams Dif.	-5.92	-8.77	Grams Dif.
Total Grams Dif.	-14.70		

Blend to Unknown Mass %

When air volumes are combined with vacuum densities to find mass (to find the total amount of absolute alcohol and the corresponding mass percent), the mass value obtained will be in excess of the intended mass:

Blend by Volume (Vac)			
Std. Batch Alc. Gallons	121.255	126.084	Water Gal.
Alc Liters	459.00	477.28	Water Liters
Density Vac	862.96	999.02	Air Den
Alc Kg.	396.099	476.810	Water Kg.
Std. Batch Alc. Kg.	395.601	476.305	Std. H20 Kg.
Alc. Grams Dif.	497.69	505.44	Grams Dif.
Total Grams Dif.	1,003.14		
Total Kg. Dif.	1.00314	Kg. of Air Displaced	
Blend Density	0.950860		
Mass * Den = Liters	0.954		

The increase will be approximately equal to the air displaced but since we are starting from volume none of the material will ever see a scale.

Blending with this atmospheric volume combined with vacuum density increases the mass and increases the proof gallons and also results in an increase in the resulting mass percent as well. The mass percent increase is not sufficient to alter the indexing of the volume percent or density values in this case since the calculated value is still within the same pigeon hole of the 2,000-place table:

Blend to Unknown Target Mass %	Concentrate		Dilute		
Proof of Alcohol 1	160.00		0.00	Proof of Alc. 2	
Mass In Kilograms	396.099		476.810	Kg. Mass	
	Vac		Vac		
Index Mass % Vac of Density at 15.556	73.517		0.000	Mass % Vac	
Mass * Mass % = Kg. of Absolute Alcohol	291.20		0.00	Kg. AA Alc. 2	
Kg. of Absolute Alcohol per PG.	1.4991	Vac Kg.	1.4991		
Kg. AA/Kg of AA per Proof Gal. = Proof Gal.	194.25		0.00		

Blend to Unknown Target Mass %	Vac			
Alc. 1 Kg. AA + Alc 2 Kg AA = Tot. Kg. AA	291.20			
Total Kg. Mass of Alcohol	872.91	1,924.43	Lbs.	
Mass of AA/Total Mass = Mass % of Blend	33.360%	33.361	Confirm w/Index	
From Mass % Index Vol. % at 15.556	40.00%	80.00	Proof at 15.556	
From Mass % Index Density % at 15.556	950.86			

Comparing to the standard batch below with 33.356 mass percent as opposed to 33.360 above—but still indexing to the 33.361 slot in the table:

Blend to Unknown Target Mass %	Concentrate		Dilute		
Proof of Alcohol 1	160.00		0.00	Proof of Alc. 2	
Mass In Kilograms	395.601		476.3050	Kg. Mass	
	Vac		Vac		
Index Mass % Vac of Density at 15.556	73.517		0.000	Mass % Vac	
Mass * Mass % = Kg. of Absolute Alcohol	290.84		0.00	Kg. AA Alc. 2	
Kg. of Absolute Alcohol per PG.	1.4991	Vac Kg.	1.4991		
Kg. AA/Kg of AA per Proof Gal. = Proof Gal.	194.01		0.00		

Blend to Unknown Target Mass %	Vac			
Alc. 1 Kg. AA + Alc 2 Kg AA = Tot. Kg. AA	290.84			
Total Kg. Mass of Alcohol	871.91	1,922.22	Lbs.	
Mass of AA/Total Mass = Mass % of Blend	33.356%	33.361	Confirm w/Index	
From Mass % Index Vol. % at 15.556	40.00%	80.00	Proof at 15.556	
From Mass % Index Density % at 15.556	950.86			

Blend to Unknown Mass %

Blend to Target Mass Percent

One cannot use vacuum mass percentages to conduct a blend on a density basis and then expect to obtain the quantities that one would obtain when using the densities under atmospheric conditions (i.e., air mass percent).

It is possible to work around this problem; the goal is to use vacuum mass percent and density values and reproduce essentially the same results as one would when using the TTB tables so that mass and volume as well as the proof gallons work out correctly under atmospheric conditions. The proof gallons serve as a very sensitive test of how well the procedures are working.

I will describe methods to accomplish these results. However, it is important that the blending equations themselves be demonstrated first in their basic form before we add the bells and whistles that will allow us to blend with vacuum values but obtain results under atmospheric conditions. I have previously described how to find reciprocal air and vacuum mass percent(s), but the density and mass/volume transformations require far more explanation and would just serve to complicate matters if I tried to cover them first.

As we have seen, most instruments provide readings as if they were in a vacuum. One thing not to get confused about is that hydrometer readings calibrated for vacuum densities are in fact accurate with respect to the demarcations on the instrument itself, since only the exposed part of the hydrometer stem is supported by the atmosphere. It is the volume and mass of the alcohol under investigation that needs to be accommodated to match real world conditions. Again, I will note that the IAT general formula as published only accommodates vacuum mass percent(s) as inputs and gives results in vacuum densities unless one changes the density of water term A_1 from 998.20123 to its air density of 997.1237 and also uses an air mass percent as an input. See the General Formula Air & Vac section, infra.

Blend to Target Mass %	Concentrate		Dilute				
Proof of Alcohol 1	160.00		0.00	Proof of Alc. 2		0.0015	Tolerance
Mass In Kilograms	395.601					0.71	Kg. Tol A2
Target Proof	80.00					0.11	Kg.Dif.
Choose "Air" or "Vac"	Vac					15.8%	% of Tol.
Result Air or Vac Target Mass % of Proof	33.361	950.86	Target Density				
	Vac		Vac				
Index Mass % Vac of Proof at 15.556	73.517		0.000	Mass %			
Kg m3 Density of Alcohol at 15.556	862.960		999.016	Density			
From Vac Density Find Mass % in Air							

Blend to Target Mass %	Vac			Air						
Target Mass %	33.361			33.352						
Mass % of Alc 1	73.517			73.508						
Target Mass % - Mass % of Alc 1 = Mass % Dif.	-40.16%			-40.16%						
Kg. Mass of Alc. 1 Restated	395.601	872.15	Lbs.	395.601	872.15	Lbs.				
Mass % Dif. * Kg Alc 1 = Result 1 (Rounded 6)	-158.86			-158.86						
Mass % of Alc 2	0.000%			0.000%						
Target Mass % Restated	33.361%			33.352%						
Mass % of Alc 2 - Target Mass % = Result 2	-0.3336	Rounded 6		-0.3335	Rounded 6					
Result 1/Result 2 = Kg. Mass of Alc. 2	476.19	1,049.83	Lbs.	476.31	1,050.07	Lbs.	0.11	Kg.Dif.		
Calculated Kg. A1 + Calck Kg. A2 = Tot. Kg.	871.79	1,921.98	Lbs.	871.91	1,922.23	Lbs.	0.11	Kg.Dif.		

Alc 1 Mass * Mass % = Kg. of Absolute Alcohol	290.84	0.00	Kg. AA Alc. 2		
Total Kg. of Absolute Alcohol	290.84	Vac	Find TTB Proof Gal. (Air Only)		
			6.60982	TTB AA Lbs. per Gal.	
Kg. of Absolute Alcohol per PG.	1.49910		641.18	Tot. Kg. AA *2.20426 Factor = Lbs. AA	
Kg. AA/Kg of AA per Proof Gal. = Proof Gal.	194.01		97.00	Lbs. /Lbs. per Gal. AA	
			194.01	Gal. AA *2 = Proof Gal.	

Blend to Target Mass %

Using our standard batch, we note that the procedure using vacuum mass percents has generated 0.11 kg difference in the mass of water (Alcohol 2) and that this amounts to 15.8% of the tolerance for that component.

By using the kilograms of absolute alcohol in vacuum rather than air we can obtain the correct number of proof gallons despite the slight mass discrepancy. This value may not be accurate if the mass error is as demonstrated. We can also calculate the kilograms per proof gallon in air but applying that value to vacuum mass percent blends is even less accurate. We will calculate these values shortly and see how effective it is.

While we have previously found the moles of alcohol I have not demonstrated how to find the molar percentage and that simple procedure is set forth below along with a confirmatory calculation of the target volume percent:

Find Moles & Molar %		
Grams of Absolute Alc.	290,836	0
Molecular Wt. per Gram	46.07	46.07
Grams/Molecular Wt.= Moles of Abs. Alc.	6,312.91	0.00
Find Moles H2O and Molar %		
(Mass Input Kg. -Kg. AA)* 1000 = Grams H2O	104,765	395,601
Molecular Wt.	18.01	18.01
Grams/Molecular Wt.=Moles of H2O	5,817.1	21,965.8
Total Moles of Alcohol & Water	12,130.0	21,965.8
Moles H20/Total Moles = Molar % Water	47.96%	100.00%
1- Molar % Molar % H20 = Molar % Alcohol	52.04%	0.00%

Find Volume %		
Density of AA	0.79303	Vac
Kg. AA/Den AA = Liters AA	366.74	
Liters of Blend	916.85	
Vol. % of Blend	40.000%	

Blend to Target Mass %

Decrease Density (Increase Proof)

Using our standard batch values, we can test the procedure by using water as the dilute and our 160-proof alcohol as the concentrate. The procedure makes a fair pass at increasing the proof to our target of 80 proof:

Decrease Density (Increase Proof)	Dilute		Concentrate				80.00	Target Proof
Proof of Alcohol 1	0.00		160.00	Proof of Alc. 2			33.361	Mass % Vac
Mass In Kilograms	476.30						33.352	Mass % Air
	Vac		Vac		Dilute Air			Conc. Air
Index Mass % Vac of Density at 15.556	0.000		73.52	Mass % Vac	0.0000	Mass %		73.51
Mass * Mass % = Kg. of Absolute Alcohol	0.00		290.90	Kg. AA Alc. 2	0.00	Kg. AA		290.87
Find Volume		290.90	Tot. kg. AA			290.87		Tot. kg. AA
Kg m3 Density of Alcohol at 15.556 (Vac)	999.02		862.96	Density	997.94	kg. m3 Air		861.86
(Mass/Den)*1000 = Vac Liter Vol. of Alc. 1	476.77		458.53	Vac Liters Alc. 2	477.29	Air Liters		459.01
Total Vac Liters	935.31		0.99	Liters Dif.	936.29	Total Air Liters		
Total Vac Gallons	247.08		0.26	Gal. Dif.	247.34	Tot. Air Gallons		
Decrease Density (Increase Proof)	Vac		Air					
Concentrate Mass % (Alc 1)	73.52%		73.51%		0.0015	Tolerance		
Dilute Mass % (Alc 2)	0.00%		0.00%		0.59	Kg. Tol A2		
Basis of 100 for calculations	100.00		100.00		0.69	Liter Tol A2		
					$13.50	Tax Rate		
Target Mass %	33.36%		33.35%		$0.00	Tax on Dilute (A1)		
Target Mass % - Dilute Mass % = Result 1	40.2%		40.2%		$2,619.71	Tax on Concentrate (A2)		
Alc 2 Mass %- Target Mass = Result 2	33.36%		33.35%		$2,619.71	Total Tax on Batch		
(Result 2/Result 1) * 100 = Kg. of Concentrate	83.08		83.06					
100 Basis + Kg. Concentrate = Tot Kg. at 100	183.08		183.06					
Scale Up to Batch Size					Calculate Vol % of Blend			
Mass of Dilute Alcohol (Alc 1)	476.30	Same Value	476.30		0.79303	Density of AA Vac		
Kg. Concentrate/100 = Scaling % to Batch Size	0.8308		0.8306		366.83	Kg. AA / Den AA = Liters		
Mass Dilute * Factor = Kg. Mass of Conc. to add	395.69		395.60		917.06	Liters of Blend		
Density of Concentrate	0.86296		0.861863		40.000%	Vol. % of Blend		
Mass/Den = Vol of Alcohol at Calib°	458.53		459.01					
Proof Gallons Already in the Batch (Dilute)	Vac		Air		Find TTB Proof Gal.			
Kg. of Absolute Alcohol per PG.	1.4991		1.49908	Kg. per PG Air	6.60982	TTB AA Lbs. per Gal.		
Dilute Kg. AA/Kg of AA per PG = Dilute PG	0.00		194.03		641.25	Tot Kg. AA * 2.204.= Lbs		
Concentrate Kg. AA/Factor = Proof Gal.	194.05		0.00		97.02	Lbs. /Lbs. per Gal. AA		
Total Proof Gallons in Batch	194.05		194.03		194.03	Gal. AA *2 = Proof Gal.		

Decrease Density (Increase Proof)

Again, we see a difference in volume and the slight difference in proof gallons, but it is more important to set forth the procedure for accomplishing the blend than to try to mitigate the differences.

Blend by Total Batch Mass & Volume

Total Batch Blend simply uses the Pearson's Square method. One does require a limit so that the target mass percent is greater than the Alcohol 2 dilute mass percent and less than the Alcohol 1 concentrate. See the Alcohol 1 and Target Mass Percent logical check boxes that each gives the "OK" for this blend:

Blend by Total Batch Mass	Alcohol 1		Alcohol 2 (Or Water)					
Proof of Alcohol 1	160.00		0.00	Proof of Alc. 2		80.00	Target Proof	
Total Batch Mass In Kilograms	871.91					33.361	Target Mass % Vac	
	Vac		Vac			Alcohol 1 Air		Alcohol 2 Air
Index Mass % Vac of Density at 15.556	73.517		0.00	Mass % Vac		73.5082	Mass %	0.00
Mass * Mass % = Kg. of Absolute Alcohol	290.87		0.00	Kg. AA Alc. 2		290.84	Kg. AA	0.00
Find Volume			290.87	Tot. kg. AA				
Kg m3 Density of Alcohol at 15.556 (Vac)	862.96		999.02	Density		861.86	kg. m3 Air	997.94
(Mass/Den)*1000 = Vac Liter Vol. of Alc. 1	458.48		476.72	Vac Liters Alc. 2		459.07	Air Liters	477.24
Total Vac Liters	935.21		1.10	Liters Dif.		936.30	Total Air Liters	
Total Vac Gallons	247.06		0.29	Gal. Dif.		247.35	Tot. Air Gallons	

Blend By Total Batch Mass Using Pearson's Square Method						OK	Alc. 1 Mass % Check
Batch Mass Restated	871.91		Parts	Ratio		OK	Target Mass % Check
Mass % of Concentrate	73.52%	Target %	0.334	0.454	395.65	Mass of Alc 1	
		33.36%					
Mass % of Dilute	0.00%		0.402	0.546	476.25	Mass of Alc 2	
		Total Parts	0.74	1.00	871.91	Total Mass	

Proof Gallons Already in the Batch (Dilute)			Find TTB Proof Gal.		
Kg. of Absolute Alcohol per PG.	1.499100		6.60982	TTB AA Lbs. per Gal.	
Alc 1 Kg. AA/Kg of AA per PG = Dilute PG	194.03		641.27	Tot. Kg. AA * Factor = Lbs. AA	
Alc 2 Kg. AA/Factor = Proof Gal. of Alc 2	0.000		97.02	Lbs. /Lbs. per Gal. AA	
Total Proof Gallons in Batch	194.03		194.03	Gal. AA *2 = Proof Gal.	

Total Mass Blend

We blended by total volume previously by calculating proof gallons necessary to accomplish our blend. In this case we will blend by total volume, using blended liters in air along with the target density in air to find the total batch mass of 871.91 kilograms (the same mass we applied in the previous worksheet), equivalent to our standard 1,922.22 pounds. 918.01 liters = 241.51 US gal., the blended volume of our standard batch:

Blend by Total Batch Net Volume Input				
Total Batch Volume in Liters (Blended Vol)	918.01		80.00	Target Proof
Target Density kg m3 Air	949.78		33.361	Target Mass %
Vol. * Den. = Kg. Mass of Batch	871.91			

Blend by Total Batch Volume	Alcohol 1		Alcohol 2 (Or Water)					
Proof of Alcohol 1	160.00		0.00	Proof of Alc. 2				
Total Batch Mass In Kilograms Restated	871.91							
	Vac		Vac			Alcohol 1 Air		Alcohol 2 Air
Index Mass % Vac of Density at 15.556	73.517		0.00	Mass % Vac		73.5082	Mass %	0.00
Mass * Mass % = Kg. of Absolute Alcohol	290.87		0.00	Kg. AA Alc. 2		290.84	Kg. AA	0.00
Find Volume			290.87	Tot. kg. AA				
Kg m3 Density of Alcohol at 15.556 (Vac)	862.96		999.02	Density		861.86	kg. m3 Air	997.94
(Mass/Den)*1000 = Vac Liter Vol. of Alc. 1	458.48		476.72	Vac Liters Alc. 2		459.07	Air Liters	477.24
Total Vac Liters	935.21		1.10	Liters Dif.		936.30	Total Air Liters	
Total Vac Gallons	247.06		0.29	Gal. Dif.		247.35	Tot. Air Gallons	

Total Volume Blend

With the total batch mass determined and the mass percents known, the batch proceeds as if it were a total mass batch as above in the first example by applying Pearson's Square.

Atmosphere

How Vacuum Densities are Determined

When doing precise work on the density of alcohol the investigations are conducted under conditions close to that of a vacuum—conditions without the buoyancy that the atmosphere provides. My initial thought was that this would be impossible since liquids boil in a vacuum. If you put a hydrometer in a beaker of alcohol, put it in a vacuum chamber, and then evacuated the air, the alcohol would simply vaporize into the chamber and the hydrometer would rest on the bottom of the beaker.

Alcohol investigators have found a way around this problem by using a tank with very low pressure (partial vacuum) and a system of balances and weights to simulate the remaining vacuum. By doing this they can obtain the most precise measurements of the density of alcohol. When these investigators publish their work, the tables and formulas they publish are "In Vacuum" with respect to mass percentage and density.

Some values for the density of water only are published as densities in air. The *Handbook of Chemistry and Physics* 89th edition lists standard densities of water at 101,325 Pa, 101.325 millibars, or 1 bar as its standard. See the *H20 HCP* worksheet for a list of the air and vacuum densities of water at different temperatures.[58]

Dry Air Buoyancy

Our atmosphere is sometimes referred to as an ocean of air. This air is densest at the bottom where we live and becomes thinner as we rise in altitude because gravity falls off with distance from the center of the Earth. Our planet's mass can only attract so much gas to itself at its current overall temperature. The weight of this air pushing on all sides of objects in the atmosphere results in a small amount of flotation and it also provides resistance to movement.

In weighing objects, we are not concerned with resistance to movement but we are concerned with the amount of suspension or buoyancy that the atmosphere provides. An easy way to explain this is to imagine yourself in a pool while stepping on a scale under the water. How much pressure or weight the scale indicated would be dependent on how much of your body was immersed in the water. In the shallow end of the pool, with half your body in the water and half your body out of it you would weigh about half your weight if standing completely outside the pool. If you were in the deep end of the pool and completely under water you would have a hard time just staying on the scale unless you let all the air out of your lungs. This is Archimedes principal: a body will experience an upward force equal to the weight of the material displaced by a given volume immersed in such liquid. This difference becomes a factor when trying to weigh an object precisely, particularly in scientific work where one is trying to control or eliminate as many variables as possible.

Because distillers are always working in the atmosphere rather than a vacuum all the tables in the Gauging Manual, except for the Specific Gravity in Vacuum values contained in Table 6, have been adjusted to reduce the vacuum density to that measured by the distiller based upon standard atmospheric conditions so one does not have to concern oneself with the difference between air and vacuum mass.

---

[58] D. R. Lide (Ed.), *The Handbook of Chemistry and Physics*, 89th edition. CRC Press, 2008. ISBN: 978-1420066791. "The weights are for dry air at the same temperature as the water up to 40° C. and at a barometric pressure corrected to 760 mm and against brass weights of 8.4 density at 0° C. Above 40° C. the temperature of the air is assumed to be 20° C., i.e., the water is allowed to cool to 20° C. prior to the weighings being made. The volumetric computations are based upon the relations that one liter = 1 dm^3 and that 1 dm^3 = 61.023744 in.^{3}. This table gives the density of water in the temperature range from 0° C. to 100° C. at a pressure of 101325 Pa (one standard atmosphere). From 0° C. to 40° C. the values are taken from the IUPAC publication in Reference 1 and refer to standard mean ocean water (SMOW), free from dissolved salts and gases. SMOW is a standard water sample of high purity and known isotopic composition. Methods of correcting for different isotopic compositions are discussed in Reference 1. The remaining values are taken from the NIST Chemistry WebBook, Reference 2. Note that the IUPAC values refer to the IPTS-68 temperature scale, while the NIST values are based on the ITS-90 scale (where the normal boiling point is 99.974° C.). The conversion between these scales can be found in Sec. 1. The difference between the scales leads to a difference in the density of water of about 20 ppm in the neighborhood of 100° C. and much less at lower temperatures." [Note, while the *HCP* talks of the density (vac) being at 1 bar it does not mean that it is measured in air. These values will have been corrected to be absolute densities but they specify the pressure because when you start measuring so accurately the pressure does start to play a role. We think of water as incompressible, but it is slightly elastic.]

Density, in terms of kilograms per meter cubed or any other unit is usually expressed on a vacuum or absolute density basis. This means that the demarcations of density units themselves, as inscribed on the hydrometer stem (or on the paper inside the stem) reflect the density as if it was in a vacuum. The hydrometer itself may be designed to be used under atmospheric conditions but the readings one will obtain from it are on a vacuum or absolute density basis. Because only the stem of the hydrometer is in the atmosphere and therefore only a small amount of the hydrometer itself is supported by air pressure it is possible to determine density in air but report it in vacuum.

Given that air pressure around us provides an upward force, let's look at just how much force this amounts to. High pressure floats us more than low pressure so the next sunny day when the barometer is above 30 inches of mercury you can feel lighter on your feet because to a small degree you are. If you feel down when it is raining part of it may be that you are in fact heavier during a low-pressure storm.

Applying a standard air pressure value of 29.92 inches of mercury or 1,013.20 millibars results in 1.2226 grams per liter of atmospheric mass displaced. This mass creates an upward force on any liter volume to that degree:

Find Dry Air Density			
Millibars	1,013.25		
Air Temp ° C.	15.556	287.058	Wikipedia Value
Constant ° C. to ° K	273.15	286.90	Engineering Toolbox SI Units (j/kg K)
Temperature (°K)	288.71	287.05	Specific gas constant dry air (287.05 J/(kg.K)
Specific gas constant dry air (287.05 J/(kg.K)	287.05		
Temp * Gas Constant Result	82,873		
Result/Millibars) * 100 = Air kg m3 (p)	1.222651	1.2226497	ABS Std. Value @ 15.556 & 1,013.20 mb
Round kg per m3 to 6 Places	1.222651	1.64E-06	Dif.

Air to Vacuum Dry Air

Relating this to a human being, the human body has a density close to that of water so we will just use zero proof as our input. A seventy-five-kilogram person (165 lbs.) occupies about 75 liters in volume (19.81 gal.).[59] Solving for these parameters gives an upward force of 80.97 grams (2.86 ounces) at standard atmospheric pressure:

Dry Air Atmospheric Mass Input to Vac Mass.			
Air Pressure	29.9213	In. of Hg. ▼	
Air Temperature ° C.	15.556	60.00	° F.
Kilograms Indicated on Scale	75.00	165.35	Lbs.
Alcohol Density (Proof @ 60° F.)	0.00		
Density in Air Kg m3	997.94	999.02	Vac kg m3
Dry Air Results			
Dry Air Den. Kg. Displaced	0.0810		
Dry Air Grams Displaced	80.97	2.86	Oz.
Indicated + Kg. Displaced = IPK Mass	75.08		

Air to Vacuum Dry Air

---

[59] Use *Gauging Drums of Alcohol* using the density (Proof) of water to make this calculation.

Recently, at my house, it was 0°F or −17.78°C and the atmospheric pressure was 30.18 inches of mercury. Solving for these values shows that the change in air pressure and temperature resulted in an increase from 81 grams to 92.33 grams of displacement by the atmosphere and thus increased buoyancy. The only time that air pressure differences will be a real factor is if one weighed out one component of a blend on a high-pressure day and then came back the next day when it was cold and rainy and weighed out its companion in the blend under low-pressure conditions.

Dry Air Atmospheric Mass Input to Vac Mass.			
Air Pressure	30.1800	In. of Hg. ▼	
Air Temperature ° C.	-17.778	(0.00)	° F.
Kilograms Indicated on Scale	75.00	165.35	Lbs.
Alcohol Density (Proof @ 60° F.)	0.00		
Density in Air Kg m3	997.94	999.02	Vac kg m3
Dry Air Results			
Dry Air Den. Kg. Displaced	0.0923		
Dry Air Grams Displaced	92.33	3.26	Oz.
Indicated + Kg. Displaced = IPK Mass	75.09		

Air to Vacuum Dry Air

Solving on a cubic meter scale shows for the density of water in air (997.94 kg/m^3 at 15.556°C), the air pressure difference in buoyancy amounts to 72.0% of the weight tolerance for blending:

Dry Air Atmospheric Mass Input to Vac Mass.				Equivalent Value		
Air Pressure	29.9213	In. of Hg. ▼		1,013.249	Millibars ▼	
Air Temperature ° C.	15.556	60.00	° F.			
Kilograms Indicated on Scale	997.94	2,200.08	Lbs.			
Alcohol Density (Proof @ 60° F.)	0.00					
Density in Air Kg m3	997.94	999.02	Vac kg m3			
Dry Air Results				Find Tolerance		
Dry Air Den. Kg. Displaced	1.0774			0.0015	Amount of Tol.	
Dry Air Grams Displaced	1,077.40	38.00	Oz.	1,496.9	Grams of Tol.	
Indicated + Kg. Displaced = IPK Mass	999.02			72.0%	Air to Vac % of Tol.	

Air to Vacuum Dry Air

By adding this air displacement to an indicated mass, one can make a good first pass at finding the vacuum density, in this case 999.02 for the density of water in a vacuum.

We saw one method of determining the density of dry air using the specific gas constant above. There is also a second method to find the density of dry air, using values derived by Regnault and Leduc/Rayleigh:[60]

Regnault Value	0.00367
Temperature	15.556
Temp * Regnalt Value =	0.057
Value	1.00
Value + Result	1.057
Leduc & Rayleigh Value	0.001293
Leduc Value/Result =	0.001
Present air Pressure (H)	760.00
mm of Hg. As Constant	760.000
Present Pres./Constant =	1.000
Result * prior result = Air Den	0.00122
As Grams per Liter	1.2232

Air to Vacuum Humid Air

Solving the equation at 15.556°C indicates that it results in essentially the same grams-per-liter displacement of mass per unit of volume.

---

[60] G. W. C. Kaye and T. H. Laby, *Tables of Physical and Chemical Constants*, Fourth Ed. London: Longmans, Green & Co., 1921, p. 27.

Air Moisture

Our atmosphere is composed of about 78% nitrogen (N_2), 21% oxygen (O_2) and 1% other gases. Nitrogen has a molecular weight of 14 so a N_2 molecule has a molecular weight of 28. Oxygen has a molecular weight of 16 so an O_2 molecule has a molecular weight of 32. Given the mixture of gases found in air, the average molecular weight of air is around 29. The molecular mass of water is 18 so the addition of water to the atmosphere, in the form of humidity, lowers its average density and decreases the buoyancy effect compared to dry air.

I thought otherwise until I looked into it. Humid air just feels closer, denser and more difficult to breathe than dry air. Also, I was initially tripped up by the phenomenon that automobiles run more efficiently with moist air than dry air. This effect is caused by the water vapor cooling the overall intake charge as it evaporates during the intake process, resulting in a denser intake charge, despite the water vapor taking up part of the vapor charge after it evaporates.

Either way, I wanted to investigate how much humidity affects the buoyancy of alcohol when measurements are taken and the result is that it has very little effect on weighings. In the following example the total discrepancy at 80% humidity is on the order of −5.49 grams per 1,000 kilograms so the joke is on me:

Dry and Humid Air Atmospheric Mass Input to Vac Mass.					Equivalent Value		
Air Pressure	29.9213	In. of Hg.	▼		1,013.249	Millibars	▼
Air Temperature ° C.	15.5560	60.00	° F.		Dew Point		
Relative Humidity	80.0%				11.6	° C.	
Kilograms Indicated on Scale	1,000.00	2,204.62	Lbs.		52.8	° F	
Alcohol Density (Proof @ 60° F.)	0.00						
Density in Air	997.94			999.02	Vac Density		
Dry Air Results				Damp Air Results			
Dry Air Den. Kg. Displaced	1.07962			1.07414	Damp Air Density kg		
Dry Air Grams Displaced	1,079.62			1,074.14	Damp Air Grams Displaced		
Indicated + Kg. Displaced = IPK Mass	1,001.080			1,001.07	Std. Kg. Mass on Scale in Damp Air		
				-5.49	Grams Dif, Damp Air v. Dry		
				99.49%	Damp Air % Dry Air		

Air to Vacuum Humid Air

I do want to report the method by which I made the calculation so the method can be replicated and any mistakes identified in the future.

Relative Humidity (RH) is the ratio of the actual water vapor pressure to the saturation water vapor pressure at the prevailing temperature.

We note below that the pressure differences for 80% humidity are noticeable in terms of atmospheric pressure and also that while the vapor pressure is calculated as a positive it needs to be subtracted from each unit since humidity decreases mass per unit volume of atmosphere:

Vapor Pressure Differences				
Start mm of Hg.	760.00		1,013.25	Start Millibars
mm of Hg. Dif. Damp to Dry	10.219		13.62	mb Dif.
Actual mm HG	749.78		999.62	Actual mb.
Expressed as a Percentage	98.66%		98.66%	%
Start Psi	14.70		29.921	Start Inches of Hg.
Psi Dif.	0.20		0.402	In. of Hg. Dif.
Actual Psi	14.50		29.519	Actual in. Hg.
Expressed as a Percentage	98.66%		98.66%	%

We will quickly run through the formulas that establish these values. First, we start with a simple dew point calculating procedure.[61] The dew point is the temperature to which the air must be cooled before dew condenses from it. At this calculated temperature the water vapor content of the air becomes equal to the saturation water vapor pressure:

Find Dew Point (Saturation Temp)	
Relative Humidity * 100	80.0
Value	100.00
Value - RH as value	20.00
Value	5.0
RH/Value = Result in ° C.	4.00
Present Air Temp Restated	15.56
Temp - Result = Dew Point ° C	11.56
° F	52.80

Air to Vacuum Humid Air

Taking the dew point and using a formula published by the US National Oceanographic and Atmospheric Administration (NOAA) one can find the millibars of pressure given the parameters of our example:[62]

Find H2O Vapor Pressure			Find Saturated Pressure	
Dew Point (Td)	11.56		15.556	Temp (T)
Value	7.50		7.50	Value
Td * Value = Result 1	86.67		116.67	Td * Value = Result 1
k Value	237.30		237.30	k Value
k Value + Td = Result 2	248.86		252.86	k Value + Td = Result 2
R 1/R2 = Result 3	0.35		0.46	R 1/R2 = Result 3
Multiplier	10.00		10.00	Multiplier
10 to the power of result = Result 4	2.2298		2.8934	10 to the power
Constant	6.11		6.11	Value
K * R4 = Millibars of Vapor Pressure	13.62		17.68	Millibars Vap. Pres.

Psi result	0.20	0.014504	Millibars to Psi
mm of HG	10.22	0.750064	millibars to mm of HG
In. of Mercury (Hg.)	0.402	0.02953	Mb to in. Hg.

Having found the vapor pressure in units of mm of mercury (Hg) one can use a formula from page 23 of *Tables of Physical and Chemical Constants* to calculate the density of humid air.

H2O Vapor Pres. Mm of Hg.	10.219		
Constant	0.3780		
Temp * Constant = Result 1	3.8628		
Barometric Height (H)	760.00		
H - Result 1 = Result 2	756.14		
Result 2/H = Result 3	0.994917		
Density of Dry Air (e'd) Restated	0.0012227	1.223	Kg m3
Dry Air * Result 3 = Den. Damp Air	0.0012164	1.216	Kg m3
Dry Air is Greater	0.000006	0.006	Kg m3
		6.21	grams/m3

The usefulness of these procedures is limited by the fact that these formulas produce 1.77 grams of difference at 0% humidity when using 1,000 kilograms. Judicious rounding might mitigate this effect. I decided not to use the procedures in the overall program because of the added complexity it would entail for the limited benefit. Also, it only applies to weighings, since density is never calculated with respect to humid air, at least in any of the published tables I have ever seen. If one is using an actual air pressure to assist in determining the mass of an

---

[61] Dew Point: http://iridl.ldeo.columbia.edu/dochelp/QA/Basic/dewpoint.html

[62] Vapor Pressure Formula //www.srh.noaa.gov/epz/?n=wxcalc_vaporpressure

object, rather than using a standard (dry) atmosphere, these tools will provide a good first approximation for the amount of difference.

Having determined that the water vapor content of air is not a significant factor in gauging alcohol, we can turn our attention back to the overall effect of the atmosphere on making precise determinations of mass.

Air to Vacuum

Reductions to Weighing in Vacuum

I've identified a few other methods to determine the mass of objects in vacuum.

The following method is reproduced from Kaye and Laby's *Tables of Physical and Chemical Constants*, page 21. It provides the following formula: "The buoyancy correction is $M\sigma (1/\Delta - 1/\rho) = Mk$. Where M is the apparent mass in grams of the body in air, σ is the density of air (= 0.0012) in grams per c.c., Δ is the density of the body, ρ is the density of the weights. The correction is true to 4% for the following limits: 740 mm. 1° to 22°; 760 mm, 8° to 29°; 780 mm, 15° to 35°. If the correction is required more accurately, multiply the value of k given below by $\sigma'/.0012$ where σ' is the true density of the air for the temperature and pressure, at the time of the weighing."

For reference the density of various weights used for determining mass or for calibration are:

Aluminum	2.65
Quartz	2.65
Steel	7.97
Brass	8.40
Lead	11.34
Platinum	21.50

In some instances, the temperature of the weights is also mentioned. For example, the density of 8.4 for brass weights is based upon them being 0°C. Brass weights seem to have become the standard by which most tables are constructed. The correction is valid across a broad range of densities and not just applicable to alcohol. Following are some calculated correction factors for various densities of materials that are reproduced from the original work:

Calculated Factor (k) in milligrams (TPCC p. 21)

Density	Brass Wt. p = 8.4	Platinum p =21.5	Aluminum p = 2.65	Density	Brass Wt. p = 8.4	Platinum p =21.5	Aluminum p = 2.65
0.5	2.26	2.34	1.95	1.6	0.61	0.69	0.30
0.55	2.04	2.13	1.73	1.7	0.56	0.65	0.25
0.60	1.86	1.94	1.55	1.8	0.52	0.62	0.21
0.65	1.70	1.79	1.39	1.9	0.49	0.58	0.18
0.70	1.57	1.66	1.26	2.0	0.46	0.54	0.15
0.75	1.46	1.55	1.15	2.5	0.34	0.43	0.03
0.80	1.36	1.44	1.05	3.0	0.26	0.34	-0.05
0.85	1.27	1.36	0.96	3.5	0.20	0.29	-0.11
0.90	1.19	1.28	0.88	4.0	0.16	0.24	-0.15
0.95	1.12	1.21	0.81	5.0	0.10	0.19	-0.21
1.0	1.06	1.14	0.75	6.0	0.06	0.14	-0.25
1.1	0.95	1.04	0.64	8.0	0.01	0.09	-0.30
1.2	0.86	0.94	0.55	10.0	-0.02	0.06	-0.33
1.3	0.78	0.87	0.47	15.0	-0.06	0.03	-0.37
1.4	0.71	0.80	0.40	20.0	-0.08	0.004	-0.39
1.5	0.66	0.75	0.35	22.0	-0.09	-0.001	-0.40

Convert to Vac

Note that the corrections decrease as the density of the body being corrected increases and that the sign becomes negative earlier based upon the density of the weights used. As the density increases, the volume of the sample decreases and becomes closer to the volume of the weights used on the other side of the balance pan, so less correction is required.

What these equations are accomplishing is to add the buoyancy force that is lifting up the volume of alcohol while simultaneously offsetting the amount of buoyancy force that is lifting up the weights.

We will start with an apparent mass of 1,000 kilograms in air at a standard dry air atmospheric pressure of 15.556°C against brass weights with a density of 8.4. We want to find the true International Prototype Kilogram (IPK) mass of this apparent mass and also calculate its volume in the atmosphere. Using 10 proof as the density of the material under investigation, we see that Kaye and Laby equation solves for an atmospheric buoyancy of 1.087 kilograms and adding that value to the apparent mass we can arrive at an IPK mass of 1,001.09 kilograms.

The reason we are adding the mass of air displaced is because we have had to overcome that amount of air buoyancy in order for the scale to indicate 1,000 kilograms in the first place.

With respect to tolerance, note that with a tolerance of 1,500 g/m^3 the air to vacuum difference is 72.5% of the weight tolerance:

Convert Air Mass to IPK Vac Mass and Air Volume			0.0015	Tolerance
True Proof of Alcohol @ 60° F.	10.00		1,500	Total Grams Tolerance
Apparent Mass In Air. Kg.	1,000.00		1,087.1	Grams of Correction
			72.5%	Air to Vac % of Tolerance
Kaye & Laby Equation				Density of Calibration Wts.
Apparent Mass in Air Kg. (M)	1,000.00	2,204.62 Lbs.	2.65	Aluminum
Air Density, Dry Air (σ') at 15.556 C.	0.00122263		7.970	Steel
Mass * Air Den = Result 1 (Mσ')	1.2226		8.40	Brass
Density of Body (Δ) Kg dm3 Vac	0.99187		11.34	Lead
Value	1.00		21.50	Platinum
1/Δ = Result 2	1.0082	7/8 Air		
Density of Weights (ρ)	8.40		0.001223	Air Density
Value	1.00		87.5%	7/8 Air
1/Density of weights = Result 3	0.1190		0.001070	Air Den * % = 7/8 Air Result
Result 2 - Result 3 = Result 4	0.8892		1,008.20	"Wrong" liters Restated
Result 4 * Result 1 = Buoyancy Cor. (K)	1.0871	Mass of Air	1.0786	Liters * 7/8 Air Den = 7/8 Air Mass
Apparent Mass + Buoyancy Cor. = Vac Mass	1,001.09	IPK Mass	1,001.08	Apparent Mass + 7/8 Air = Vac Mass
Find Volume			8.53	7/8 Grams Dif. From K&L Grams
Index Kg m3 Vac	991.87		0.57%	Corrected % of Tolerance
(Vac IPK Mass/ Vac Den) = IPK Mass Liter Vol.	1,009.30			
Correction/Air Mass = % of Air Applied	88.92%			
Calculate wrong volume			Calculate Correct Volume.	
Kg m3 Vac	991.87		990.79	Air Density
(Indicated Mass/ Vac Den)= Wrong Volume	1,008.20		1,009.30	Indicated/Air Den = Vol.
Volume Dif. From Vac mass liters	-1.096		0.001	Dif. From IPK Vol.

Convert to Vac

Once the true IPK mass is found one can use the vacuum density to calculate the correct number of liters. What does not work is to use the apparent mass and the vacuum density which will calculate a lesser volume than is actually present in air. This can be remedied by obtaining the air density and using the indicated mass on the scale.

There is a simple shortcut to using the Kaye and Laby (K&L) equation. It's called the *7/8 air rule*. Take the Standard Atmospheric air pressure in kilograms per liter and multiply it by 87.5%, then use this reduced value of air per liter and multiply by the "wrong" liters (which without the K&L equation is the only volume that can be calculated since one does not actually know the IPK mass). This correction will arrive at a close approximation to the actual air buoyancy. It will also apply to density correction as well. One should not add full air to a density correction since the density measured is the atmospheric density just the same as the indicated mass is also the atmospheric mass overcoming the buoyancy of the atmosphere. Accordingly, density corrections should not add the full mass of atmosphere displaced since they are beginning with an air density.

If one knows the true vacuum density, then the full displacement of the atmosphere can be applied. We will examine this in more detail in the section on density correction.

Notice that for this low-proof alcohol the percentage of air actually applied is only 88.92% of the actual air pressure, either against the volume or the weights. Using aluminum weights with a density of 2.65 the percentage drops to 63% of the air density and the 7/8 rule no longer works:

Convert Air Mass to IPK Vac Mass and Air Volume			0.0015	Tolerance
True Proof of Alcohol @ 60° F.	10.00		1,500	Total Grams Tolerance
Apparent Mass In Air. Kg.	1,000.00		771.3	Grams of Correction
			51.4%	Air to Vac % of Tolerance
Kaye & Laby Equation				Density of Calibration Wts.
Apparent Mass in Air Kg. (M)	1,000.00	2,204.62 Lbs.	2.65	Aluminum
Air Density, Dry Air (σ') at 15.556 C.	0.00122263		7.970	Steel
Mass * Air Den = Result 1 (Mσ')	1.2226		8.40	Brass
Density of Body (Δ) Kg dm3 Vac	0.99187		11.34	Lead
Value	1.00		21.50	Platinum
1/Δ = Result 2	1.0082	7/8 Air		
Density of Weights (ρ)	2.65	0.001223	Air Density	
Value	1.00	87.5%	7/8 Air	
1/Density of weights = Result 3	0.3774	0.001070	Air Den * % = 7/8 Air Result	
Result 2 - Result 3 = Result 4	0.6308	1,008.20	"Wrong" liters Restated	
Result 4 * Result 1 = Buoyancy Cor. (K)	0.7713	Mass of Air 1.0786	Liters * 7/8 Air Den = 7/8 Air Mass	
Apparent Mass + Buoyancy Cor. = Vac Mass	1,000.77	IPK Mass 1,001.08	Apparent Mass + 7/8 Air = Vac Mass	
Find Volume		-307.29	7/8 Grams Dif. From K&L Grams	
Index Kg m3 Vac	991.87	-20.49%	Corrected % of Tolerance	
(Vac IPK Mass/ Vac Den) = IPK Mass Liter Vol.	1,008.98			
Correction/Air Mass = % of Air Applied	63.08%			

Changing the proof to 190 and the weights back to brass with a density of 8.4 shows that the air percentage is now 110.77% of the density of air at the time of the measurement:

Convert Air Mass to IPK Vac Mass and Air Volume			0.0015	Tolerance
True Proof of Alcohol @ 60° F.	190.00		1,500	Total Grams Tolerance
Apparent Mass In Air. Kg.	1,000.00		1,354.3	Grams of Correction
			90.3%	Air to Vac % of Tolerance
Kaye & Laby Equation				Density of Calibration Wts.
Apparent Mass in Air Kg. (M)	1,000.00	2,204.62 Lbs.	2.65	Aluminum
Air Density, Dry Air (σ') at 15.556 C.	0.00122263		7.970	Steel
Mass * Air Den = Result 1 (Mσ')	1.2226		8.40	Brass
Density of Body (Δ) Kg dm3 Vac	0.81519		11.34	Lead
Value	1.00		21.50	Platinum
1/Δ = Result 2	1.2267	7/8 Air		
Density of Weights (ρ)	8.40	0.001223	Air Density	
Value	1.00	87.5%	7/8 Air	
1/Density of weights = Result 3	0.1190	0.001070	Air Den * % = 7/8 Air Result	
Result 2 - Result 3 = Result 4	1.1077	1,226.71	"Wrong" liters Restated	
Result 4 * Result 1 = Buoyancy Cor. (K)	1.3543	Mass of Air 1.3123	Liters * 7/8 Air Den = 7/8 Air Mass	
Apparent Mass + Buoyancy Cor. = Vac Mass	1,001.35	IPK Mass 1,001.31	Apparent Mass + 7/8 Air = Vac Mass	
Find Volume		41.92	7/8 Grams Dif. From K&L Grams	
Index Kg m3 Vac	815.19	2.79%	Corrected % of Tolerance	
(Vac IPK Mass/ Vac Den) = IPK Mass Liter Vol.	1,228.37			
Correction/Air Mass = % of Air Applied	110.77%			

This is because, as a substance becomes less dense, the air has a proportionately larger effect on the overall buoyancy since it is a larger proportion of the total mass. Even with this discrepancy between the K&L equation and the 7/8 rule, the outage for the 7/8 rule is less than 3% of the tolerance.

Ratio of Fractions Air/Vac

Let's look at how density of a material changes its relationship to the weight of the atmosphere as proof increases. When the table compilers determined the density of 1-proof alcohol they took 1 liter of alcohol or 1 liter of water and weighed them. For convenience let's assume they did this under atmospheric conditions:

	1 Proof Alc (Air)		Water (Air)			
Density of One Proof Alcohol	997.17	Kg m3	997.94	Water Den	0.7661	Density Dif.
Mass of Displaced Air	1.223		1.223	Mass of Displaced Air		
Mass of Air/Den of Water = Fraction	0.001226		0.001225	Mass of Air/Den of Water = Fraction		

	200 Proof (Air)		Water (Air)			
Density of 200 Proof Alcohol	791.92		997.94	Water Den	206.016	Density Dif.
Mass of Displaced Air	1.223		1.223	Mass of Displaced Air		
Mass of Air/Den of Water = Fraction	0.001544		0.001225	Very Different Fractions		

Convert to Vac

The mass of 1,000 liters of 1-proof alcohol would be 997.17 kg and the mass of the water would be 997.94 kg. These masses are so similar that when we add the mass of the displaced air of 1.223 kg the mass of the air is virtually the same fraction of the mass of the alcohol as it is of the mass of the water (0.001226 v. 0.001225).

When we perform the same calculation with respect to 200-proof alcohol we see that the density difference is much greater (206 total kg) and results in very different fractions between the two calculations.

This same effect is seen when one uses vacuum densities as well:

	1 Proof Alc (Vac)		Water Vac			
Density of One Proof Alcohol	998.25		999.02	Water Density	0.7660	Density Dif.
Mass of Displaced Air	1.223		1.223	Mass of Displaced Air		
Mass of Air/Den of Water = Fraction	0.001225		0.001224	Mass of Air/Den of Water = Fraction		

	200 Proof Alc (Vac)		Water Vac			
Density of One Proof Alcohol	793.03		999.02	Water Density	205.986	Density Dif.
Mass of Displaced Air	1.223		1.223	Mass of Displaced Air		
Mass of Air/Den of Water = Fraction	0.001542		0.001224	Very Different Fractions		

Convert to Vac

What is happening is that the mass of air with respect to 200 proof is making up a larger proportion of the mass of alcohol. Therefore, by using vacuum and vacuum one can eliminate a variable with respect to weight of air making up a larger and larger fraction of the relative specific gravity as proof increases.

The same effect, in essentially the same proportions, is seen when using SGs as well. What has happened is that the SGs have been calculated against a basis of 1.0 as the base unit. The difference is so small it has been ignored and the specific gravity result one obtains is close to the same values that one would obtain if they had determined the weight of water in a vacuum as well. Some confusion has also entered the picture with respect to SGs because the TTB SG vacuum values were calculated against the density of water in air rather than that of a vacuum.

The atmospheric pressure standard used by the Tables is most likely that of 1 standard atmosphere at sea level (which was the standard in effect in 1909 when the original calculations were made by the Bureau of Standards) rather than the value of 25 meters above sea level which is currently used. It also used the definition of the liter in effect at that time which we will discuss infra.

Table 6 SG Values.

The Bureau of Standards data on alcohol density was compiled using the weight of alcohol in a simulated vacuum. The values listed in Table 6 under the column Specific Gravity in Vacuum are the values compiled from the original work. The specific gravity in Air values are derived from the vacuum values by subtracting units of specific gravity from the vacuum values along a curve that begins at zero with one proof and then climbs to a maximum of 0.00024 SG units at 200 proof.

Proof	SG in Air	SG in Vac.	Difference
1	0.99925	0.99925	0.00000
2	0.99850	0.99850	0.00000
3	0.99776	0.99776	0.00000
4	0.99703	0.99703	0.00000
5	0.99630	0.99630	0.00000
6	0.99559	0.99559	0.00000
7	0.99487	0.99488	-0.00001
8	0.99418	0.99419	-0.00001
9	0.99349	0.99350	-0.00001
194	0.80770	0.80792	-0.00022
195	0.80555	0.80577	-0.00022
196	0.80333	0.80356	-0.00023
197	0.80104	0.80127	-0.00023
198	0.79866	0.79889	-0.00023
199	0.79620	0.79643	-0.00023
200	0.79365	0.79389	-0.00024

The reason that this is done is because, with alcohol and water mixtures, as one moves up the proof scale the alcohol, which is less dense than water, becomes a larger proportion of the mass.

It displaces the same amount of volume but higher proof alcohol is proportionally more affected by the atmospheric buoyancy because more of the volume is represented by less-dense material.

The air has a larger (relative) effect on a cubic meter of alcohol than on a cubic meter of water because the buoyancy due to the air is the same for both (both being 1 cubic meter) but the alcohol weighs less and so the effect of the air is a bigger fraction of the alcohol weight than it is of the water weight.

Another way to examine this relationship is to use the *Strength v. Density* worksheet which demonstrates the difference in terms of grams per m^3 between the modified ABS values and the traditional TTB values:

Strength v. Density

Proof	1.00		200.00	Proof		
	Kg m3 Vac		Kg m3 Vac	%	1-%	
Kg m3 Vac	998.25		793.03	79.44%	20.56%	
Total Kg m3 ∆	205.2			Value	1.00	
			1/% = Ratio of Change		4.86	

	Vac	Vac				Air	Air
ABS Density Values	Tden SG Vac	Tden SG Vac				Tden SG Air	Tden SG Air
SG Vac	0.99923	0.79381	SG Vac		SG Air	0.99923	0.79356	SG Air
Vac H2O Density 15.56	0.999016	0.999016	Vac H2O Density 15.556		Air H2O 15.556	0.997939	0.997939	Air H2O 15.556
Density Kg m3	998.25	793.03	Density Kg m3		Density Kg m3	997.17	791.93	Density Kg m3
Tden Vac Density	998.25	793.03	Tden Vac Density		Tden Air Value	997.17	791.93	Tden Air Value
Grams per m3 Dif.	0.275	0.138	Grams per m3 Dif.	Air Vac SG Dif.	g per m3 Dif.	-0.189	-0.153	g per m3 Dif.
				-0.00025				

	Vac	Vac				Air	Air
TTB SG Values	TTB SG Vac	TTB SG Vac				TTB SG Air	TTB SG Air
TTB SG Vac	0.99925	0.79389	SG Vac		SG Air	0.99925	0.79365	SG Air
Vac H2O Density 15.56	0.999016	0.999016	Vac H2O Density 15.556		Air H2O 15.556	0.997939	0.997939	Air H2O 15.556
Density Kg m3	998.27	793.11	Density Kg m3		Density Kg m3	997.190	792.014	Density Kg m3
Grams m3 Dif.	19.98	76.92	Grams per m3 Dif.	Air Vac SG Dif.	g per m3 Dif.	20.96	88.82	g per m3 Dif.
				-0.00024				

Strength v. Den

We see a very small difference at 1 proof amounting to ~20 g/m^3 but a much larger difference amounting to ~70–90 g/m^3 at 200 proof. Notice that the air to vacuum difference in SG is similar at ~24 at 200 proof.

Note the change in the difference between air and vacuum over the scale of proof:

Proof	S.G. Air	S.G. Vac	Differenece		Dif/Vacuum value	
Zero Proof	1.000000	1.000000	0.000000		0.000000	
1 Proof	0.999250	0.999250	0.000000	One Part in	0.000000	One Part in
7 Proof	0.994870	0.994880	0.000010	100,000.0	0.00001005	99,488.0
20 Proof	0.986600	0.986610	0.000010	100,000.0	0.00001014	98,661.0
40 Proof	0.975940	0.975960	0.000020	50,000.0	0.0000205	48,798.0
60 Proof	0.965300	0.965340	0.000040	25,000.0	0.000041	24,133.5
80 Proof	0.951720	0.951780	0.000060	16,666.7	0.000063	15,863.0
100 Proof	0.934180	0.934260	0.000080	12,500.0	0.000086	11,678.3
120 Proof	0.913330	0.913440	0.000110	9,090.9	0.000120	8,304.0
140 Proof	0.889860	0.889990	0.000130	7,692.3	0.000146	6,846.1
160 Proof	0.863640	0.863800	0.000160	6,250.0	0.000185	5,398.7
180 Proof	0.833620	0.833820	0.000200	5,000.0	0.000240	4,169.1
200 Proof	0.793650	0.793890	0.000240	4,166.7	0.000302	3,307.9

The lighter a given volume is the more it is supported by the atmosphere. More precisely, the greater the proportion made up by the atmospheric weight of a given volume with respect to its total weight, the greater the density difference between air and vacuum.

True SG

I have found two methods of correcting apparent density to true density. Apparent density is what I refer to as SG Air and true density is equivalent to SG Vac. Apparent SGs are given by pycnometer or weighing bottle. In such cases one needs to convert to True SG (Vacuum SG).

Even though we use them in air, hydrometers (except TTB hydrometers) and volume percent hydrometers are calibrated for true density (vacuum) so that is what you read off of them. When using a hydrometer there is no need to convert since the hydrometer reads the true density.

The first method of converting SG Air to SG Vac or vice versa is sourced from *Tyco's Tables*, page 51,[63] which sets forth the algorithm as follows:

$$True\ Density_{Vac} = \big((1 - Apparent\ Density) * 0.00120\big) + Apparent\ Density$$

$$Apparent\ Density = \big((True\ Density - 1) * 0.00120\big) + True\ Density$$

This method assumes 50% humidity and 760 mm of barometric pressure. It also assumes the measurement is made against brass weights with a density of 8.4.

---

[63] *Tyco's Tables*, Rochester, NY: Taylor Instrument Companies, 1918.

I have set up the worksheet to work in both directions, Vac to Air and Air to Vac, in order to check the reciprocity of transformation since, starting from the TTB standpoint of atmospheric density, I needed to calculate vacuum densities:

Convert SG from Vac to Air & Vice Versa				
True Proof of Alcohol @ 60° F.	160.00		0.0015	Tolerance
Apparent Mass in Air Kg.	1,000.00		1,500	Total Grams Tolerance
	Vac		Air	
SG Vac @ 60° F.	0.86381		0.86364	SG Air @ 60° F.
H2O Density at 60° F. Vac	0.99902		0.99795	H2O Den
(SG * H2O Density)*100 = Kg m3	862.96		861.87	Kg m3

Tyco's Tables Version	Vac to Air		Air To Vac	
SG Vac = True Density	0.863810		0.863643	Apparent Density (SG Air)
The Value of 1	1.00		1.00	The Value of 1
True Density -1 = Result	-0.14		0.13636	1 - Apparent Density Result
Constant	0.0012251753		0.0012236761	Constant
Result * Constant = Result 2	-0.00017		0.00017	Result * Constant = Result 2
T.D. + Result 2 = Apparent Den. (SG Air)	0.86364		0.86381	Apparent Den + Result = SG Vac
SG Dif. From input SG Air	1.43E-07		-1.43E-07	Ending SG Dif.

H2O Density at 60° F. (Air)	0.997950		0.999016	H2O Density at 60° F. (Vac)
(SG * H2O Density)*100 = Kg m3	861.87		862.96	Kg m3
Grams per m3 of Density Dif.	0.1		-3.63	Grams of Density Dif.
% of Tolerance	0.0095%		-0.24%	% of Tolerance

Tuned Constant Vac to Air	0.0012251753
Tuned Constant Air to Vac	0.0012236761
Original Tycos Constant	0.00120

SG Air & Vac

I have tuned the original *Tyco's* constant of 0.00120 (which I'm assuming stood in for the mass of air in kg/dm^3) to individual values that produce more reciprocal results for each transformation. In order to achieve this degree of reciprocity (10–7 decimal places) individual constants for each transformation were empirically derived. We will see the need for different constants when going from air to vacuum or vacuum to air again when we modify the density of the brass weights used in translating weighings from air to vacuum and vice versa.

Note that after correction when going from SG Vac to SG Air one uses the density of water in air to find the density; and contrary-wise going from SG Air to the density in vacuum one uses the density of water in a vacuum to find the density. The formula applies equally well to SGs and densities.

A second method that accomplishes the same task is a modified version of the Kaye and Laby equation published in *Tables of Physical and Chemical Constants*. It also assumes the measurement is made against brass weights with a density of 8.4 and an air pressure of 760 mm Hg.[64]

---

[64] Atmospheric Pressure - Standard Atmosphere - Sea Level in Millibars 1013.20 Millibar = 1000 Dynes per square centimeter - Millibars to Dynes Converter 1000.00 Standard Air Pressure in Inches of Mercury 29.92 Standard Air Pressure in Pounds Per Square Inch 14.70

In order to calculate the mass air based upon the standard pressure we use the following procedure:

Dry Air Density Equation			Convert to other pressure units
Present Air Pressure in mm Hg	760.00	0.039370079	mm Hg to In Hg
mm Hg. To Pascals	133.3220	29.9213	Inches Hg
Pascals	101,325	0.019336721	mm Hg to PSI
Air Temp ° C.	15.5560	14.696	PSI
Kelvin Constant	273.15	33.863788	mm Hg to PSI
Temperature (°K)	288.71	1,013.247	Millibars
Specific gas constant dry air (287.05 J/(kg.	287.05	0.75006376	millibars to mm of HG
Temp * Gas Constant Result	82,873	760.00	mm of HG
Result/Pascals = Air kg m3 (p)	1.2226	0.029530069	millibars to inches of HG
Expressed as Kg. per Liter	0.001222650	29.921	Inches of Mercury
		0.49115272	Inches of HG to PSI
		14.696	PSI
		0.014503774	Milibars to PSI
		14.696	PSI

Water and Alcohol Analysis

The mass of dry air at this pressure works out to 0.00122265 kilograms per liter at 15.556°C.

We will use this air mass value to run a slightly-modified Kaye and Laby equation to solve for SG and density from vacuum to air, and also the converse, from air to vacuum, to check the reciprocity of the procedure:

Convert SG from Vac to Air & Vice Versa				
True Proof of Alcohol @ 60° F.	160.00	0.0015	Tolerance	
Apparent Mass in Air Kg.	1,000.00	1,500	Total Grams Tolerance	
	Vac	Air		
SG Vac @ 60° F.	0.86381	0.86364	SG Air @ 60° F.	
H2O Density at 60° F. Vac	0.99902	0.99795	H2O Den	
(SG * H2O Density)*100 = Kg m3	862.96	861.87	Kg m3	
Modified Kaye & Laby Equation Method	Vac to Air	Air to Vac		
SG Vac Restated	0.86381	0.86364	SG Air Restated	
Density of Air Kg. per Liter	0.00122263	0.00122263	Density of Air Kg. per Liter	
Density of Water Kg. per Liter (Air)	0.997939	0.999016	Density of Water Kg. per Liter Vac	
Air Density / Water Density = Result 1	0.001225	0.001224	Air Density / Water Density R1	
The value of 1	1.0	1.0	The value of 1	
SG Vac minus 1 = Result 2	-0.13619	-0.13636	SG Apparent in air minus 1 = R2	
Result 1 * Result 2 = Result 3	-0.000167	-0.000167	Result * Result	
SG Vac + Result 3 = SG Air	0.86364	0.86381	SG Air - Result = SG Vac	
Calc SG Air Dif. From Start SG Air	1.5E-07	-1.21E-07	SG Dif.	
(SG * H2O Density) * 100 = Kg m3 (Air)	861.86	862.96	Density	
Grams per m3 of Density Dif.	-9.725	-3.605	Grams of Density Dif.	
% of Tolerance	-0.6%	-0.24%	% of Tolerance	

SG Air & Vac

This formula uses the density of air/the density of water in order to find Result 1, where the original equation uses the mass times the air density with the SG standing in for the mass component.

Note that when I calculated my SGs from density and populated my SG table I did not round them and Excel defaults to 15 places as we have seen; that is why one can see such small discrepancies (10–7) between the resulting SGs.

Using the full Kaye and Laby equation one applies the density of air against the density of the weights and leaves out the density of water. In both cases the density of the weights is set to 1.0 since we are just converting SG back and forth between Air and Vacuum values and not conducting a weighing.

Origional Kaye and Laby Equation	Vac to Air		Air to Vac	
SG Vac Restated	0.86381		0.86364	SG Air Restated
Air Density, Dry Air (σ')	0.001223		0.001223	Air Density, Dry Air (σ')
Mass * Air Den = Result 1 (Mσ')	0.0010561		0.001056	Mass * Air Den = Result 1 (Mσ')
Value	1.000		1.00	Value
1/ Value (Δ) = Result 2	1.157662		1.157886	1/ Value (Δ) = Result 2
Density of Wts. (ρ) (Set Density to 1.0)	1.000		1.000	Density of Weights (ρ)
Value	1.00		1.00	Value
1/Density of weights = Result 3	1.0000		1.0000	1/Density of weights = Result 3
Result 2 - Result 3 = Result 4	0.157662		0.15789	Result 2 - Result 3 = Result 4
Result 4 * Result 1 = Buoyancy Cor. (K)	0.00016651		0.0001667	Result 4 * Result 1 = Buoyancy Cor. (K)
Vac SG - Air Cor. = Air SG	0.863643		0.86381	Air SG + Air Cor = Vac SG
SG Dif.	4.90E-07	Dif.	-2.85E-07	SG Dif.

SG Air & Vac

As you can see the equation also operates at a high degree of reciprocity (10–7) but not as well as the modified equation.

Interpolated 15.556°C Water Density.

The TTB tables rely on several different values for the density of water under atmospheric conditions. Most scientific work relies on vacuum densities for their calculations. Harmonizing these two regimens requires several steps.

I am using values from the *Handbook of Chemistry and Physics* (*HCP*) 89th edition for this example.[65] Using vacuum densities and transitioning to air values requires that the 60°F (15.556°C) values be interpolated from the vacuum values listed for 15.5°C and 16.6°C. None of the reference books I have found lists 15.556°C as a recognized interval in their work:

Find Water Value @ 15.556 using HCP Table
Using HCP 89th Ed. Water Mass per dm3

Water Temp H2O Restated	15.50		15.5556		15.60
Index g/cm3 In Vac using HCP Table	0.99902470	Interp. Val.	0.999015971	Rounded 9	0.9990090
Interpolate 15.556 Value for HCP 89th			-8.73E-06 Less than 15.5 Value		
Kilograms Dif. Between Temps	0.0000157		6.97E-06 Greater Than 15.6 Value		
Temp Dif. Between 15 & 15.556	0.5560				
Kg. Dif. * Temp Dif. = Density Change	0.00000873				
Subtract Dif. From Larger 15° C. Value	0.99901597080	Interpolated Density 15.556 for HCP 89th			

Water and Alcohol Analysis

To correctly interpolate, first find the difference in density between 15.5°C and 16.6°C (0.0000157 kg) and then multiply it by 0.556 to find the change in density as 0.00000873 kg. This change corresponds to a move 55.56% of the "distance" from 15.5°C to 15.6°C. Since density is decreasing as temperature is increasing one must subtract the value of 0.00000873 from the 15.5°C density value.

---

[65] Section 6-4 of *HCP* 89th, "This table gives the density of water in the temperature range from 0° C. to 100° C. at a pressure of 1013.25 Pa (one standard atmosphere). From 0° C. to 40° C. the values are taken from the IUPAC publication in Reference 1 and refer to standard mean ocean water (SMOW), free from dissolved salts and gases. SMOW is a standard water sample of high purity and known isotopic composition. Methods of correcting for different isotopic compositions are discussed in Reference 1. The remaining values are taken from the NIST Chemistry WebBook, Reference 2. Note that the IUPAC values refer to the IPTS-68 temperature scale, while the NIST values are based on the ITS-90 scale (where the normal boiling point is 99.974° C.). The conversion between these scales can be found in Sec. 1. The difference between the scales leads to a difference in the density of water of about 20 ppm in the neighborhood of 100° C. and much less at lower temperatures."

After interpolating the vacuum density, it is necessary to calculate the density of air in order to find the density of water in air. The following dry air equation was used to make that determination:

Dry Air Density Equation	
Present Air Pressure in mm Hg	760.00
mm Hg. To Pascals	133.3220
Pascals	101,325
Air Temp ° C.	15.5560
Kelvin Constant	273.15
Temperature (°K)	288.71
Specific gas constant dry air (287.05 J/(kg.K)	287.05
Temp * Gas Constant Result	82,873
Result/Pascals = Air kg m3 (p)	1.2226
Expressed as Kg. per Liter	0.0012226497
Kg. per Liter Rounded 9	0.001222650

Convert Vac to Air

Having found (interpolated) the vacuum density of water at 15.556°C, we need to translate it to an air density in order to use it with the TTB tables and generally to find air densities from vacuum ones. To do this we will use the Kaye and Laby equation with our standard atmosphere and a modified density of the weights employed:

Change density of weights to obtain reciprocality from Vacuum Direct Input To Air				
HCP Interpolated H20 (Vac)	0.9990159708		0.99793857	Air Density
Kaye and Laby Equation	Vac to Air		Air to Vac	
Air Density, Dry Air (e')	0.001222650		0.001222650	
Mass * Air Den = Result 1 (Me')	0.0012214		0.0012201	
Value	1.00		1.00	
1/ HCP Value (Delta) = Result 2	1.0010		1.00206568	
Density of Weights (p)	8.409068823		8.40000	
Value	1.00		1.00	
1/Density of weights = Result 3	0.1189		0.1190	
Result 2 - Result 3 = Result 4	0.8821		0.8830	
Result 4 * Result 1 = Buoyancy Cor. (K)	0.00107740		0.00107740	
Vac Mass - Air Den = Air Density	0.997938574		0.99901597	Air Den + Air = Vac
Lbs per Gal. Air	8.328201013	Lbs. Gal. Air	0.0000000	Dif. From Start
Wine Gal. Per Lb. Air	0.12007395		3.65E-14	As Scientific
Old ABS Wine Gallons per Lb. Table 4	0.12007395		1.077396	Kg. m3 Air mass
Dif. From Old ABS Value	-7.76E-10		8.337192322	Lbs. per Gal. Vac
	Vol 1st		Mass 1st	
Liters per Gal.	0.26417205		2.2046226	Kg. to Lbs.
Vac Den/.26417205	3.7816869		2.202453187	Den* K = Lbs.
Kilograms to Lbs.	2.20462260		3.7854118	Liters per Gal.
Lbs. per Gal. Vac	8.33719232		8.33719228	Lbs * K = Lbs per Gal.
Tyco's Tables Vac Value	8.3372200		0.000000039	Dif. Each Other
Difference from Tyco's Value	2.77E-05		3.92E-08	Dif. Each other

Water and Alcohol Analysis

Note that in the above diagram that the density of the weights used to transition from vacuum to air density has been modified from 8.4 to 8.40906882. The original equation was designed for reductions to weighings in vacuum, not for going from vacuum density to atmospheric density. In order to make the equation reciprocal I modified the density of the weights when transitioning from vacuum to air densities so that the equation operated reciprocally.

The conversion back to English units can be accomplished in two different ways: either by using the volume converter first or by using the mass converter first. This choice to some degree affects the resulting output but not to the degree that it matters with respect to the value that *Tyco's Tables* published (page 49) for the density of one US gallon of water in vacuum at 15.556°C (8.33722).

The density of water in units of pounds per US gallon in air calculated for the program are those interpolated from the *HCP* 89th values using the Kaye and Laby equation modified to be reciprocal. They are shown below in comparison to other published values:

Air Values for water density	Lbs. per WG	WG per Lb.	
HCP 72nd Lbs. per Gal. Air 15.556	8.328264	0.12007304	
Lbs. per Gal § 30.66 Old §186.66 & Tyco's	8.328230	0.12007353	
HCP 89th & Kaye & Laby Eq. Vac to Air	8.328201	0.12007395	ABS Std.
Lbs. per Wine Gal. §30.41	8.328200	0.12007400	

Water and Alcohol Analysis

Converting the air density values shows minimal to no variation in the resulting values.

Convert back to metric density	Air to Metric	Air Metric to Lbs.	
HCP 89th Lbs. per Gal. (Air)	8.328201015	0.997938575	ABS Air Den H20
Lbs. to Kg.	0.45359237	2.2046226	Kg. to Lbs.
Kilograms	3.77760844	2.20007793	H2O Den * K = Lbs.
Liters per Gal.	3.78541180	0.26417205	Gallons per Liter
kg/L per Gal.	0.997938569	8.3282010	Lbs * K = Lbs per Gal
K&L Calculated Value (Starting Value)	0.997938575	8.3282010	Starting Air Value
Dif. From Calculated Value	5.20E-09	0.00E+00	Difference

Water and Alcohol Analysis

Gallon Discrepancy

Since the TTB Pounds per Gallon and Gallons per Pound values were all determined using the pre-1930 value of the inch and the pre-1964 value assigned to the liter I wanted to see if I could untangle the differences to see how it would affect the density and overall mass of each defined unit.

Prior to 1930 when the modern Industrial Inch was first adopted, the number of cubic inches per liter (as then defined) was set at 61.025. Modern volumetric computations are based upon the relations that one liter = 1 dm^3 and that 1 dm^3 = 61.023744 in.^{3} Using the liter to gallon converter then in effect (0.264178) resulted in a cubic-inches-per-gallon value that was less than the defined value of 231 cubic inches by 0.00045 cubic inches. Using the modern defined values for a decimeter cubed results in a gallon volume that is slightly greater than 231 inches cubed.[66]

Inch to Meter Conversion	1866-1930 Value	1930 "Industrial Inch"	
Value of 1 to Represent Meter Length	1.0	25.4	mm per Inch
Divisor for fraction of Inch per Meter	39.97	1,000	mm per Meter
Fraction of meter per inch	0.0250188	0.02540	% meters per inch
Old Version is Less than Modern	-0.00038	0.00038	Industrial Inch is Greater

	Modern		
Industrial Yard Fraction of Meter	0.9144		
Inches per Yard	36.00	Pre 1930	
mm per Inch	25.40	25.4000508	Pre 1930 mm per In.
Inches * mm per In. = mm per yard.	914.400	36.00	Inches per Yard
Modern Yard is Shorter	-0.001829	914.401829	mm per Yard

Water and Alcohol Analysis

Fortunately, the Industrial Yard works out to the same 25.4 mm per inch.

Apart from the TTB values, the most complete source I have found for converters in effect when the BOS work was undertaken is a book titled *Tyco's Tables*. See the *Constants* section for a listing of those values in comparison to the modern values.

Examining the differences between the new and old values we see that it is a simple matter to divide the difference of 0.0000017 gallons into one gallon. This, while sometimes expressed as 1.7 parts per million, is not in fact an accurate description of the amount of outage. 1/0.0000017 = 587,049 gallons per cubic inch of outage which, when multiplied by 231 cubic inches, equals 135,608,319; this represents the number of gallons required to accumulate a one-gallon discrepancy in volume when using the 0.26417205 converter. Using different converters results in different amounts of accumulated outage as we will see. After finding the gallons per one gallon of outage, it is fairly straightforward to find the cubic size of the volume represented as containing the one additional gallon of volume:

---

[66] "In 1930, the British Standards Institution adopted an inch of exactly 25.4 mm. The American Standards Association followed suit in 1933. By 1935, industry in 16 countries had adopted the 'industrial inch' as it came to be known. In 1946, the Commonwealth Science Congress recommended a yard of exactly 0.9144 meters for adoption throughout the British Commonwealth. This was adopted by the United States on 1 July 1959, and the United Kingdom in 1963, effective on 1 January 1964. The new standards gave an inch of exactly 25.4 mm, 1.7 millionths of an inch longer than the old imperial inch and 2 millionths of an inch shorter than the old US inch." The US Geological Survey still uses the old inch because it makes a difference based upon the scale of the maps they are creating.

Compare Post 1964 to Pre 1964 (Tycos) Cubic Inches per US Gal.

	HCP 72nd Post 1964		Tycos	
1 dm3 vol. in cubic inches	61.0237440		61.025	Cubic Inches per dm3
Liters to Gallons Post 1964	0.26417205		0.264178	Pre 1964 (Tycos)
Cubic In./Gal. per liter Gal. = Cubic In. per Gal.	231.0000017		230.99955	Cubic Inched per Gal.
Standard Definition of US Gal. (Cubic Inches)		231.00		
Post 1964 volume Dif. (Cubic Inches)	1.703E-06		-4.47E-04	Pre 1964 Difference
Cu In.Dif/231 Cu In. = Gallons per 1 Gal Outage	135,608,319.13		3.78541180	Liters per Gal.
			513,333,331	Gal. * L per 1 Gal. = Liters per 1 Gal. Out
Cube Root of Gal. Outage = Cu Gal. per Side	513.76		800.69	Cube Root = Decimeters per 1 Gal. Out
Cubic Inches per Gal.	231.00		80.07	Decimeters/10 = Meters per 1 Ga. Out
Cube Root of Cu In = Gal. Inches per Side	6.1358			
Cubic Gal per side * Cu In per =Tot. Inches	3,152.34		3.2808399	Meters to Feet
Total Inches/12 = Ft. per side per 1 gal outage	262.69		262.69	Feet per Side per 1 Gal. Outage
Feet to Meters	0.30480			
Ft. per Side * K = meters per side per 1 gal. out	80.07			

Find Volume per 1 Liter Outage			Prove Gallon Outage at Scale	
Liters per Gal.	3.78541180		1,000,000	Scale Value
Cube Root of Liters per Gal = Result	1.56		1.703	Cubic In. Dif * Scale = PPM per Cu In.
m per side/Result = Ft. per side per 1 liter out	51.38		587,049	1/Cu In. Dif = Cu In per Cubic In Dif.
Feet to Meters	0.30480		231.00	Standard Definition Cu In per Gal.
Meters per side per 1 liter outage	15.66		135,608,319.1	Cu In Dif. * 231 = Gal. per 1 Gal. Out.
Find Volume				
m per side * 10 = Decimeters Length per side	156.59		231.0000017	Calc Cubic In
Decimeters Cubed = Liters per 1 Liter Outage	3,840,000		135,608,320.1	Cu In Dif. * 231 = Gal. per 1 Gal. Out.
Ratio of liters out to "Gal Liters out"	133.68		1.0000000	Gallon Difference.

Volume Analysis

Converting to a procedure, we can calculate the English and metric volume of discrepancy:

Find Gallons per 1 Liter Outage			Sample Values	
Gallons per 1 US Gallon Outage	135,608,319.1		35,823,927.83	Gal. per 1 Liter Outage
Liters per Gal.	3.78541180		17,911,963.92	1/2 = Gal. per 500 ml
Gallons/ liters per Gal. = Gal. per 1 Liter Outage	35,823,927.83		8,955,981.96	1/4 = Gal. per 250 ml

Find Gallons and Liters Outage from Gallon Input Volume			Easy Method to Find Dimension of Vol. using Constant	
US Gallons at 231 Cubic Inches Per Gal.	8,955,982.0		7.4805195	Gallons per Cubic Foot
Gallons per 1 US Gallon Outage	135,608,319.13		1,197,241	Gallons /Gal per Cubic Foot = Cubic Ft.
Gal. Input/Gallons per 1 Outage = Gal. Out %	6.604%		106.18	Cube Root = Ft. per Side of % Gal. Out
Liters per Gal. Post 1964	3.78541180		0.30480	Feet to Meters
% outage * Liters per Gal = Liters of Out (dm3)	0.25000		32.36	Meters per Side per % Gallon Outage
dm3 * 1,000 = cm3 (ml) of Outage	250.00			
Cubic Inches per Gal.	231.00		61.023744	dm3 to Cubic Inches
% Gal. * Cu In. per Gal. = Cubic In. of Outage	15.26		15.256	ml out 8 K = Cubic Inches of Outage

Volume Analysis

Applying the gallons per 1 gallon outage value to a volume equal to 250 ml confirms that the outage will be 250 ml and amount to 15.26 cubic inches per ~ 9,000,000.00 gallons.

As mentioned above, this is all tied to the value of the converter used. It is possible to derive values to greater decimal places. Using the modern Industrial Inch definition of 2.54 centimeters per inch and the defined 231 cubic inches per gallon one can generate essentially "perfect" converters. My version of Excel only provides 15 decimal places, but that is sufficient to make the point:

Derive dm3 per gallon		
Centimeters per Inch (By Definition)	2.54	Industrial Inch
Cube of 2.54 = Cubic cm per Cubic In.	16.38706400000	Terminates to Zero
Cubic Inches per US Gallon (By Definition)	231.0	
Cubic cm per Cu In. * Cu In per Gal = cm3 per Gal.	3,785.4117840000	Terminates to Zero
one cm^3 fraction of dm3	0.001	
cm^3 per US Gal. * fraction of dm3 = dm3 per Gal.	3.7854117840000	Terminates to Zero
one cm^3 fraction of dm3	0.001	
Cubic Cm per Cubic In * .001 = Inch cubed fraction of a liter	0.0163870640	Terminates to Zero
1/ In Cubed to Liters = Cu In per Liter=	61.0237440947323	Non Terminating at 15 places
Derive Gallons per dm3 and		
1/dm3 per Gal. = Gallons per dm3 (15 places)	0.264172052358148	Non Terminating at 15 places

Derive Values

While the derived values will produce perfectly symmetrical results when used in conjunction with each other, they do not work as well when used independently. Through trial and error, I have arrived at values that are rounded to 9 places that seem to give the best overall results. The in.^{3}/dm^3 value, when rounded to 10 or 11 places, was generating values that were less than 231 cubic inches by more than the rounded 9-place value was over that amount. Also, my preference is to meet the whole number value one is targeting and be slightly over rather than knowing one is always shorting the outcome when using any converter.

HCP 72nd Post 1964	61.023744000
Derived Value	61.0237440947323
Round 9	61.0237440950

Liters to Gallons Post 1964	0.264172050
Derived Value	0.264172052358148
Round 9	0.2641720520

Liters per Gallon Post 1964	3.78541180
Derived Value (Self Terminating)	3.7854117840

Derive Values

Applying the "rounded 9" values shows that we have improved the calculation for both converters and that the 3.785… value is in the billions of gallons and liters range before 1 unit of either volume in outage is achieved.

1 dm3 vol. in cubic inches	61.023744095	61.023744095
Liters to Gallons	0.2641720520	3.7854117840
Cubic In./Gal. per liter Gal. = Cubic In. per Gal.	231.000000314	231.0000000010
Standard Definition of US Gal. (Cubic Inches)	231.00	231.00
Post 1964 volume Dif. (Cubic Inches)	3.142E-07	1.013E-09
Cu In.Dif/231 Cu In. = Gal per 1 gal. Outage	**735,225,829**	**227,944,524,134**
Liters per Gal.	3.7854117840	3.7854117840
Gal. * L per 1 Gal. = Liters per 1 Gal. Out	2,783,132,517	862,863,887,755
Cube Root = Decimeters per 1 Gal. Out	1,406.62	9,520.23
Dm3/10 = Meters per Side per 1 Gal. Out	140.66	952.02
Find Volume per 1 Liter Outage		
Liters per Gal.	3.785411784	3.785411784
Cube Root of Liters per Gal = Result	1.56	1.56
m per side/Result = Ft. per side per 1 liter out	90.26	610.86
Feet to Meters	0.30480	0.30480
Meters per side per 1 liter outage	27.51	186.19
Find Volume		
m per side * 10 = Decimeters Length per side	275.10	1,861.91
Decimeters Cubed = Liters per 1 Liter Outage	**20,819,277**	**6,454,670,121**
Ratio of liters out to "Gal Liters out"	133.68	133.68

Ratio	310.03
Percentage	0.32%

Volume Analysis

The ratio of 310 indicates the relative accuracy of one converter value to another. My conclusion is that one should not use the liters-to-gallons value of 0.264172… when making volume conversions. One should always use the liters-per-gallon value of 3.785…, either with multiplication or division, to transform volumes. Note that the ratio of "liters out" to "gallons per liter out" has remained constant at 133.68 and this value can be used as a divisor to gallons or a multiplicand to liters to find the equivalent value in the other volume unit.

The consequence of this decision is that, when employing multiplication, the converter as designed is set to exceed the target volume unit and will operate to result in overages in the volume being converted into; this in effect underestimates (underdetermines), after conversion, the actual original volume since the original volume was assumed known.

When using the reciprocal value 0.264… in multiplication, the opposite effect will occur. The resulting volume will be less than the original volume which means that more volume is present than underdetermined by the converter.

Examining these errors induced by converters a little further we see that either the ABS converter or the modern converter of 3.7854… will work to obtain reciprocal results. Applying the 0.2641…. value using multiplication in either ABS or modern form results in −1,000 and −4,000 liters difference respectively: an underdetermined result from an assumed original value:

Conversion Procedure	ABS		Modern	
Gallons per 1 gal. Outage (Restated)	227,944,524,134	Same	227,944,524,134	
Liters per Gallon	3.7854117840		3.78541180	
Gal. * Converter (K) = Liters	862,863,887,755		862,863,891,402	
Reverse Procedure				
ABS Liters per Gal.	3.785411784		3.78541180	
Liters/Converter = Gallons	227,944,524,134		227,944,524,134	
Difference	0.000000		0.000000	
Apply .264 Converter	ABS		Modern	
1st Conversion Liters Restated	862,863,887,755		862,863,891,402	Liters
Gallons per Gal. Constant (K)	0.2641720520		0.264172050	K.
Liters * K = Resulting Gal.	227,944,523,825		227,944,523,063	Gal.
Gallon Dif.	-309.0		-1,071.3	Gal Dif.
Liters Dif.	-1,169.82		-4,055.30	Liters Dif.
Liters per Gal. Constant (K)	0.264172052		0.264172050	K
Original Gal./K = Liters	862,863,888,925		862,863,895,457	Liters
Liters Dif. From 3.785 Converter	1,169.8		4,055.3	Liters Dif.

Applying the 0.264… value using division reverses the error to over-determine liter volume from assumed gallon volume as the source.

This means that one should always store the original volume unit entered so as to minimize the number of transformations of volume that take place. This will minimize cumulative volume transformation discrepancies.

Another way to examine this issue is to deliberately define a converter that is one part in 1,000 off in a positive and negative direction and see that the "short" converter will underestimate original volume and the "long" converter will overestimate volume. We will use this approach when we examine the history of the liter and the changes that have occurred to it over time:

US Gallons at 231 Cubic Inches Per Gal.	37,854.11784			
Liters per Gal. Standard Converter	3.785411784			
Gal. * Converter = Liters Result	143,293.42			
Liters per Gallon	3.78541178			
1/1000	0.0010			
Standard Converter * .001 = Result	0.0037854		Converter + .001 Dif = .001 Oversized Converter	
Converter - .001 Dif. = Undersized Converter	3.78163	3.7891972	Over Sized Converter	
Gal. * .001 off Converter = Liters Result	143,150.13	143,436.72	Gal. * .001 off K = Liters Result	
Liters Difference	-143.29	143.29	Liters Difference	
Amount of Difference	0.001	-0.001	Amount of Difference	
Gallons	37,816.26	37,891.97	Gallons	
Gallon Dif.	-37.85	37.85	Gallon Dif.	

Volume Analysis

I have not gone to the trouble to convert old liters to old gallons but, in volume 2 we will make an attempt to harmonize the densities derived and published in older data sets with modern ones by accounting for the differing definitions of volume in use at the time the researchers conducted their work. By altering the volume we will in effect revise the mass contained in that volume; which will in effect alter the density with respect to mass per unit of volume.

Using A 15°C Hydrometer

In order to use the TTB tables with a hydrometer calibrated for a temperature other than 60°F it is necessary to adjust for the difference in buoyancy that the hydrometer exhibits at its calibration temperature compared to what the tables expect as an input. Many hydrometers are calibrated to 15°C (59°F). Translating the values obtained from these hydrometers into a proof scale based on 60°F (15.56°C) is often done directly and ignores the temperature calibration standard difference. To be as accurate as the instruments and tables will allow involves an additional temperature correction of the difference between the calibration temperature of the hydrometer involved and the standard Temperature/Density of the Table one is referencing. For a first example let's use our tables calibrated for 60°F (15.56°C) and use a hydrometer calibrated for 15°C.

Comparing the hydrometer calibration temperature to the Table temperature results in a difference in 0.556°C (15.556–15.00 = 0.556°C). This 0.556°C is conveniently almost exactly 1.0°F (1.0008°F to be exact).

The conversion from degrees F. to degrees C. actually results in a number that repeats indefinitely but is rounded up at the last place to 6 as 0.55555...6 and this makes it inconvenient and imprecise to accomplish the conversion; but three decimal places as 15.556 will obtain 60.0008°F—only 8/10,000 different from a truly 60°F calibrated Table:

Fahrenheit To Celsius Formula °F -32 * 1.8 = °C				
° F	minus	Result	Constant	Result/Constant
60.000	32.0	28.0000	1.8	15.555556

° C	Constant	° F
0.556	1.8	1.000800

Celsius To Fahrenheit Formula °C * 1.8 +32 = °F				
° C	Constant	°C * Constant =	Plus	Result + 32 = ° F
15.5560	1.8	28.0008	32.0	60.00080

° F	Constant	° C
1.000	1.8	0.555556

Converters

As to the calibration temperature of 59°F (15.00°C is indeed exactly 59°F), the 15°C instrument is calibrated to read 1.0 in a solution of water at 15°C (59°F). This is equivalent to the initial mark one would make when the hydrometer exhibited its buoyancy in a solution at its calibration temperature.

We will call this the Zero Point. The Table also has a Zero Point. It is the temperature it is calibrated for. When using a 60°F hydrometer no correction will occur when using the hydrometer or the TTB tables when that temperature is used.

The difference in temperature between the Table Zero Point and the hydrometer calibration temperature yields the value necessary to connect the instrument to the Table Zero Point.

In applying this offset value, the order of operations is important:

Table Temp – Hydrometer Calibration Temp °F yields temperature offset value. In this case 60 – 59 = 1.

Obtaining the offset value is only the first step. The second step is to apply the offset value to the temperature of the liquid being sampled:

Alcohol Temp. + Dif. from 60°F = Table Entry Temp.

In this case a solution of 59° (15°C) fluid would have the Table difference of 1°F added to it to achieve a match with the Table of 60°F. This will induce the table to perceive the fluid as less dense than the hydrometer indicates. To make this a little easier to follow I'm including a proof lookup to SG Vac at 60°F so we are in effect testing a hydrometer demarcated in both proof and SG at 15.556°C:

Courtesy Proof to SG Vac Input (Proof Input)	160.00			0.00150	Blending Tolerance
SG Vac at 60° F.	0.86381			0.300	Degree of Proof
				1,294.55	grams/m3 Tolerance
Apparent Specific Gravity of Alcohol (Vac)	0.86381	0.0000	Hyd. Cor.	32.4%	% of Tolerance
Density of H20 at Calib. Temp	0.99910				
SG * H2O Den = Density Kg m3 Vac	863.034				
Temperature of Alcohol in ° F	59.00	15.000	Celsius Equivalent		
Temp of Hydrometer Calibration in ° C.	15.000	59.000	Fahrenheit Equivalent		
Find Celsius To Fahrenheit Hydrometer Offset.					
60° F. Table Dif. from Hyd. Calibration Temp.	1.00	0.556	° C.	True Proof using a 60° F. Hydrometer	
Temp. + Dif. from 60°F. = Tbl. Entry Temp.	60.00	0.000	Therm. Cor.	160.3	60° F. Proof
Apparent SG Vac Proof Equivalent	160.00			0.3	Proof Difference
				0.86339	SG Vac Equivalent
Hydrometer Corrected True Proof	160.0			-0.00042	SG Dif.
True SG Vac @ 60°	0.86381			0.99902	H2O at 15.556 C. Vac
H2O at 15.556 C. Vac (Use H2O Den of Table)	0.99902			862.540	Density Kg m3 Vac
Density Kg m3 Vac	862.960			420.0	Grams per m3 Dif.

Hydrometer Correction SG Blend by Wt.

Applying 1°F of temperature correction has compensated for the 1° difference between the hydrometer's calibration temperature and that of the TTB Table 1 to arrive at a true proof of 160 at 59° F rather than at 60°F. We can also see that if one was using a 60°F hydrometer in a bath of 59°F alcohol that the true proof would be 160.3 because the alcohol would in fact be denser with respect to the hydrometer's perception of it.

It is also possible to perform this correction by using the built-in thermometer correction in the TTB Table algorithm. However, usually this is reserved to input the correction factor associated with the instrument itself so that it reads the right temperature. No two thermometers ever read the same in my experience no matter how much you pay for them.

Now, we will raise the temperature by 1°F and change the apparent proof to 160.3. The true proof result is now modified by an increase of 0.3 proof. This shows the effect of the Tables operating to obtain true proof over a 1°F temperature change:

Courtesy Proof to SG Vac Input (Proof Input)	160.30			0.00150	Blending Tolerance
SG Vac at 60° F.	0.86339			0.300	Degree of Proof
				1,293.92	grams/m3 Tolerance
Apparent Specific Gravity of Alcohol (Vac)	0.86339	0.0000	Hyd. Cor.	32.5%	% of Tolerance
Density of H20 at Calib. Temp	0.99910				
SG * H2O Den = Density Kg m3 Vac	862.614				
Temperature of Alcohol in ° F	60.00	15.556	Celsius Equivalent		
Temp of Hydrometer Calibration in ° C.	15.000	59.000	Fahrenheit Equivalent		
Find Celsius To Fahrenheit Hydrometer Offset.					
60° F. Table Dif. from Hyd. Calibration Temp.	1.00	0.556	° C.	True Proof using a 60° F. Hydrometer	
Temp. + Dif. from 60°F. = Tbl. Entry Temp.	61.00	0.000	Therm. Cor.	160.3	60° F. Proof
Apparent SG Vac Proof Equivalent	160.30			0.3	Proof Difference
				0.86339	SG Vac Equivalent
Hydrometer Corrected True Proof	160.0			-0.00042	SG Dif.
True SG Vac @ 60°	0.86381			0.99902	H2O at 15.556 C. Vac
H2O at 15.556 C. Vac (Use H2O Den of Table)	0.99902			862.540	Density Kg m3 Vac
Density Kg m3 Vac	862.960			420.0	Grams per m3 Dif.

Hydrometer Correction SG Blend by Wt.

Change the apparent proof back to 160 but leave the temperature at 60° and the true proof drops by 0.3 proof:

Courtesy Proof to SG Vac Input (Proof Input)	160.00			0.00150	Blending Tolerance
SG Vac at 60° F.	0.86381			0.300	Degree of Proof
				1,294.55	grams/m3 Tolerance
Apparent Specific Gravity of Alcohol (Vac)	0.86381	0.0000	Hyd. Cor.	32.4%	% of Tolerance
Density of H20 at Calib. Temp	0.99910				
SG * H2O Den = Density Kg m3 Vac	863.034				
Temperature of Alcohol in ° F	60.00	15.556	Celsius Equivalent		
Temp of Hydrometer Calibration in ° C.	15.000	59.000	Fahrenheit Equivalent		
Find Celsius To Fahrenheit Hydrometer Offset.					
60° F. Table Dif. from Hyd. Calibration Temp.	1.00	0.556	° C.	True Proof using a 60° F. Hydrometer	
Temp. + Dif. from 60°F. = Tbl. Entry Temp.	61.00	0.000	Therm. Cor.	160.0	60° F. Proof
Apparent SG Vac Proof Equivalent	160.00			0.3	Proof Difference
				0.86381	SG Vac Equivalent
Hydrometer Corrected True Proof	159.7			-0.00042	SG Dif.
True SG Vac @ 60°	0.86423			0.99902	H2O at 15.556 C. Vac
H2O at 15.556 C. Vac (Use H2O Den of Table)	0.99902			862.960	Density Kg m3 Vac
Density Kg m3 Vac	863.380			420.0	Grams per m3 Dif.

Hydrometer Correction SG Blend by Wt.

The same true proof difference of 0.3 proof shows up. What we are seeing is the amount of change we are applying to the hydrometer to correct it for use with this Table.

Now raise the temperature to 84°F and note the result:

Using a Hydrometer Calibrated in SG Vac at 15 ° C. . Find Offset True Proof and Blend 1 Alcohol to a target proof by Weig

Courtesy Proof to SG Vac Input (Proof Input)	160.00			0.00150	Blending Tolerance
SG Vac at 60° F.	0.86381			0.300	Degree of Proof
				1,294.55	grams/m3 Tolerance
Apparent Specific Gravity of Alcohol (Vac)	0.86381	0.0000	Hyd. Cor.	30.9%	% of Tolerance
Density of H20 at Calib. Temp	0.99910				
SG * H2O Den = Density Kg m3 Vac	863.034				
Temperature of Alcohol in ° F	84.00	28.889	Celsius Equivalent		
Temp of Hydrometer Calibration in ° C.	15.000	59.000	Fahrenheit Equivalent		
Find Celsius To Fahrenheit Hydrometer Offset.					
60° F. Table Dif. from Hyd. Calibration Temp.	1.00	0.556	° C.	True Proof using a 60° F. Hydrometer	
Temp. + Dif. from 60°F. = Tbl. Entry Temp.	85.00	0.000	Therm. Cor.	151.8	60° F. Proof
Apparent SG Vac Proof Equivalent	160.00			0.3	Proof Difference
				0.87491	SG Vac Equivalent
Hydrometer Corrected True Proof	151.5			-0.00040	SG Dif.
True SG Vac @ 60°	0.87531			0.99902	H2O at 15.556 C. Vac
H2O at 15.556 C. Vac (Use H2O Den of Table)	0.99902			874.050	Density Kg m3 Vac
Density Kg m3 Vac	874.450			400.0	Grams per m3 Dif.

Hydrometer Correction SG Blend by Wt.

The same difference of 0.3 proof is maintained even when the temperature is increased significantly to 84°F.

We will test this against a blend to see if it really will work. Returning to our standard batch and blending one alcohol by weight we see that with a temperature of 84° the total batch weight is 1,800.64 lbs. and the proof gallons are 181.74.

Blend One Alcohol by Weight using Table 4 then add components by weight or volume

Temp of Alcohol	84.00	0.000	Thermometer Cor.
Apparent Proof of Alcohol	160.00	0.000	Hydrometer Cor.
True Proof Result	151.80	0.00150	Blending Tolerance
Target Proof	80.00	0.300	Tolerance as ° of Proof
Weight of Alcohol	872.15	395.60	Kilograms (Air)
Volume of Alcohol at 60° F.	119.716	453.17	Liters
Table 7 Volume Correction Factor to be applied	1.0130		
Volume of Alcohol corrected to Present Volume	121.272	$13.50	Tax per Proof Gallon
Present Volume Difference from 60° F. Volume	1.56	750.00	ML per Bottle
60° F. Volume compared to present volume	-1.56	12.00	Bottles per Case

Results

Proof Gallons of Alcohol	181.74	$2,453.47	Tax
Gallons of Water @ 60° F. to Add	111.49	90.87	Absolute Alcohol Gallons = P.G./2
Gross Volume of Batch @60°F.	231.20	227.17	Net Blended Wine Gallons
Volume Reduction due to Blending	-4.03	40.000%	Absolute Alc. Gal./Net. Gal. = % Alc
Actual "net" 60° F. Volume available for Bottling	227.17	-15.25	Liters
Weight of Water to Add	928.49	-20.34	# of 750 ML bottles "Short"
Weight of Alcohol	872.15	1.74%	Vol Difference/Gross Gallons
Total Weight of Batch	1,800.64	1.77%	(Gross Gal/Net Gal)-1

Blend One Alcohol by Weight

This simply establishes what a 60°F hydrometer would generate from a true proof of 151.8 proof with our standard weight of alcohol of 872.15 lbs.

Blending with a 15°C hydrometer shows exactly the same results for the blend so long as we add 1°F (0.556°C) and 0.3 proof to the input proof to compensate for the additional temperature we are including. This is correcting both sides of the procedure but is necessary to see that it is producing accurate results:

Courtesy Proof to SG Vac Input (Proof Input)	160.30				0.00150	Blending Tolerance	
SG Vac at 60° F.	0.86339				0.300	Degree of Proof	
					1,293.92	grams/m3 Tolerance	
Apparent Specific Gravity of Alcohol (Vac)	0.86339	0.0000	Hyd. Cor.		30.1%	% of Tolerance	
Density of H20 at Calib. Temp	0.99910						
SG * H2O Den = Density Kg m3 Vac	862.614						
Temperature of Alcohol in ° F	84.00	28.889	Celsius Equivalent				
Temp of Hydrometer Calibration in ° C.	15.000	59.000	Fahrenheit Equivalent				
Find Celsius To Fahrenheit Hydrometer Offset.							
60° F. Table Dif. from Hyd. Calibration Temp.	1.00	0.556	° C.		True Proof using a 60° F. Hydrometer		
Temp. + Dif. from 60°F. = Tbl. Entry Temp.	85.00	0.000	Therm. Cor.	152.1	60° F. Proof		
Apparent SG Vac Proof Equivalent	160.30			0.3	Proof Difference		
				0.87452	SG Vac Equivalent		
Hydrometer Corrected True Proof	151.8			-0.00039	SG Dif.		
True SG Vac @ 60°	0.87491			0.99902	H2O at 15.556 C. Vac		
H2O at 15.556 C. Vac (Use H2O Den of Table)	0.99902			873.660	Density Kg m3 Vac		
Density Kg m3 Vac	874.050			390.0	Grams per m3 Dif.		
Blend with Hydrometer Corrected True Proof							
Weight of Alcohol In Pounds	872.15	395.61	Kilograms				
Target Proof	80.00	0.951797	Target Proof SG Vac				
Results							
Proof Gallons of Alcohol	181.74		$2,453.47	Tax			
Gallons of Water @ 60° F. to Add	111.49		90.87	Absolute Alcohol Gallons = P.G./2			
Gross Volume of Batch @60°F.	231.20		227.17	Net Blended Wine Gallons			
Volume Reduction due to Blending	-4.03		40.00	Target % Alcohol By Volume			
Actual "net" 60° F. Volume available for Bottli	227.17		-15.25	Liters			
Weight of Water to Add	928.49		-20.34	# of 750 ML bottles "Short"			
Weight of Alcohol	872.15		1.74%	Vol Difference/Gross Gallons			
Total Weight of Batch	1,800.64		1.77%	(Gross Gal/Net Gal)-1			

Hydrometer Correction SG Blend by Wt.

The procedure works just as well at temperatures below 60°F.

The question is whether we can use this procedure to correct for a hydrometer calibrated for 20°C and allow it to use the TTB tables. The answer is, just barely.

First, we set the calibration temperature to 20°C and to compensate for the calibration temperature difference, then we subtract 8°F so that the TTB tables do not perform any correction. We note that now there is a difference of 2.7 degrees of proof in a positive direction for the 20° C hydrometer, rather than the negative one when the calibration temp was below 60°F. So far, so good.

Courtesy Proof to SG Vac Input (Proof Input)	160.30			0.00150	Blending Tolerance
SG Vac at 60° F.	0.86339			0.300	Degree of Proof
				1,292.77	grams/m3 Tolerance
Apparent Specific Gravity of Alcohol (Vac)	0.86339	0.0000	Hyd. Cor.	-287.8%	% of Tolerance
Density of H20 at Calib. Temp	0.99821				
SG * H2O Den = Density Kg m3 Vac	861.844				
Temperature of Alcohol in ° F	68.00	20.000	Celsius Equivalent		
Temp of Hydrometer Calibration in ° C.	20.000	68.000	Fahrenheit Equivalent		
Find Celsius To Fahrenheit Hydrometer Offset.					
60° F. Table Dif. from Hyd. Calibration Temp.	-8.00	-4.444	° C.	True Proof using a 60° F. Hydrometer	
Temp. + Dif. from 60°F. = Tbl. Entry Temp.	60.00	0.000	Therm. Cor.	157.6	60° F. Proof
Apparent SG Vac Proof Equivalent	160.30			-2.7	Proof Difference
				0.86712	SG Vac Equivalent
Hydrometer Corrected True Proof	160.3			0.00372	SG Dif.
True SG Vac @ 60°	0.86339			0.99902	H2O at 15.556 C. Vac
H2O at 15.556 C. Vac (Use H2O Den of Table)	0.99902			866.264	Density Kg m3 Vac
Density Kg m3 Vac	862.543			-3720.3	Grams per m3 Dif.

Even though the proof difference is not a positive number we still must add it, as a positive number, to the apparent proof because it is the hydrometer we are offsetting:

Courtesy Proof to SG Vac Input (Proof Input)	162.70			0.00150	Blending Tolerance
SG Vac at 60° F.	0.86003			0.300	Degree of Proof
				1,287.73	grams/m3 Tolerance
Apparent Specific Gravity of Alcohol (Vac)	0.86003	0.0000	Hyd. Cor.	-282.7%	% of Tolerance
Density of H20 at Calib. Temp	0.99821				
SG * H2O Den = Density Kg m3 Vac	858.484				
Temperature of Alcohol in ° F	68.00	20.000	Celsius Equivalent		
Temp of Hydrometer Calibration in ° C.	20.000	68.000	Fahrenheit Equivalent		
Find Celsius To Fahrenheit Hydrometer Offset.					
60° F. Table Dif. from Hyd. Calibration Temp.	-8.00	-4.444	° C.	True Proof using a 60° F. Hydrometer	
Temp. + Dif. from 60°F. = Tbl. Entry Temp.	60.00	0.000	Therm. Cor.	160.1	60° F. Proof
Apparent SG Vac Proof Equivalent	162.70			-2.6	Proof Difference
				0.86367	SG Vac Equivalent
Hydrometer Corrected True Proof	162.7			0.00364	SG Dif.
True SG Vac @ 60°	0.86003			0.99902	H2O at 15.556 C. Vac
H2O at 15.556 C. Vac (Use H2O Den of Table)	0.99902			862.821	Density Kg m3 Vac
Density Kg m3 Vac	859.181			-3640.4	Grams per m3 Dif.

Hydrometer Correction SG Blend by Wt.

We see that the true proof with a 60°F hydrometer is not 160.0 but 160.1. What we are seeing is that 8°F is sufficient for the TTB tables to include 0.1° of proof of hydrometer thermal correction in a positive direction.

Changing the temperature to 44°F shows that instead of 2.7° of offset we have 2.4 (when we should have 2.7°) and at 0.3 proof this is the limit of blending tolerance: This despite the fact that because of the lesser density of 170.2 proof alcohol, the blending tolerance outage is less, as a % than in the prior example.

Courtesy Proof to SG Vac Input (Proof Input)	162.70			0.00150	Blending Tolerance
SG Vac at 60° F.	0.86003			0.300	Degree of Proof
				1,287.73	grams/m3 Tolerance
Apparent Specific Gravity of Alcohol (Vac)	0.86003	0.0000	Hyd. Cor.	-275.6%	% of Tolerance
Density of H20 at Calib. Temp	0.99821				
SG * H2O Den = Density Kg m3 Vac	858.484				
Temperature of Alcohol in ° F	44.00	6.667	Celsius Equivalent		
Temp of Hydrometer Calibration in ° C.	20.000	68.000	Fahrenheit Equivalent		
Find Celsius To Fahrenheit Hydrometer Offset.					
60° F. Table Dif. from Hyd. Calibration Temp.	-8.00	-4.444	° C.	True Proof using a 60° F. Hydrometer	
Temp. + Dif. from 60°F. = Tbl. Entry Temp.	36.00	0.000	Therm. Cor.	167.8	60° F. Proof
Apparent SG Vac Proof Equivalent	162.70			-2.4	Proof Difference
				0.85269	SG Vac Equivalent
Hydrometer Corrected True Proof	170.2			0.00355	SG Dif.
True SG Vac @ 60°	0.84914			0.99902	H2O at 15.556 C. Vac
H2O at 15.556 C. Vac (Use H2O Den of Table)	0.99902			851.850	Density Kg m3 Vac
Density Kg m3 Vac	848.301			-3548.5	Grams per m3 Dif.

Hydrometer Correction SG Blend by Wt.

From these examples we can determine that a 15°C-calibrated hydrometer can easily adapt to using the TTB tables throughout their range while a 20°C hydrometer will maintain blending tolerance but is not as accurate in all cases.

The consequence of this offset is that, because we are adding or subtracting temperature to our readings, we will run out of retrievable table values at the high or low temperature end of the table. For instance, if the 60°F table has a lower limit of 31°F then at 38°F we would be unable to subtract 8°F and still obtain a temperature value.

This same effect of hydrometer calibration temperature offset can also be employed by changing the amount of thermometer correction in the right direction. So, one can use the plain SG blending worksheet procedures to blend by volume or weight in Air or Vacuum SG.

The open question is to determine whether a 15°C hydrometer marked in SG (Air) as opposed to SG (Vac) relative density would be better or worse in its performance. The short answer is that it always makes it worse. SG Air is always less than SG Vac. Consequently, below 15.556°C calibration temperatures the result is towards lesser density and higher proof than intended and when correcting to calibration temperatures above 15.556°C the correction reverses and the result are densities that are greater than intended or lower proof.

As to hydrometer thermal correction; that feature is built into the TTB tables and should be applied to the calibration temperature of the hydrometer rather than to the table temperature when using a hydrometer calibrated for a temperature other than that of the table in use.

Notes

Density & Volume Correction

It is possible to use the TTB table 7 volume correction algorithm to determine the density of alcohol at different calibration temperatures after a true proof correction has been performed and the density determined at 60°F (15.556°C). The converse of the question is whether one can determine true density at another calibration temperature and then use Table 7 volume correction to find the density at 15.556°C. The TTB regulations have a section dealing with both sides of this question, so we will spend some time covering it.

Section 30.24 & 30.25.

While almost every distiller in the US uses proof hydrometers, the TTB used these regulations to define how to use a set of hydrometers that indicate alcohol density using specific gravities with values above 1.0 rather than below 1.0. The regulations describe how these hydrometers are to be used to correct specific gravity determined with respect to 60°F to different temperatures.

While they do accomplish this task, the same procedure can be employed when using hydrometers calibrated at calibration temperatures other than 60°F to find 60°F densities. In order to reach this conclusion, we must first work through the regulation example.

Section 30.24 first defines the unique SG Hydrometers that mark decreasing density as an increase in the value indicated by the instrument.

§30.24 Specific gravity hydrometers. (a) The specific gravity hydrometers furnished by proprietors to appropriate TTB officers shall conform to the standard specifications of the American Society for Testing and Materials (ASTM)[67] for such instruments. Such specific gravity hydrometers shall be of a precision grade, standardization temperature 60°/60°F, and provided in the following ranges and subdivisions:

Range	Subdivision
1.0000 to 1.0500	0.0005
1.0500 to 1.1000	0.0005
1.1000 to 1.1500	0.0005
1.1500 to 1.2000	0.0005
1.2000 to 1.2500	0.0005

Note that the regulation specifies that these hydrometers are calibrated at 60°/60°F. After defining hydrometers, the TTB then describes how to apply Table 7 values to correct SG to 60°F. We are assuming that the value of 1.0 is that of water and 1.25 is close to 200-proof absolute alcohol; since 0.79192 kg/dm³ (Air) is the density of 200-proof alcohol and that value is 0.20808 or 20.81% less than 1 it would make sense for this set of hydrometers to top out at 1.25, which is simply taking density decreasing as an increase in values. This is the same way that proof hydrometers indicate density.

The next section describes how to use the instruments.

§30.25 "Use of precision specific gravity hydrometers. Because of temperature density relationships and the selection of the standardization temperature of 60°/60°F, the specific gravity readings will be greater at temperatures below 60 degrees Fahrenheit and less at temperatures above 60 degrees Fahrenheit. Hence, correction of the specific gravity readings will be made for temperature other than 60 degrees Fahrenheit. Such correction may be ascertained by dividing the specific gravity hydrometer reading by

---

[67] "No instrument shall be in error by more than 0.0005 specific gravity. (b) A certificate of accuracy prepared by the instrument manufacturer for the instrument shall be furnished to the appropriate TTB officer. (c) Incorporation by reference. The 'Standard Specification for ASTM Hydrometers,' (E 100–72 (1978)), published in the *1980 Annual Book of ASTM Standards* (STP 25 1062 (1980)), is incorporated by reference in this part. This incorporation by reference was approved by the Director of the Federal Register on March 23, 1981. This publication may be inspected at the National Archives and Records Administration (NARA), and is available from the American Society for Testing and Materials, 1916 Race Street, Philadelphia, Pennsylvania 19103. For information on the availability of this material at NARA, call 202–741–6030, or go here. (http://www.archives.gov/federal_register/code_of_federal_regulations /ibr_locations.html (Sec. 201. Pub. L. 85–859, 72 Stat. 1358, as amended (26 U.S.C. 5204); 80 Stat. 383, as amended (5 U.S.C. 552(a))) [T.D. ATF–198, 50 FR 8535, Mar. 1, 1985, as amended by T.D. ATF–381, 61 FR 37004, July 16, 1996; 69 FR 18803, Apr. 9, 2004]"

the applicable correction factor in Table 7. *Example*: The specific gravity hydrometer reading is 1.1525, the thermometer reading is 68 degrees Fahrenheit, and the true proof of the spirits is 115 degrees. The correct specific gravity reading will be ascertained as follows: (a) From Table 7, the correction factor for 115° proof at 68°F is 0.996. (b) 1.1525 divided by 0.996=1.1571, the corrected specific gravity. (Sec. 201, Pub. L. 85–859, 72 Stat. 1358, as amended (26 U.S.C. 5204)). The assumptions in this example are that true proof is already known at 115 proof, with a hydrometer that is already calibrated at 60°F and the temperature is now 68° and one wants to correct it back to 60°F."

Working the example shows that the Table 7 value of 0.996 does mathematically result in the numbers set forth in the regulation. Note that we must know the true density (true proof) in order to find the correct Table 7 value to apply. With only an apparent temperature and an apparent proof the Table 7 algorithm would not work properly. Solving the regulation shows the following results:

§ 30.25 Correction	§30.24 Hyd.	Proof
True Proof at 60° F.	1.1525	115.0
Table 7 Correction Factor to 68° F. (20° C.)	0.9960	0.9960
Reading/T7 Factor = Result	1.1571	115.462
Units of Change	0.00463	0.462

30.25 Density Correction

When using this set of hydrometers, because the starting point is 1.0 and higher proofs result in higher values (lesser densities), the division set out in the regulation is required to achieve the proper amount of correction. This is similar to hydrometers demarcated in proof, where decreasing density is reported as increasing proof.

The correction for proof is not accurate in terms of the change in proof, but shows that both hydrometer demarcation regimens move in the same direction towards an increase in value when subjected to the procedure from 60°F to 68°F. Now that we know what direction the hydrometers run we can better determine what the regulation is telling us.

Restating the regulation language in terms of proof:

"The (Proof) readings will be greater at temperatures below 60 degrees Fahrenheit and less at temperatures above 60 degrees Fahrenheit."

With respect to apparent proof this statement is incorrect as to "readings" since colder alcohol will always be more dense than warmer alcohol and consequently read a lower proof. With respect to True Proof, it is a correct statement, since starting with a denser alcohol and the same apparent proof "reading" the true proof will be higher (less dense) if the starting temperature is below 60°F after it is corrected to 60°F and less, lower proof and greater density, when the alcohol is above that temperature.

What we can say is that from a known true density at 15.556°C, we can find density at any other reasonable calibration temperature, say 20°C, by using the procedure. Conversely, we can also posit that from a known true density at 20°C, or any reasonable calibration temperature, one can find the density at 60°F by using the procedure.

In changing the volume or density of true proof alcohol from 60°F to 68°F the density would decrease and volume would increase as temperature increased. We use this procedure in our Table 7 volume correction when we want to change volume from 60°F to some other temperature. Applying this procedure to find the volume at 60°F, when one multiplies any volume by 0.9960, the result will be a smaller volume, which is to be expected when the temperature changes from 68°F to 60°F. The volume change to a smaller volume also results in a greater density:

Reciprocal, Apply Factor from 20° C. to 15.556.

"Proof" at 20° C.	1.1571	115.462
Table 7 Correction Factor	0.9960	0.9960
Reading * T7 Factor = Result	1.1525	115.00
Units of Change	-0.00463	-0.462

Hydrometer Reading	1.1571	115.00
(1- Factor)+1 = Opposite Factor	1.004	1.004
Factor/ reading =SG	1.1525	115.46
Ending Dif. From Starting	-0.00461	0.4600

The two methods demonstrated above are almost mathematically equivalent, but not quite. Both are serviceable but I have found that when faced with this issue it is better to use the (Reading * Factor) procedure rather than the (1 + Opposite Factor/reading) procedure.

What the regulation is mathematically describing is a method whereby the Table 7 volume correction factors can be used as a density correction factor to determine the density change from 20°C (68°F) to 15.556°C (60°F), and, by implication, to do the converse from 15.556°C back to 20°C. Before we can explore that possibility further we have to introduce two better ways to accomplish the same purpose and then come back and see if Table 7 is sufficient to do the same job by itself.

Table 37a Density

A very accurate method of density correction and volume correction was created as a result of original research undertaken by N. S. Osborne, E. C. McKelvy and H. W. Bearce on behalf of the United States Bureau of Standards.[68] This work was used to develop the Official Methods of Analysis tables of 1919 that are still in use. Both of those works were used to develop the more user-friendly TTB tables.

The formula developed by the research team was a least-squares reduction that related to a table of alpha, beta and gamma values which are reported in Table 37a (page 402 of the original publication). The formula is designed to apply those values in order to determine the density of alcohol at 25°C:

$$D\frac{t}{4} = D\frac{25}{4} + \alpha(t - 25) + \beta(t - 25)^2 + \gamma(t - 25)^3$$

The investigators used this formula to calculate the values in Table 49 (on page 424–425) which reports the density of alcohol at various temperatures based upon the percentage mass of alcohol (although they call it "percent alcohol by weight"). The results are reported for temperatures between 10°C and 40°C. I found an example at the bottom of page 396 in order to check the operation of the formula and we will work through it now as an introduction to the formula.

The full text of the Table 37a density values and the 37b volume values, along with Table 49, are reproduced in the Tables section.

---

[68] N. S. Osborne, E. C. McKelvy, and H. W. Bearce, "Density and Thermal Expansion of Ethyl Alcohol and of its Mixtures with Water." Bulletin of the [US] Bureau of Standards 9 (1913): 327–474, http://nvlpubs.nist.gov/nistpubs/bulletin/09/nbsbulletinv9n3p327_A2b.pdf.

The original research on alcohol and water mixtures from which the U.S. Government developed its alcoholometric tables was conducted Osborne, McKelvy and Bearce on behalf of the United States National Bureau of Standards in 1910 and 1911. Their results were published in 1913, in five consecutive parts (see Bibliography, infra).

TABLE XXXIII
Determination of Thermal Expansion of Alcohol-Water Mixture Containing 80.036 Per Cent Alcohol by Weight
Reduction and Adjustment of Experimental Results

396

Bulletin of the Bureau of Standards

[Vol. 9

Determination number	Observed temperature t'	Observed density $D^{t'}_4$	Reduction to integral temperature		Apparent permanent change between series $\Delta'\times10^6$	Adjusted correction for permanent change $\frac{\Delta}{2}\times10^6$	Density adjusted to mean concentration D^t_4	Mean observed density D^t_4	Calculated D^t_4	Observed D^t_4 minus calculated $D^t_4\times10^6$
			t	D^t_4						
1	10.032	0.851856	10	0.851883	−1	−1	0.851882	0.851882	0.851882	0
14	10.026	.851860		.851882		+1	.851883			+1
2	15.028	.847619	15	.847643	−1	0	.847643	.847642	.847644	−1
13	15.019	.847626		.847642		0	.847642			−2
3	20.025	.843342	20	.843364	−2	+1	.843365	.843363	.843362	+3
12	20.029	.843337		.843362		−1	.843361			−1
4	25.034	.838997	25	.839027	5	+2	.839029	.839030	.839031	−2
11	25.025	.839010		.839032		−2	.839030			−1
5	30.006	.834636	30	.834641	11	+3	.834644	.834646	.834646	−2
10	30.020	.834634		.834652		−3	.834649			+3
6	34.988	.830210	35	.830199	6	+4	.830203	.830202	.830202	+1
9	34.994	.830210		.830205		−4	.830201			−1
7	39.966	.825722	40	.825691	7	+5	.825696	.825694	.825694	+2
8	39.981	.825715		.825698		−5	.825693			−1

Equation for adjustment of Δ: $\Delta=9.1-1.821\,(7-n)$.

Probable value of D^t_4 from above determinations: $D^t_4=0.839031-[871.4\,(t-25)+1.08\,(t-25)^2+0.0069\,(t-25)^3]\,10^{-6}$.

The data on page 396 indicates that that alcohol-water mixture under investigation consisted of 80.036 percent alcohol by weight. The published table of alpha, beta and gamma values only contains values for whole percentage alcohol by weight values but this is as close to 80 percent alcohol by weight as we will get in order to check the formula that was created from the investigation. Using that value, we lookup the three values we will need to operate the formula:

Alpha, Beta, Gamma Lookups					
% by Wt.	80.00	% Alcohol By Weight			
Exact Match % Weight Row	81			Absolute Values	
Alpha	-8714	-871.4	Alpha/10	871.40	Alpha
Beta	-108	-1.08	Beta/100	1.08	Beta
Gamma	-5.00	-0.0050	Gamma/1000	0.005	Gamma

Table 37a Density

Since we need to check the table as well as the formula, we retrieve the published alpha, beta and gamma values and see that they are negative numbers. We also reduce them by dividing each by 10, 100 and 1,000 respectively. Because the example is using positive values rather than negative ones we also need to reverse the sign and use the absolute value of each. Then we can compare what we can discern from the published table with what the article example indicates:

	Table 37a Values	Table 33 Calculations		Reverse	
Start Density D 25/4	0.839031	0.839031		0.843362	
Alpha	871.4	871.4		-871.4	Table Value
T = Observed Temperature	20	20		20	
Formula Start Temp 25	25	25		25	
Tepm - Formula Start Temp t-25 =	-5.00	-5.00		-5.00	
D 25/4 + Alpha * t-25 Result	-4,357.00	-4,357.00		4,357.00	
Beta	1.08	1.08		-1.08	Table Value
t-25 Squared	25.00	25.00		25.00	
Beta * t-25 Squared	27.00	27.00		-27.00	
Gamma	0.00500	0.00690	***	-0.00500	Table Value
t-25 Cubed	-125.00	-125.00		-125.00	
Gamma * t-25 Cubed	-0.62500	-0.86250		0.6	
Alpha + Beta + Gamma Results =	-4,330.6	-4,330.9		4,331	
The value of 1 to the minus 6	1.0E-06	1.0E-06		1.0E-06	
Multiply Result by 1 to minus 6 = Den Change	-0.004331	-0.004331		0.004331	
Start Den.- Reduction = Den @ Observed t/4	0.843362	0.843362		0.839031	
Difference Between Each Run of Equation	0.0000002			0.00000000	Difference
Difference as Scientific Notation	2.4E-07			0.0E+00	

Table 37a Den

Using the value of 20°C for the reduction in value in temperature from 25°C as the beginning temperature, we see that the resulting density value of 0.843362 matches the published calculated D t/4 value to six decimal places. This small difference is caused by a difference in gamma values used in Table 33, which applies 0.0069 and rather than the published value of the absolute value of –0.005. From these results we can draw the conclusion that the equation works satisfactorily in conjunction with the published values

Reversing the equation involves using the published values in Table 37a, some of which are negative and others positive. In this case they are all negative values, and we see that this results in a completely reciprocal solution of the equation.

The next task is to apply the formula to reproduce the values in Table 49 and to arrange it so one can translate from any start temperature to any destination temperature. We will accomplish this by running the equation twice: once to go from the observed temperature to 25°C. and the second time from 25°C. to the chosen destination temperature. We will start with an example of 50% mass of alcohol by weight, changing from 10°C to 40°C. We will use the table value at 10°C as the input, then run the formula, then compare the result to the table value at the destination temperature.

Diagram follows on the next page.

Make Equation Variable for Start and Destination Temperatures						
Start Density D 25/4	0.92162		Destination		Table 49 Lookup	
T = Observed Temperature ° C.	10.00		40.000		Alpha, Beta, Gamma Lookups	
Formula Start Temp 25° C.	25.0		25.0	Fixed Value	50.000	Mass % Input
Temp - Formula Start Temp t-25 =	-15.00		15.00		50.00	Round % Mass Whole
Alpha	-803.5		-803.50		51	Exact Match % Row
D 25/4 + Alpha * t-25 Result	12,052.50		-12,052.50		-803.50	Alpha/10
Beta	-1.28		-1.28		-1.2800	Beta/100
t-25 Squared	225.00		225.00		-0.0020	Gamma/1,000
Beta * t-25 Squared	-288.00		-288.00			
Gamma	-0.00200		-0.0020			
t-25 Cubed	-3,375.00		3,375.00			
Gamma * t-25 Cubed	6.75000		-6.75			
Alpha + Beta + Gamma Results =	11,771.3		-12,347.3			
The value of 1 to the minus 6	1.0E-06		1.0E-06			
Multiply Result by 1 to minus 6 = Den Change	0.011771		-0.01235			
Find Density Values at Temp	Vac					
Observed Temp to 25° C. Value Restated	0.011771					
Density Kg. per Liter Restated (Vac)	0.921620					
Density - Formula Change = 25° Density	0.90985					
25° to Destination ° C. Value	-0.01235					
25° Den + Formula Change = Destination Den.	0.89750		0.0015	Tolerence		
Kg. per m3 Vac	897.502		1.35	Kg per m3 Tolerance		
Table 49 Value	897.50		1.50	Grams per m3 Dif.		
Kg. per m3 Difference from Table 49 Value	0.0015		0.11%	% of Tolerance		

Table 49 Values							
Temperature ° C.	10	15	20	25	30	35	40
BOS Alc. Density D t/4	921.620	917.760	913.840	909.850	905.800	901.680	897.500

Table 37a Den

The results show that the difference in density is only 1.5 g/m^3 from the Table 49 value. This represents only the rounding required to provide 5 decimal places in the table itself.

The first thing to note is that in order to obtain these results we have been required to use the table values themselves regardless of whether they are positive or negative (in this case all negative), and not the absolute values that we have been using up to this point. The second thing to note is that besides having to run the equation twice (once to 25°C and then from 25°C to the destination temp of 40°C), one never knows the actual numerical value used to accomplish the entire transformation: it is a combination of both runs of the equation and so is hidden. (The density at 25°C is part of the transformation to which the second formula run is applied.) I have not found any means of reversing the subtraction or the order in which the two corrections are offset against 25°C that will suffice to allow one to offset the table values or absolute values and obtain reliable results. One simply has to run the equation twice.

I have not found any discrepancy greater than 10 g/m^3 within the parameters of Table 49. In practice I have had very little difficulty using the formula down to 4°C, while comparing to tables using that calibration temperature. I have not had much experience with applying it at temperatures above 40°C.

The equation relies upon knowing the mass percent of the alcohol under investigation. The TTB tables allow us to calculate the mass percent at 15.556° (in Air) with a fair degree of certainty as we have seen. Because the formula only has intervals of whole mass percents for different alpha, beta and gamma values, the small difference between air and vacuum mass percentages can be ignored for the moment; we can use the formula to calculate the density of alcohol at temperatures other than 15.556°C and compare the results to table 49.

This is required because we need to know with certainty what SG a vacuum density hydrometer would read at any given calibration temperature (20°C for starters) in order to see if it can be adapted to use the TTB tables directly with the proper offset.

Finally, while this example is set up to have density as an input, I've found that the equation also works when specific gravity is used as an input and this avoids the necessity of finding density; but my preference is to work in densities.

One advantage of Table 37a is that it will accommodate air densities or SGs as input as well as vacuum ones and conduct its transformation to a result in the same air or vacuum regimen. As we will see, the General Formula of the IAT tables can be made to accomplish this as well but it requires additional manipulation.

Table 37b Volume

In order to calculate the volume of alcohol at various temperatures, Osborne, McKelvy and Bearce developed a similar least-squares reduction method along with a set of A, B, and C, values which are reported in Table 37b (page 403):

$$V_t = V_{25}[1 + A(t-25) + B(t-25)^2 + C(t-25)^3]$$

Like the density formula, it is designed to translate volume to 25°C and in order to use it to find another destination temperature one must run the equation from 25°C to that temperature. The example is set up to run from 10.0°C to 38°C which is the highest Celsius temperature at which we can compare the results to a Table 7 volume correction. We will start with an assumed 1,000 units of volume at 10°C. (In this case volume can be liters or gallons since the units themselves are dimensionless; you can consider them any volume unit you like). I've also chosen 125.7 proof since this happens to be a TTB proof with a mass percent very close to a whole number at 55.004%. The results show that even at this large difference in temperature Table 7 and the 37b formula results agree with one another to 17.5% of the tolerance allowed. For less extreme temperature changes the agreement is even better and only limited by the fact that Table 7 provides only three decimal places:

Present Volume (Any Unit)	1,000.00		125.70	True Proof
Temperature of Liquid ° C.	38.000		1258	Match Proof Row
Destination Temperature ° C.	10.000		54.996	Mass %
37b Two Column Vol to Vol Result	974.64			
Table 7 Vol. Result	974.90		0.0015	Tolerance
Liters Dif. From 37b	0.26		1.50	Liter Tol.
% of Tol.	17.5%			

Table 37b Vol. Cor. Two Col. Vol to Vol	Start °	Destination °		
Temperature of Liquid ° C.	38.0000	10.000	Destination °	
Calibration Temp of BOS Work	25.000	25.000	BOS Temp	
Start ° C. - BOS Calib ° = Temp Dif.	13.000	-15.00	Destination - BOS	
A	91.40	91.40		
A * Temp Dif. = Result	1,188.20	-1,371.00		
Temp Dif. Squared	169.00	225.00		
B	0.2200	0.2200		
B * Temp Dif. Squared	37.180	49.500		
Temp Dif. Cubed	2,197.00	-3,375.00		
C	0.00100	0.0010		
C * Temp Dif. Cubed	2.19700	-3.37500		
A + B + C Results =	1,227.5770	-1,324.88		
1 to the minus 5	1.0E-05	1.0E-05		
Reduce by 1 to the -5 to Output Value	0.012276	-0.013249		
The Value of One	1.0	1.0		
1 Minus Correction Value = Net Factor	0.98772	0.98675	One Plus Factor	
Present Liters Restated	1,000.00	987.72	25° C Vol.	
Net Factor * Present Liters = 25° C. Vol.	987.72	974.64	Vol. @ Target	

Table 37b Volume

Again, it is required that the equation be run to 25°C volume and then that volume used to determine the volume at the destination temperature. I was misled by the fact that with this equation it is possible to offset the values and arrive at a net factor that works most of the time but which suffers from certain deficiencies such that when the liquid temperature is below 25°C the results are out of tolerance.

We see that Table 7 and the Table 37b two-column method agree quite well with each other. Note that offsetting the correction factors for 37b creates truly anomalous results. This was disappointing since offsetting the factor provided a means of having no correction at all when none was required. However, doing it the "right" way shows 0.02 liters difference at 30/30 where none would be required. This is annoying but there is nothing that can be done about it:

Present Volume (Any Unit)	1,000.00		125.70	True Proof
Temperature of Liquid ° C.	30.000		1258	Match Proof Row
Destination Temperature ° C.	30.000		55.004	Mass %
37b Two Column Vol to Vol Result	999.98		0.0015	Tolerance
			1.50	Liter Tol.
Table 7 Vol. Result	999.83			
Liters Dif. From 37b	-0.15			
% of Tol.	-9.8%			

Table 37b Volume

It was also disappointing because I was hoping to be able to obtain a numerical factor to apply and compare or use as a density conversion factor as well. Table 7 also suffers from the same requirement that the intermediate volume, in that case at 15.556°, must be calculated before moving on to the destination temperature.

Despite being off by up to 0.15 liters (150 ml) when the calibration temperature and sample are the same temperature, the two-column method is superior to the offset method because one never is out of tolerance in the transformation. I traced this problem to the occasionally negative A, B and C values which throw off any attempt to offset values. One can use a single-column method and find a factor value if one is staying below 25° C. or going from 25°C upward, but it needs to be reversed to operate when calibration temp and sample temp are reversed.

Begin 30.25 Analysis

Now that we have introduced the Table 37a (density) and 37b (volume) methods and compared them to Table 7, we can apply them to change the density of alcohol to different temperatures.

We will start at 60°F densities and use all three methods to change the density to different calibration temperatures and compare those densities to the Table 49 values. Then we will turn it around and allow hydrometers calibrated at other temperatures to use the TTB tables to find density, mass percent and volume percent at 60°F and at the calibration temperature of the hydrometer.

We will assume that we know the true proof and mass percent at 15.556°C and that we wish to change the density to 20°C. The only actual numerical factor we obtain to accomplish this is provided by Table 7 and that value is 0.996. We apply that factor directly to both the density in air and the SG vac at 15.556°. The total kg per meter cubed change is a decrease of 3.62 kg/m^3 and the difference between the Table 7 method and the Table 37a method is only ~75 g/m^3. Comparing 37a as a dedicated density converter to 37b which is a volume converter shows that all three results arrive at roughly the same resulting density value and are only ~22 g/m^3 different from each other. This difference is more related to vagaries in the tables and the values used for the density of water at the time the tables were created than the overall accuracy of each method:

§ 30.25 Density Correction

Temp of Density Desired ° C.	20.00	68.000	° F
True Proof at 60° F.	125.70		Using SG Vac & Table 7
Index Mass % for 37a formula	54.996	0.90701	SG Vac at 60° F.
Index kg m3 Vac at 15.556	906.12	0.9960	Table 7 Factor
Table 7 Correction Factor to Apply	0.9960	0.90338	SG Vac * T7 Factor = SG
Reading * T7 Factor = Air Den at Calib. °	902.49	0.99902	H2O Den @ 15.556 Vac
Kg. m3 Change in Density	-3.62	902.49	SG*H20 = kg. m3 Vac at Calib.
Resulting Apparent Proof at Temp	127.90	0.47	Grams Dif. From Den. Result
37a Results	Vac		
37a Kg. m3 Result from Known Mass %	902.54	902.52	37b Results kgm3 Vac
Grams per m3 Dif. From T7	52.02	-21.60	Grams Dif. 37a Result

Find Mass per m3 Dif.	TP -.3	TP -.2	TP -.1	Proof @ °	TP +.1	TP +.2	TP +.3	
Proof Values	127.6	127.7	127.8	127.9	128.0	128.1	128.2	
Exact Match of Proof Row	1277	1278	1279	1280	1281	1282	1283	
Kg. m3 (Vac) at Proof	903.93	903.82	903.70	903.59	903.47	903.35	903.24	
Grams Dif. Each .1 Proof	116.0	116.0	115.0		-116.0	-116.0	-115.0	
Total Grams per .3 Proof (.0015) Change	347.0			902.58	Table 49 Density (Vac)		-347.0	Grams Dif.
T37a Grams Dif. % of .3 proof Tol.	-10.4%			-35.97	37a Grams Dif. From T49		10.4%	37a % of Tol.
T37b Grams Dif. % of .3 Proof Tol.	-16.6%			-57.57	37b Grams Dif. From T49		16.6%	37b % of Tol.
Table 7 Grams Dif. % of .3 proof Tol.	-25.4%			-87.99	T7 Grams Dif. From T49		25.4%	T7 % of Tol.

30.25 Density Correction

I've also included a method of determining the total grams per meter cubed change for 0.3 proof on either side of the true proof assumed. Since 0.3 proof is equal to the tolerance of 0.0015 this allows us to compare the density results to that standard. Note that the grams per meter cubed values vary between individual 0.1 proof increments. All of the values are sufficiently close to one another to retain close tolerance but the Table 7 value performs the least well.

I have applied all of these methods to both density and SG and obtained reliable results. The Table 49 values are simply looked up in an array based upon the mass percent as an input. SGs are calculated based upon the density of water at each temperature. I've also identified those proof values that come closest to whole proof values in order to obtain the best results possible for this test:

Closest Proof to Whole Mass % at 15.556

Retreive Table 49 Values at Whole Mass %

					Proof	TTB Mass %
55.0	Round Mass % Whole				24.8	10.004
56	Match % Wt. Column				53.7	21.997
Temp	Density	H2O Den.	SG (Vac)		79.2	33.003
10	910.55	999.70	0.91082		101.1	42.997
15	906.59	999.10	0.90741		125.7	55.004
20	902.58	998.21	0.90420		135.4	60.004
25	898.50	997.05	0.90116		155.6	71.003
30	894.37	995.65	0.89828		169.3	79.011
35	890.16	994.04	0.89550		185.0	88.9780
40	885.89	992.22	0.89284		190.8	92.9900
T49 Relay	902.58	T49 Relay	0.90420		197.5	97.9870

The reason that so few proof increments align with whole mass percent values is that proof is a species of volume percent and volume percent and mass percent are two different regimens for determining alcohol content. One other caveat is that in this test we are having to use the mass percent in air as opposed to vacuum but since the formula only has values at whole mass percents there is no real difference in outcome.

Just to be sure this all works as we intend it to, we will raise the temperature at which we would like to obtain the density to 35°C and note that the overall correction itself has changed to −15.86 fewer kilograms per meter cubed but the differences from each method are still well within tolerance. However, Table 37a, which is really designed specifically for the purpose, is performing the best with only 30.35 grams per m³ of difference between its result and the Table 49 published values.

§ 30.25 Density Correction			
Temp of Density Desired ° C.	35.00	95.000	°F
True Proof at 60° F.	125.70		Using SG Vac & Table 7
Index Mass % for 37a formula	54.996	0.90701	SG Vac at 60° F.
Index kg m3 Vac at 15.556	906.12	0.9825	Table 7 Factor
Table 7 Correction Factor to Apply	0.9825	0.89114	SG Vac * T7 Factor = SG
Reading * T7 Factor = Air Den at Calib. °	890.26	0.99902	H2O Den @ 15.556 Vac
Kg. m3 Change in Density	-15.86	890.26	SG*H20 = kg. m3 Vac at Calib.
Resulting Apparent Proof at Temp	138.20	0.47	Grams Dif. From Den. Result
37a Results	Vac		
37a Kg. m3 Result from Known Mass %	890.13	890.05	37b Results kgm3 Vac
Grams per m3 Dif. From T7	-129.78	-83.53	Grams Dif. 37a Result

Find Mass per m3 Dif.	TP -.3	TP -.2	TP -.1	Proof @ °	TP +.1	TP +.2	TP +.3	
Proof Values	137.9	138.0	138.1	138.2	138.3	138.4	138.5	
Exact Match of Proof Row	1380	1381	1382	1383	1384	1385	1386	
Kg. m3 (Vac) at Proof	891.70	891.58	891.46	891.33	891.21	891.09	890.97	
Grams Dif. Each .1 Proof	122.0	122.0	123.0		-122.0	-122.0	-123.0	
Total Grams per .3 Proof (.0015) Change	367.0			890.16	Table 49 Density (Vac)		-367.0	Grams Dif.
T37a Grams Dif. % of .3 proof Tol.	-8.3%			-30.35	37a Grams Dif. From T49		8.3%	37a % of Tol.
T37b Grams Dif. % of .3 Proof Tol.	-31.0%			-113.88	37b Grams Dif. From T49		31.0%	37b % of Tol.
Table 7 Grams Dif. % of .3 proof tol.	27.1%			99.44	T7 Grams Dif. From T49		-27.1%	T7 % of Tol.

30.25 Density Correction

In Volume 2 we will revise the Table 49 values to accommodate the changes in the definition of the liter since publication of the table values: the differences will be even smaller.

There are a few other things that can be learned from Section 30.25:

Find Air Density						
SG Air at 60° at Start Proof	0.90690		From SG to Density use Table Temp H2O			
H2O Den @ 15.556 (60° F. Air)	0.99794	0.90690	SG Air Result Restated	0.90690	SG Air Result Restated
SG*H20 = kg. m3 (Air) at Calib.	905.03	0.120074	Wine Gal. per Lb. of Water in Air	8.32820	Lbs. per WG of Water in Air
		7.55283	SG/WG per lb H2O = Alc. Lbs. per Gal.	7.55283	SG* Lbs. per WG H2O = Lbs. per Gal.
Find Lbs. per Gal. (Air)		7.55283	Index WG per Lb. at Start ° Proof.	7.55283	Index WG per Lb. at Start ° Proof.
Den (Air) at Temp Restated	905.03	0.00000	Lbs. per Gal. Dif.	0.00000	Lbs. per Gal. Dif.
Kilograms to Pounds Avdp	2.2046226			T37a Find Lbs. per Gal. (Vac) at Calib °	
Kg. per m3 * Factor = Lbs. m3	1,995.25		Find SG from Density use Calib ° Water	902.55	37a Den. Vac Restated
Gallons per m3 value	264.17205	0.90255	37a Den. Kg dm3 at Calib	2.2046226	Kg. to Lbs.
Lbs./Gal.m3 = Lbs. per Gal.	7.55283	0.99821	H2O Den (Vac) at Calib Temp	1,989.78	Lbs. Vac
Lbs. per Gal. Dif. From Calculated Value	0.00000	0.90417	Den/H2O = SG H2O at Calib°	264.17205	Gal per m3
		0.00000?	Dif. From T49	7.53213	Lbs. per Gal. Vac
				0.132765	Gal. per Lb. Vac

30.25 Density Correction

The first calculation simply multiplies SG air by the value in kilograms per decimeter cubed for water in air at 15.556° to find the density at that temperature. When using a table value to find density it is important to use the density of water at that temperature and also the same type of density, either air or vacuum. In the second column, again using SG Air and the value provided in section 30.25 for the density of water (air) at 15.556° of 0.120074 wine gallons per pound, we divide the SG by that value to find the pounds per wine gallon; this results in a value identical to the Table 4 value (my Table 4 has been modified to harmonize with SG air and vacuum values based upon modern values for the definition of water and the volume of a liter which is not quite a dm³). Using multiplication, one can also take the pounds per wine gallon and also find the same value.

It is also possible to find the pounds per wine gallon by using a kilograms-to-pounds converter coupled with a gallons-per-meter-cubed value. By using the vacuum density as an input, it is also possible to find the pounds per wine gallon in absolute density (vacuum) as well.

General Formula

The International Organization of Legal Metrology, (Organization Internationale de Metrologie Legale or *OMIL*) published their International Alcohol Tables (IAT) in 1973 to "harmonize ... data relating to the density and to the alcoholic strengths by mass and by volume of mixtures of water and ethanol."[69] The General Formula, derived from a least-squares analysis of the harmonized data, is a reliable and very accurate method that can be used to find the density of alcohol at different temperatures based upon the mass percentage of the alcohol.[70]

The Greek *sigma* (Σ) indicates summations. As each "Σ" is expanded, coefficients are retrieved from lists (A_k, B_k, and $C_{i,k}$). Inputs are mass percent (p, expressed as decimal form) and sample temperature in degrees Celsius (t). The output is the corresponding density (ρ).

The value for term A_1 of 998.20123 is the OIML-defined density of water in vacuum at 20°C. In the previous volume we observed that this value can be changed based upon whether the air or vacuum mass percent of alcohol is used find air densities as well. We will cover that topic in more detail in this volume.

$$\rho = A_1 + \sum_{k=2}^{12} A_k\, p^{k-1} + \sum_{k=1}^{6} B_k\,(t-20°C)^k + \sum_{i=1}^{n}\sum_{k=1}^{m_i} C_{ik}\, p^k\,(t-20°C)^i$$

Where $n = 5$, $m_1 = 11$, $m_2 = 10$, $m_3 = 9$, $m_4 = 4$, $m_5 = 2$, with the numerical coefficients

$A_1 = 9{,}982\,012\,300 \cdot 10^2$ (This is essentially the IAT determination of the density of water in vacuum at 20 C.)

A_k [kg/m³]

2	$- 1{,}929\,769\,495 \cdot 10^2$
3	$3{,}891\,238\,958 \cdot 10^2$
4	$- 1{,}668\,103\,923 \cdot 10^3$
5	$1{,}352\,215\,441 \cdot 10^4$
6	$- 8{,}829\,278\,388 \cdot 10^4$
7	$3{,}062\,874\,042 \cdot 10^5$
8	$- 6{,}138\,381\,234 \cdot 10^5$
9	$7{,}470\,172\,998 \cdot 10^5$
10	$- 5{,}478\,461\,354 \cdot 10^5$
11	$2{,}234\,460\,334 \cdot 10^5$
12	$- 3{,}903\,285\,426 \cdot 10^4$

B_k

1	$- 2{,}061\,851\,3 \cdot 10^{-1}$ kg/ (m³ ·°C)
2	$- 5{,}268\,254\,2 \cdot 10^{-3}$ kg/ (m³ ·°C²)
3	$3{,}613\,001\,3 \cdot 10^{-5}$ kg/ (m³ ·°C³)
4	$- 3{,}895\,770\,2 \cdot 10^{-7}$ kg/ (m³ · °C⁴)
5	$7{,}169\,354\,0 \cdot 10^{-9}$ kg/ (m³ ·°C⁵)
6	$- 9{,}973\,923\,1 \cdot 10^{-11}$ kg/ (m³ ·°C⁶)

C_{1k}
kg/ (m³ · °C)

1	$1{,}693\,443\,461\,530\,087 \cdot 10^{-1}$
2	$- 1{,}046\,914\,743\,455\,169 \cdot 10^{1}$
3	$7{,}196\,353\,469\,546\,523 \cdot 10^{1}$
4	$- 7{,}047\,478\,054\,272\,792 \cdot 10^{2}$
5	$3{,}924\,090\,430\,035\,045 \cdot 10^{3}$
6	$- 1{,}210\,164\,659\,068\,747 \cdot 10^{4}$
7	$2{,}248\,646\,550\,400\,788 \cdot 10^{4}$
8	$- 2{,}605\,562\,982\,188\,164 \cdot 10^{4}$
9	$1{,}852\,373\,922\,069\,467 \cdot 10^{4}$
10	$- 7{,}420\,201\,433\,430\,137 \cdot 10^{3}$
11	$1{,}285\,617\,841\,998\,974 \cdot 10^{3}$

C_{2k}
kg/ (m³ · °C²)

1	$- 1{,}193\,013\,005\,057\,010 \cdot 10^{-2}$
2	$2{,}517\,399\,633\,803\,461 \cdot 10^{-1}$
3	$- 2{,}170\,575\,700\,536\,993$
4	$1{,}353\,034\,988\,843\,029 \cdot 10^{1}$
5	$- 5{,}029\,988\,758\,547\,014 \cdot 10^{1}$
6	$1{,}096\,355\,666\,577\,570 \cdot 10^{2}$
7	$- 1{,}422\,753\,946\,421\,155 \cdot 10^{2}$
8	$1{,}080\,435\,942\,856\,230 \cdot 10^{2}$
9	$- 4{,}414\,153\,236\,817\,392 \cdot 10^{1}$
10	$7{,}442\,971\,530\,188\,783$

C_{3k}
k kg/ (m³ · °C³)

1	$- 6{,}802\,995\,733\,503\,803 \cdot 10^{-4}$
2	$1{,}876\,837\,790\,289\,664 \cdot 10^{-2}$
3	$- 2{,}002\,561\,813\,734\,156 \cdot 10^{-1}$
4	$1{,}022\,992\,996\,719\,220$
5	$- 2{,}895\,696\,483\,903\,638$
6	$4{,}810\,060\,584\,300\,675$
7	$- 4{,}672\,147\,440\,794\,683$
8	$2{,}458\,043\,105\,903\,461$
9	$- 5{,}411\,227\,621\,436\,812 \cdot 10^{-1}$

C_{4k}
kg/ (m³ · °C⁴)

1	$4{,}075\,376\,675\,622\,027 \cdot 10^{-6}$
2	$- 8{,}763\,058\,573\,471\,110 \cdot 10^{-6}$
3	$6{,}515\,031\,360\,099\,368 \cdot 10^{-6}$
4	$- 1{,}515\,784\,836\,987\,210 \cdot 10^{-6}$

C_{5k}
kg/ (m³ · °C⁵)

1	$- 2{,}788\,074\,354\,782\,409 \cdot 10^{-8}$
2	$1{,}345\,612\,883\,493\,354 \cdot 10^{-8}$

---

[69] Sources to be "harmonized": research from Tokyo, Japan (Kawasaki, et al.); Montpellier, France (Jaulme, et al.); Russia (Mendeleev); Braunschweig, Germany (Wagenbreth); and the United States (Osborne, Kelvey, and Bearce for the Bureau of Standards.

[70] https://www.oiml.org/en/ The tables can be found at https://www.oiml.org/en/files/pdf_r/r022-e75.pdf/view For some reason the file name is RO22e75.This formula is also reproduced in the European Union's Annex V Federal Law on Weights and Measures - Alcohol Tables - Federal Regulations of 2004, which holds for values of temperature ranging between –20°C and +40°C.

In practice the formula is solved by arranging it as follows (in this case we are assuming that we know the mass percent at 20°C, which is 55%, and we want to know the density at 35°C):

General Formula Find Density @ Temp (-20 to 40)

	55.000%	Known Mass % (Vac)
	35.000	Temperature ° C
	890.110	Density Result at Temp

A1		Sum 3 (Cik)
998.20123		-8.267

k	Ak	p^(k-1)	Ak*pk-1	Sum 1 (Ak)
2	-1.930E+02	0.550	-106.137	-95.653
3	3.891E+02	0.303	117.710	
4	-1.668E+03	0.166	-277.531	
5	1.352E+04	0.092	1,237.362	
6	-8.829E+04	0.050	-4,443.638	
7	3.063E+05	0.028	8,478.232	
8	-6.138E+05	0.015	-9,345.288	
9	7.470E+05	0.008	6,255.070	
10	-5.478E+05	0.005	-2,523.032	
11	2.234E+05	0.003	565.978	
12	-3.903E+04	0.001	-54.378	

k2	C2k	C*p^k*(t-20)^	Sum C2k
1	-1.193E-02	-1.476E+00	9.102E-01
2	2.517E-01	1.713E+01	
3	-2.171E+00	-8.125E+01	
4	1.353E+01	2.786E+02	
5	-5.030E+01	-5.696E+02	
6	1.096E+02	6.828E+02	
7	-1.423E+02	-4.874E+02	
8	1.080E+02	2.036E+02	
9	-4.414E+01	-4.574E+01	
10	7.443E+00	4.242E+00	

AK just deals with Mass %

k	Bk	(t-20)^k	Bk*(t-20)^k	Sum 2 (Bk)
1	-2.062E-01	15.000	-3.093E+00	-4.172E+00
2	-5.268E-03	225.000	-1.185E+00	
3	3.613E-05	3,375.000	1.219E-01	
4	-3.896E-07	50,625.000	-1.972E-02	
5	7.169E-09	759,375.000	5.444E-03	
6	-9.974E-11	11,390,625.00	-1.136E-03	

k3	C3k	C*p^k*(t-20)^	Sum C3k
1	-6.803E-04	-1.263E+00	-1.173E-01
2	1.877E-02	1.916E+01	
3	-2.003E-01	-1.124E+02	
4	1.023E+00	3.159E+02	
5	-2.896E+00	-4.919E+02	
6	4.810E+00	4.494E+02	
7	-4.672E+00	-2.401E+02	
8	2.458E+00	6.946E+01	
9	-5.411E-01	-8.411E+00	

i	k1	C1k	C*p^k*(t-20)^	Sum C1k
1	1	0.169	1.397	-9.078
2	2	-10.469	-47.504	
3	3	71.964	179.594	
4	4	-704.748	-967.332	
5	5	3,924.090	2,962.400	
	6	-12,101.647	-5,024.720	
	7	22,486.466	5,135.128	
	8	-26,055.630	-3,272.611	
	9	18,523.739	1,279.629	
	10	-7,420.201	-281.925	
	11	1,285.618	26.865	

k4	C4k	C*p^k*(t-20)^	Sum C4k
1	4.075E-06	1.135E-01	2.713E-02
2	-8.763E-06	-1.342E-01	
3	6.515E-06	5.487E-02	
4	-1.516E-06	-7.022E-03	

k5	C5k	C*p^k*(t-20)^	Sum C5k
1	-2.788E-08	-1.164E-02	-8.554E-03
2	1.346E-08	3.091E-03	

General Formula

One advantage of this formula over Table 37a is that it will accommodate fractional mass percent values. It only solves to vacuum density and one must be careful to use *mass percent in vacuum* values rather than TTB values in air. We will cover how to change mass percent from air to vacuum and vice versa shortly.

General Formula Air and Vac

The IAT General Formula can also be configured to change density on an air or vacuum basis.

The general formula is very sensitive to the value of the density of water in vacuum that is built into the equation as published. This value is ~5 g/m³ different from that published in the *Handbook of Chemistry and Physics* (*HCP*) which is the basis of my program. The *HCP* reports values for temperatures beyond 40°C, which is the limit for the General Formula's reliability—although I've found the GF to be reliable up to about 55°C.

IAT Vac H2O @ 20	998.20123
HCP 89th Vac H2O	998.20630
Grams per m3 Dif.	5.07

IAT Air H2O	997.1237
HCP Air H2O	997.1288
Grams per m3 Dif.	5.10

Additionally, it is required that the density of water in air be acquired by using the Kaye & Laby equation to run the IAT density of water to that of air:

IAT Value for A1 is H2O at 20° C.	998.20123
Vac to Air Algorithm	Vac to Air
Vac Density kg/dm3	0.9982012
Air Density, Dry Air (σ')	0.00122265
Mass * Vac Den = Result 1 (Mσ')	0.001220
Value	1.0
1/ HCP Value (Δ) = Result 2	1.0018
Density of Weights (ρ)	8.4090688
Value	1.00
1/Density of weights = Result 3	0.1189
Result 2 - Result 3 = Result 4	0.8829
Result 4 * Result 1 = Buoyancy Cor. (K)	0.0011
Vac Den - Air Mass = Air Den	0.99712
Result Kg m3 Air	997.1237

Begin with mass percent on a density basis: in this regimen, which also includes SGs and pounds per gallon, etc., the mass percent changes between air and vacuum, but the density remains constant. First, we run a vacuum mass percentage of 33.85% through the general formula to find a density very close to 950 kg/m³, while using the IAT density of water in vacuum as the reference (998.20123) to yield 950.001 kg/m³ (left diagram):

General Formula Find Density @ Temp (-20 to 40)

Known Mass % (Vac)	33.850%
Temperature ° C	15.556
Density Result at Temp	950.001

A1
998.20123

General Formula Find Density @ Temp (-20 to 40)

Known Mass %	33.227%
Temperature ° C	15.556
Density Result at Temp	950.017

	A1
IAT Air H2O	997.12370

Then we change the mass percent and the density of water to their air values and note the difference in results (right diagram).[71]

Density

Vac Mass % to Den	950.001
Air Mass % to Den	950.017
Grams per m3 Dif.	16.00

---

[71] The procedures for finding Air mass % from the TTB pounds per wine gallon or pounds per proof gallon were set out in the Analysis of Obscuration section and Modules and Blending section and are also found in Volume 2. On a metric basis it essentially involves boosting the vacuum density by the density of air to the same degree that we reduce it to find air density; and then retrieving the vacuum mass % at that density.

The resulting air and vacuum densities are within 16 g/m³ of each other. But we can do better: we can use our air to vacuum mass percent procedure and apply those results in addition to the straight mass percent in air values. In this way we convert the mass percent in air to a value that will yield the same density as the vacuum mass percent does.

Find Modified Mass % in Air	Vac Values	Air Values
Corrected Density of Sample	950.001	948.924
Density of AA @ 15.556 C.	789.239	788.131
Vac Den/Air Den = Calc Air Mass %	100.11%	
Vac Density AA/Air Density AA	100.14%	
AA Density Ratio/Alc Den Ratio	100.03%	
Mass % Air Restated	33.227	
Mass % * % Change = Vac mass %	33.236	
Rounded to 3 Places	33.236%	

General Formula Find Density @ Temp (-20 to 40)	
Known Mass % (Vac)	33.236%
Temperature ° C	15.556
Density Result at Temp	950.002

	A1
IAT Air H2O	997.12370

General Formula

The result is a very favorable 950.002 kg/m³ which is 1 g/m³ different from our starting density. One gram per meter cubed is one part out of one million.

It does not mean that the original air mass percent was wrong. Running a Solver[72] on 950 kg/m³ at 15.556° yields a mass percent of 33.225%, compared to the 33.227% from our tables.

The Solver is a method to reduce either density or volume percent to a mass percent. The differential equation Solver feature is installable in Excel as an add-in; it takes the density of alcohol and iteratively guesses the mass percent until the equation converges upon a solution using the GRG Nonlinear or Simplex LP modes:

Density Output Units (Rounded)	950.00000
Mass %	33.225
Vol % at Calib. Temp	39.856
Proof at 60° F. (15.556° C.)	79.71

Calibibration Temp Density Output	950.001
Mass %	33.227
Vol % at Calib. Temp	39.859
Proof at 60° F. (15.556° C.)	79.72

Solver 1.7   *Blending 21.8*

This procedure is a manipulation that allows the general formula to generate the correct density when used in conjunction with the density of water in air.

I have also found this useful when entering volume percents in air, such as in proof, to run the equation in the opposite direction for vacuum to air.

Find Mass % Air for Vol % Input	Vac Values	Air Values
Density of Sample (Vac)	950.001	948.924
Density of AA @ Calib Temp	789.239	788.131
Air Density/Vac Density	99.887%	
Air Density AA/Vac Density AA	99.960%	
AA Density Ratio/Alc Den Ratio	99.973%	
Mass % Vac Restated	33.236	
Vac Mass % * % Change = Air mass %	33.227	
Rounded to 3 Places	33.227	

General Formula

The result is the air mass % that we started with.

---

[72] From Microsoft Office Online Help: "Solver is a Microsoft Excel add-in program you can use for what-if analysis. Use Solver to find an optimal (maximum or minimum) value for a formula in one cell, called the objective cell, subject to constraints, or limits, on the values of other formula cells on a worksheet. Solver works with a group of cells, called decision variables or simply variable cells that are used in computing the formulas in the objective and constraint cells. Solver adjusts the values in the decision variable cells to satisfy the limits on constraint cells and produce the result you want for the objective cell." Frontline Systems, Inc. P.O. Box 4288 Incline Village, NV 89450-4288 (775) 831-0300 Web site: http://www.solver.comE-mail: info@solver.com Solver Help at www.solver.com.

Turning our attention to volume percent, except for the small amount of variation noted above, in the volume % regimen the mass percent does not fluctuate between air and vacuum to the degree that it does with respect to densty measurements. This is because in volume % the density does vary. When using mass percents derived from volume percents, one can substitute the density of water in air at 20°C for the vacuum value built into the General Formula and obtain the correct density at any given temperature and the result is only 8 g/m³ different.

First, we run a mass percent of 33.85% through the general formula to find very close to 950 kg/m³ using the IAT density of water in vacuum as we did above, to yield 950.001 kg/m³:

General Formula Find Density @ Temp (-20 to 40)		General Formula Find Density @ Temp (-20 to 40)	
Known Mass % (Vac)	33.850%	Known Mass % (Vac)	33.850%
Temperature ° C	15.556	Temperature ° C	15.556
Density Result at Temp	950.001	Density Result at Temp	948.924

A1		A1
998.20123	IAT Air H2O	997.12370

Then we change just the density of water to an air density in kg/m³ (997.1237) in air.

The resulting density is compared to the IAT Tables and found to be only 8 grams/m³ different:

Vol %	
Vac Mass % w/Air H2O	948.916
Vac Mass % w/Air H2O	948.924
Grams per m3 Dif.	8.0

Of course, all of this is very sensitive to the density of air used by the K&L equation as well as the modified density of the weights that are applied so that the vacuum to air and air to vacuum versions of the equation are reciprocal.

It would be possible to use the General Formula to generate the density of water at any temperature within the range of the formula (up to 40°C) and use those values rather than the HCP or some other set of values. I've used the HCP because they provide values to higher temperatures than the IAT formula will generate. There is a ~6 g/m³ difference in values between the two methods that is pretty consistent throughout the range of temperatures.

It will also be helpful if we can find a way to calculate the mass percentage at different calibration temperatures. To do that we will need an accurate fix on the volume percent at other temperatures so we will examine the Table 37b volume correction procedure next.

Test Density From 15.556°C

With this additional IAT General Formula we can continue investigating how accurately we can find densities at various temperatures. Using Table 49 as our standard we can compare Tables 7, 37a, 37b and the IAT formula to each other.

Looking at our standard transformation to 20°C we see that the results are very similar to those obtained when we calculated density previously at only 36 g/m^3 difference between the Table 37a calculation and Table 49 and 21.5 g/m^3 difference between the IAT general formula result and Table 49 values. This amounts to only 10.5% and 6.3% of the tolerance totaling 343 g/m^3 for 0.3 proof of change:

Test Density Change from 15.556 against Table 49 and IAT at Whole Mass % Values

True Proof at 15.556 C.	125.70		Find Density from SG	
Mass % @ 15.556 (Air)	54.996		0.90701	SG (Vac) at Calib. Temp Proof
Lookup Density at 15.556 (Vac)	906.12		0.999016	H2O Den. at 15.556 C. (60° F. Vac)
			906.12	SG*H20 = kg. m3 (Vac) at Calib.
Temp at which Density is to be Found	20.000	68.00	° F.	
Table 49 Value at Temp & Whole Mass %	902.58	3.54	TTB Kg per m3 Dif. than T49 Value (Start Dif.)	
Use Table 37a to Find Calibration Temp. Density Using 15.556 Mass			IAT General Formula from 15.556 Mass %	
T37a Den. at Calib	902.54		902.56	IAT Result Mass % Input
Grams Different from Table 49 Value	-36.0		-21.5	Grams Dif. T49
% of .3 proof Tol.	-10.5%		-6.3%	% of .3 proof Tol.
			From 37a Calcs Find SG and Difference	
			0.90254	Density as kg. per dm3
37b Den Using Vol. Cor. To Correct Den.	902.52		0.99821	Calib H2O Density on Ouput Side.
Grams Dif. T49	-57.6		0.90417	Den per dm3/H20 at Calib. = SG Vac
% of .3 proof Tol.	-16.8%		-0.00004	SG Dif. From T49

Find Mass per m3 Dif.	TP -.3	TP -.2	TP -.1	True Proof	TP +.1	TP +.2	TP +.3	
Proof Values	125.4	125.5	125.6	125.7	125.8	125.9	126.0	
Exact Match of Proof Row	1255	1256	1257	1258	1259	1260	1261	
Kg. m3 (Vac) at Proof	906.46	906.35	906.23	906.12	906.00	905.89	905.77	
Grams Different from Surrounding Proofs	114.0	115.0	114.0	0.0	-114.0	-115.0	-114.0	
Total Grams per .3 Proof (.0015) Change	343.0						-343.0	Grams Dif.
T37a Grams Dif. % of .3 proof Tol.	-10.5%						10.5%	37a % of Tol.
IAT Grams Dif. % of .3 Proof Tol.	-6.3%						6.28%	IAT % of Tol.
T37b Grams Dif. % of .3 Proof Tol.	-16.8%						16.79%	37b % of Tol.
Table 7 Grams Dif. % of .3 proof Tol.	-25.8%						25.8%	T7 % of Tol.

Test Density Change from 15.556

When calculating density from SG on the input side one must use the density of water at the input calibration temperature, in this case at 15.556°C. When calculating the SG on the output side it is necessary to use the water density at that temperature. It is also possible to calculate an SG with respect to a different temperature (for instance, 4°C) by using the density of water at that temperature. Many tables are now constructed against that densest water value.

The mass percent value is rounded to a whole number for use by 37a. The IAT formula will accept fractional mass percent values. It is very sensitive to the mass percent input and generally works best using vacuum-based mass percentages. In this instance, for some reason that I can't adequately explain, it was found that using air mass percentages in conjunction with the general formula produced the best results in comparison to Table 49 regardless of the calibration temperature. Since no hydrometer is in use, there is no hydrometer thermal correction and this is probably not the cause of the necessity/desirability of using air mass percentage values in this case.

Stress testing the conversion to 35°C shows that Table 49 was probably compiled using the 37a formula (the difference is only -30.3 g/m^3) but that the IAT formula also performs well at only -39.7 g/m^3 difference:

Test Density Change from 15.556 against Table 49 and IAT at Whole Mass % Values

True Proof at 15.556 C.	125.70		Find Density from SG
Mass % @ 15.556 (Air)	54.996	0.90701	SG (Vac) at Calib. Temp Proof
Lookup Density at 15.556 (Vac)	906.12	0.999016	H2O Den. at 15.556 C. (60° F. Vac)
		906.12	SG*H2O = kg. m3 (Vac) at Calib.
Temp at which Density is to be Found	35.000	95.00	° F.
Table 49 Value at Temp & Whole Mass %	890.16	15.96	TTB Kg per m3 Dif. than T49 Value (Start Dif.)
Use Table 37a to Find Calibration Temp. Density Using 15.556 Mass		IAT General Formula from 15.556 Mass %	
T37a Den. at Calib	890.13	890.12	IAT Result Mass % Input
Grams Different from Table 49 Value	-30.3	-39.7	Grams Dif. T49
% of .3 proof Tol.	-8.8%	-11.6%	% of .3 proof Tol.
		From 37a Calcs Find SG and Difference	
		0.89013	Density as kg. per dm3
37b Den Using Vol. Cor. To Correct Den.	890.05	0.99404	Calib H2O Density on Ouput Side.
Grams Dif. T49	-113.9	0.89547	Den per dm3/H2O at Calib. = SG Vac
% of .3 proof Tol.	-33.2%	-0.00003	SG Dif. From T49

Test Density Change from 15.556

Now that we know that we can accurately predict density change with temperature, the question becomes, can we find volume percent after we have changed the density to another calibration temperature.

Find Volume Percent

The first feature of the procedure we need to look at is the difference in liters generated if one uses the same assumed mass of 1,000 kg. for both the air and vacuum mass inputs. Because of the difference in density per unit of volume between air and vacuum regimens this assumption leads to the calculation of different volumes: in this case 1.33 liters fewer in total with respect to the vacuum liters.

The problem is that the air liters calculated, while matching in both density and wine gallons per pound calculations, are the actual liters that are being employed based upon an assumed mass of 1,000 kg. But, if the mass is actually weighed on a scale *and the atmospheric buoyancy is not taken into account*, the volume will not be based upon actual IPK kilograms. This air and vacuum volume difference along with the mass to volume difference will flow down to the calculation of volume percent and we can avoid that by making correct assumptions.

Of course, the TTB tables are in air so when using TTB table values like wine gallons per pound to calculate volume one will always calculate atmospheric volumes, based upon atmospheric densities. This is fine if one is blending by volume, but not when one is blending by mass. The same problem also exists when calculating proof gallons using proof gallons per pound and this is the real issue because it affects taxation and company income. One part out of a thousand is worth investigating and keeping track of:

Find Density and Vol % at Temp, Using 15.556 Values				30.000	Destination ° C.		
Proof Input	125.70			1,000.000	Kilograms	Liters Air at 60° F. via WGPP	
Closest Mass % of Proof Input (Air)	54.996			1.10698	Kg. 7/8 Air	1,000.00	Kg. for TTB
				1,000.00	Kg. for Air Input	2,204.62	Lbs.
		Vac	Air			0.13240	WG per Lb. Air
Kg. per m3 Vac at 60°	906.12	Liter Dif.	905.03	Kg. per m3 Air		291.89	WGPP * Lbs. = Gal.
Mass/Density = dm3 Volume @ 60° F.	1,103.61	1.33	1,104.94	Liter Volume Air		1,104.94	Liters at 60° F.
Vac Mass % at 60° F. (15.556 C.)	55.007		54.996	Air Mass % at 60° F.		-1.96	milliliters Dif.
Volume % Vac. At 15.556	62.851%		62.850%	Vol % Air			

Find Vol % from 15.556

Note that even if one was using a calibrated scale, one that has been set to zero after compensating for the atmospheric buoyancy, the assumption that both air and vacuum mass are the same will retain the difference in calculated liters based upon the vacuum density.

The volume percent issue can be addressed by assigning the correct air or vacuum mass percent to the calculation to arrive at essentially identical volume percents.

The mass to volume issue can be addressed by subtracting the atmospheric buoyancy so that the indicated mass on a scale is reduced to the true vacuum mass. For reasons I will explain, that value is about 7/8 of the standard atmospheric density amounting to 1.06982 kg/m^3. By subtracting this value in kilograms from the input mass, the volume values become essentially the same at only 0.11 liters difference (110 ml) per meter cubed. We have also adjusted the kilogram input which applies to the TTB calculations as well:

Find Density and Vol % at Temp, Using 15.556 Values				30.000	Destination ° C.		
Proof Input	125.70			1,000.000	Kilograms	Liters Air at 60° F. via WGPP	
Closest Mass % of Proof Input (Air)	54.996			1.10698	Kg. 7/8 Air	998.89	Kg. for TTB
				998.89	Kg. for Air Input	2,202.18	Lbs.
		Vac			Air	0.13240	WG per Lb. Air
Kg. per m3 Vac at 60°	906.12	Liter Dif.		905.03	Kg. per m3 Air	291.57	WGPP * Lbs. = Gal.
Mass/Density = dm3 Volume @ 60° F.	1,103.61	0.11		1,103.72	Liter Volume Air	1,103.72	Liters at 60° F.
Vac Mass % at 60° F. (15.556 C.)	55.007			54.996	Air Mass % at 60° F.	-1.96	milliliters Dif.
Volume % Vac. At 15.556	62.851%			62.850%	Vol % Air		

Find Vol % from 15.556

As a practical matter, if one weighs out 1,000 kilograms in air—and assumes it is a true 1,000 kg—one will always have, depending on atmospheric conditions, only 998.89 kilograms on the scale and that volume will in fact be the vacuum volume of ~1,103.7 liters.

The converse solution is to load the scale until the indicated mass includes the atmospheric buoyancy and to subtract that buoyancy:

Find Density and Vol % at Temp, Using 15.556 Values				30.000	Destination ° C.
Proof Input	125.70			1,001.107	Kilograms
Closest Mass % of Proof Input (Air)	54.996			1.10698	Kg. 7/8 Air
				1,000.00	Kg. for Air Input
		Vac			Air
Kg. per m3 Vac at 60°	906.12	Liter Dif.		905.03	Kg. per m3 Air
Mass/Density = dm3 Volume @ 60° F.	1,104.83	0.11		1,104.94	Liter Volume Air
Vac Mass % at 60° F. (15.556 C.)	55.007			54.996	Air Mass % at 60° F.
Volume % Vac. At 15.556	62.851%			62.850%	Vol % Air

Note that if a true 1,000 kilograms are measured the volume will indeed be 1,104.94 liters but in order to achieve this volume using vacuum densities one will be required to add the air buoyancy to the measured kilograms.

Returning to the question as to whether we can change the temperature and find the density and volume percentage we will again start with our standard 55 % mass percentage alcohol:

Find Density and Vol % at Temp, Using 15.556 Values			30.000	Destination ° C.
Proof Input	125.70		1,000.000	Kilograms
Closest Mass % of Proof Input (Air)	54.996		1.10698	Kg. 7/8 Air
			998.89	Kg. for Air Input
	Vac		Air	
Kg. per m3 Vac at 60°	906.12	Liter Dif.	905.03	Kg. per m3 Air
Mass/Density = dm3 Volume @ 60° F.	1,103.61	0.11	1,103.72	Liter Volume Air
Vac Mass % at 60° F. (15.556 C.)	55.007		54.996	Air Mass % at 60° F.
Volume % Vac. At 15.556	62.851%		62.850%	Vol % Air
Find 3b Vol % & Proof at 15.556	Vac		Air	
Restate Density Value at 15.556	906.12		905.03	
AA Den at 60° F.	793.032		791.925	
Ratio	1.143		1.143	
Mass % at 15.556	55.007%		54.996%	
Ratio * Mass % = Vol % @ Temp	62.851%		62.850%	
Vol % *2 * 100 = Proof at 15.556	125.70	Proof	125.70	
	Vac		Air	
37a Density at Destintation Temp	894.33		893.24	
Mass/Density = Calc Liter Vol. at Temp	1,118.15	Liter Dif.	1,118.28	Volume at Temp
37b Liter Volume at Destination Temp	1,118.08	0.11	1,118.19	Vol. at Temp °C
37b Liter Dif.	-0.07		-0.09	Liter Dif.
Vol. % at Calib. ° Using Mass % @ 15.556	63.017%		63.017%	
Find Vol % Using Den. of AA at Calib Temp	Vac		Air	
Restate Density Value at Calib °	894.33		893.24	
AA Den at Calibration Temp.	780.65		779.54	
Ratio	1.146		1.146	
Mass % at 15.556	55.007%		54.996%	
Ratio * Mass % = Vol % @ Temp	63.017%		63.017%	
Vol % *2 * 100 = Find Proof at Temp	126.03	Wrong!		
Index Proof closest to Calib Den.	135.70	Actual "Proof" based upon Density at Temp		
Index Mass % of Calib Den.	60.16	Mass % at Calib Temp on Vol % Basis		
Vac Density on a Volume % Basis	882.37	Second Run of T37a equation		

Find Vol % from 15.556

Starting with density only and using the density of absolute alcohol at 15.556° along with the IAT Table 3b method of finding volume percentage shows that we can reproduce the starting true proof at 125.70 proof at 15.556°. We also note that for volume percent there is essentially no difference in the calculated air or vacuum values since neither volume percent or proof have this as an attribute (they are dimensionless as to air and vacuum). Subsequently we reproduce the same procedure for finding volume percentage at the calibration temperature (30°C) using the density of absolute alcohol at 30°C along with the air and vacuum mass percent values we have calculated. While the volume percent values work out properly, we can no longer divide volume percent by 2 and multiply by 100 to obtain the "proof" at the calibration temperature. Because of the temperature increase the density has decreased by ~12 kg/m^3 from 906 down to 894 and consequently the corresponding "proof" to the density at 30°C has increased from 125.7 to 135.7. One interesting feature of the calculation is that, because proof is a species of volume percent it is possible to find the mass percent on a volume-percent basis by indexing the mass percent of 135.7 proof. This value of 60.17 is accurate at least as an initial starting point to investigate further. We have also been able to find the density on a volume percent basis which I will explain shortly. What we haven't been able to do is to find the mass percent on a density basis at the destination temperature of 30°C and we will come back to this problem later.

One additional computation that can be made with this same information is to calculate the grams per 100 cc of both alcohol and water based upon the density and mass percent at 15.556°:

Find Grams per 100 CC at 15.556	Vac		Air	
Vac Kg. per Liter	0.90612		0.90503	Air Kg. per Liter
Alc Mass %	55.007%		54.996%	Alc Mass %
(Alc % * Kg. per liter)*100 = Grams per 100 CC	49.84		49.77	(% Alc * Kg. per liter)*100 = Grams per 100 CC
Grams per 100 CC*10 = Grams per liter (Kg m3)	498.42		497.72	Grams per 100 CC*10 = Grams per liter (Kg m3)
Alc Kg. per Liter * 1000 =Vac Kg. per m3 of Alc.	906.12		905.03	Air Kg. per m3 of Alc.
Kg. per m3 Alc -Alc Kg = Kg/m3 H2O	407.69		407.30	Kg. per m3 Alc -Alc Kg = Kg/m3 H2O
Grams per 100 CC of H2O	40.77		40.73	Grams per 100 CC of H2O

The point to note is that once mass percent, on a density basis, is known for a particular calibration temperature, it never changes. The volume of alcohol and water change at different rates and accordingly the volume percent of alcohol is a different ratio to the whole volume based on the temperature of the entire mixture. While we can find the density at different calibration temperatures and the volume percent, finding the mass percentage is a much harder matter.

One means of confirming the volume percent at the calibration temperature, while still using essentially a version of the mass percentage at 15.556°, is to use the parts alcohol and parts water values from Table 6 and then calculate the change in volume for each component to arrive at a composite volume. I call this a method of finding the "volume sharing" between alcohol and water. It is not as accurate as the 3b method but gives some idea of what is going on "under the hood" when alcohol and water mixtures change temperature:

Find Volume Sharing	Vac/Vac & 60/60		Air /Air & 60/60	
Kilograms	1,001.11		1,000.00	
% Ethanol of Alc.	55.007%		54.996%	% Ethanol of Alc.
Kg. * Alc % Ethanol = Kg. Ethanol	550.68		549.35	Kg. * Alc % Ethanol = Kg. Ethanol
AA Kg. m3 Vac at Calib °	780.65		779.54	AA Kg. m3 Air at Calib °
Alc 1 Mass/Mass per m3 = m3 at Temp	0.705		0.705	Alc 1 Mass/Mass per m3 = m3 at Temp
m3*1000 = liters of Absolute Alc. at temp	705.41		704.71	m3*1000 = liters of Absolute Alc. at temp
Kg. Mass - Alc Mass = Mass H2O	450.43		450.65	Kg. Mass - Alc Mass = Mass H2O
H2O Mass per m3 at Temp	995.65		994.57	H2O Mass per m3 at Temp
H2O Mass/Mass per m3 = m3 H2O at Temp	0.452		0.453	H2O Mass/Mass per m3 = m3 H2O at Temp
m3 * 100 = Liters of H2O at Temp	452.40		453.11	m3 * 100 = Liters of H2O at Temp
Alc Liters + H2O Liters = Total Liters Mixed	1,157.81		1,157.82	Total Liters Mixed Together
Find Volume Sharing %				
Parts Alcohol At True Proof (15.556)	62.85			
Parts Water	40.81		30.00	Destination ° C.
Parts Total	103.66			
Total Parts Dif. from 100 = Shared Volume	3.66			
Vol. Share/Total Parts = Vol. Share %	3.53%			
Total Liters * Share % = Liters of Sharing	40.88		40.88	Liters of Sharing
Total Liters - Sharing Liters = Calc Liters	1,116.93		1,116.94	Total Liters - Sharing Liters = Calc Liters
37b Volume Corrected Liters Restated	1,118.19		1,118.19	37b Volume Corrected Liters Restated
37b Liters Dif. From Sharing Liters	1.26		1.25	37b Liters Dif. From Sharing Liters

Find Vol % from 15.556

In this case we need to assume 1,000 kilograms in air and add the air buoyancy to the vacuum input in order to obtain the best results for both air and vacuum. With these assumed masses and the density of absolute alcohol at the destination temperature and the mass percent at 15.556° one can find the liter volume of that absolute alcohol at the destination temperature. Doing the same thing using the density of water and the mass of water based upon the mass percent of alcohol one finds the volume of water at the destination temperature as well. Summing these values provides the total volume of each component as if they were separated. Then using the parts alcohol and parts water for the mass percent of the alcohol at the true proof at 15.556° (in air, since we needed air kilograms for the primary input) one finds the difference between the total parts and 100; this is the "shared volume of the alcohol and water on a parts basis. This value is then expressed as a percentage. Then one multiplies the total liters by the sharing percentage to find the total liters of shared volume and subtracts it from the total to arrive at the net liters at the calibration temperature. Finally, we can compare our results to a Table 37b volume correction and determine that we are within ~1.2 liters of what that calculation yielded as a result. This is just barely within tolerance but is sufficiently informative to report that one can dis-assemble alcohol and water in this manner yet it is not accurate enough to base volume correction or volume percent findings upon.

Finding Density and Mass Percent

Continuing with our previous example that begins at 127.5 proof and finds the density and "proof" at 30° C, it is possible to take that apparent proof (in this case, 135.7), and the apparent density of 894.37 kg/m^3 and in conjunction with the apparent mass percent, 60.152, employ the Table 37a density correction formula to alter the temperature again from 15.556° to 30°C in order to find the resulting density at that temperature. In this case that density is 882.37 on a vacuum density basis. This density is the correct density of alcohol with a mass percent of 60.152 percent at a calibration temperature of 30°C. The procedure operates in both air and vacuum and also finds the mass of air between the two regimens:

Find Density on a Volume % Basis					
Table 37a Density Module					
Calibration Temperature Proof	135.70		30.000	Destination °C.	
Exact Match Proof Row	1358		60.152	Mass %	
15.556 ° C. Kg. per m3 Vac	894.37		893.28	Kg. per m3 Air	
37a Formula to Find Density from 60° F.					
D t/4 = D25/4 + α(t -25) + β(t -25)2 + γ(t -25)3					
Start Temp = Table Temp ° C.	15.556		25.00	Formula Start Temp	
Formula Destination Temp ° C.	25.00		30.00	Destination ° C	
Alpha	-835.90		-835.90	Alpha	
Table Temp - Formula Temp = Temp Dif.	-9.44		5.00		
Alpha * Temp Dif. = Result	7,894.24		-4,179.50		
Beta	-1.21		-1.21	Beta	
Temp Dif. Squared	89.19		25.00		
Beta * Temp Dif. Squared	-107.92		-30.25		
Gamma	-0.004		-0.004	Gamma	
Temp Dif. Cubed	-842.30		125.00		
Gamma * Temp Dif. Cubed	3.36921		-0.5000		
Alpha + Beta + Gamma Results =	7,789.69		-4,210.25		
1 to the minus 6	1.0E-06		1.0E-06		
Reduce Result by Value = Net Change	0.00779		-0.00421		
Find Density Values at Temp	Vac		Air		
Net Change Restated	0.00779		0.00779		
60° F. Density Kg. per Liter (Vac)	0.89437		0.89328	Air Kg dm3	
60° Den - Formula Change = 25° Density	0.88658		0.88549		
25° to Destination ° C. Value	-0.00421		-0.00421		
25° Den + Change = Target ° C. Kg dm3	0.88237	Air Kg.	0.88128		
Kg. per m3 Vac Unrounded	882.37	1.0930	881.28	Kg. m3 Air	

Find Vol % from 15.556

End Volume 1

This concludes the first volume of this work which has focused mainly on TTB blending procedures and an introduction to the general formula and IAT tables. It should be sufficient for most distillers, who use the TTB system, to understand and conduct accurate gauging and blending operations.

The second volume will examine a roughly equal number of advanced topics and conduct a full examination of the IAT tables and the procedures for building blending applying them at various other calibration temperatures using air or vacuum densities.

Notes

Constants and Converters

The following identify some of the constants and converters that have been used in this book along with the derived values that are also employed.

Constants	Standard Values	ABS Derived Values	
Liters per US Gallon	3.78541180	3.785411784	Liters per US Gal. ABS
Gallons per Liter	0.264172050	0.264172052	Gallons per Liter ABS
Pounds Avdp to Kilograms	0.453592370	61.02374400	Cubic Inches per Liter HCP 72nd Post 1964
Kilograms to Pounds Avdp	2.20462260	61.02374410	Cubic Inches per Liter ABS Derived Value
Gallon = U.S. Gal. of 231 Cubic In.	231.00	16.387064	Cubic Inches to Cubic CC (mm)
Fluid Ounces to Liters	0.02957350	0.016387064	Cubic Inches to dm3
Fluid Ounces to Milliliters (CC's)	29.573530	0.061023744	Cubic Centimeters to Cubic Inches
		ABS Derived Value	
Wt. Oz. Avoirdupois To Grams	28.3495230	7.48051948	Gallons per Cubic Foot
Grams to Weight Ounces Converter	0.0352739620	7.48051950	Std. Value Gallons per Cubic Foot
Grams per Lb.	453.592370	3,785.4118	cc's per US Gallon

Constants

Water Values in Air at 15.556	Lbs. per US Gal.	Gal. per Lb.		
HCP 72nd Lbs. per Gal. Air 15.556	8.328310	0.12007238	Lbs. per US Gallon H2O @ 15.556 (Vacuum)	
Lbs. per Gal § 30.66 Old §186.66 & Tyco's	8.328230	0.12007353	8.337266	HCP 72nd Lbs. per Gal. Vac 15.556
HCP 89th & K& L Eq. Vac to Air (ABS Std.)	8.328201	0.12007395	8.337220	Tycos Tables Value
Lbs. per Wine Galon §30.41	8.328200	0.12007400	8.337192	ABS K&L Value

Kg m3 Air Density of H2O @ 15.556	Air		Kg m3 Vac	
HCP 89th & K& L Eq. Vac to Air (ABS Std.)	997.93857		999.0159708	Vac H2O Density @ 15.556 HCP 89th
Lbs. per WG §30.41 & Tycos Tables	997.93856	-0.01 g m3 Dif.	999.0101372	General Formula H2O Density @ 15.556
Den 60° F. § 30.66 Old §186.66 & T5	997.94204	3.5 g m3 Dif.	-5.834	HCP Grams per m3 Dif.

20° C. Water Values	Vac		Vac	
			998.230	BOS @ 20
IAT Coefficient is Water Density at 20° C.	998.20123		0.999972	Factor
HCP 89th Water at 20° C.	998.20630		998.20205	BOS * Factor = Den.
HCP Grams per m3 Dif.	5.070		4.3	HCP Grams per m3 Dif.

	Vac	Air
Absolute Alcohol at 20 C kg m3	789.239	788.131

Kilograms per Proof Gallon	Kg AA per PG Air		Kg AA per PG Vac
	1.498900	Rounded 4	1.501000
Moles per Proof Gal.	Moles per Proof Gal. Air		Moles per Proof Gal. Vac
	32.535270	Rounded 5	32.580850

	Vac	Air
Absolute Alcohol at 15.556 kg m3	793.0320	791.925
GF Raw Result AA at 15.556	793.0321462	

Standard Atmosphere Values		Pressure Converters	
Kg. per Liter 100 % Std. Atm. Full Air	0.00122265	0.039370079	mm Hg to In Hg
1/8 Air Kg. per dm3	0.00015283	0.019336721	mm Hg to PSI
1/8 Air grams per m3	152.83125	33.863788	mm Hg to PSI
Kg. per Liter 87.5 % Std. Atm. (7/8 Air)	0.00106982	0.75006376	millibars to mm of HG
Kg. per m3 87.5 % Std. Atm. (7/8 Air)	1.06982	0.029530069	millibars to inches of HG
Std. Atm. 7/8 Lb. per Gal. @60° F.	0.008928090	0.49115272	Inches of HG to PSI
		0.014503774	Milibars to PSI

Some older values before the adoption of the modern SI definitions.

Tycos Tables Values 1901-1964 Stds.	
Gal. per Liter Tycos	0.2641780
Liters per Gal. Tycos	3.785332
Pounds Avdp to Kilograms	0.45359243
Kilograms to Pounds Avdp Tycos	2.204622341
Grains per kilogram Tycos	15,432.356
Grams per Kilogram Tycos	1,000.00
Pounds Troy per Kilogram Tycos	2.26792285
Cubic Inches per Liter Tycos	61.025
Cubic Centimeters per Liter Tycos	1,000.027
milliliters per liter (CC's)	1,000.00
Kilograms of Water per Gallon Tycos	3.777623
Cubic Inches per Gallon Tycos	231
Imperial Gallons per US Gal	0.8237
Cubic Inches per Imperial Gal.	277.274
Liters per Imperial Gal.	4.54346
US Gal per Imperial Gal.	1.20032
Lbs. H20 per Imperial Gal. @ 60° F.	9.996

Some representative volume conversion values.

While we are on the topic of unit conversions the following are the volume converters used by the program.

Volume Translation from US Gallons to Other Units					
Gallon Input	1,000.000				
Converters from US Gal. to Unit	To Unit	Results		To US Gal	US Gallons
m3	0.0037854118	3.785	Cubic Meters	264.17205	1,000.00
dm3 (= liter)	3.78541180	3,785.41	Liters	0.26417205	1,000.00
Imp. Gal.	0.832674180	832.67	Imperial Gal.	1.2009499	1,000.00
yd3	0.0049511317	4.951	Cubic Yards	201.97403	1,000.00
ft3	0.1336805600	133.681	Cubic Feet	7.4805195	1,000.00
in3	231.0	231,000.0	Cubic Inches	0.004329	1,000.00
Pint UK	6.6613935	6,661.39	Pints UK	0.15011874	1,000.00
Pint US	8.00	8,000.00	Pints US	0.125	1,000.00
Barrel (Oil)	0.023809524	23.81	Barrels Oil	42.0	1,000.00
US fl. Oz.	128.00	128,000.00	Fluid Oz. US	0.0078125	1,000.00
UK fl. Oz.	133.22787	133,227.87	Fluid Oz. UK	0.007505937	1,000.00

From Liters to Other Units					
Liters to Unit	1,000.000				
Converters from Liters to Unit	To Unit	Results		To Liters	Liters
Liters to US Gallons	0.264172	264.172	US Gallons	3.7854118	1,000.00
Liters to UK Gal.	0.219969	219.969	Gal. UK	4.54609	1,000.00
Liters to ft3	0.035315	35.315	Ft. Cubed	28.316847	1,000.00
Liters to yd3	0.001308	1.308	yd3	764.55486	1,000.00
Liters to Barrel (Oil)	0.006290	6.290	Barrel (Oil)	158.98729	1,000.00
Liters to m3	0.001	1.000	m3	1,000.0	1,000.00
Liters to US fl. Oz.	33.814023	33,814.0	US fl. Oz.	0.02957353	1,000.00

Constants

Tables

Introduction

These tables are published to document the values compiled and calculated for use by the ABS program and discussed in this book on Alcohol Blending and Accounting.

The tables are published because without a verifiable source, authentication and transparency no one else would really be able to independently determine the accuracy of the entire data set or to make improvements or correct errors.

The IAT calibration temperature tables are 200-place abridgements of 8,000-place tables that actually serve as the reference tool for ABS calculations. The tables are set forth so that individuals that want to reproduce or improve upon them, using the procedures outlined in the text, will have a sufficient number of data points to compare to if they decide to do so.

Several TTB and IAT tables have been omitted. The most useful of these tables are TTB Table 1 which is just too large to republish and is readily available online.

Web address of TTB Table 1: https://www.ttb.gov/foia/Gauging_Manual_Tables/Table_1.pdf

I have taken the step of publishing an abridged table of mass percents and densities at 15.556°C at 2,000 places from which the remaining TTB values can be derived.

The other tables are short enough and add sufficiently to the usefulness of the work that they are produced or reproduced in this volume.

This volume has also served as a convenient place to archive the current TTB gauging manual regulations and various other information that may change in the future or no longer be available.

Table 37a Density

Bulletin of the Bureau of Standards Volume 9 Page 402.

% by Wt.	Alpha	Beta	Gamma	% by Wt.	Alpha	Beta	Gamma
0	-2565	-484	32	50	-8035	-128	-2
1	-2574	-490	32	51	-8074	-127	-3
2	-2591	-496	32	52	-8111	-126	-3
3	-2613	-499	31	53	-8147	-126	-3
4	-2646	-501	31	54	-8181	-125	-3
5	-2689	-502	30	55	-8212	-124	-3
6	-2745	-501	30	56	-8244	-124	-3
7	-2816	-498	29	57	-8274	-123	-4
8	-2901	-495	29	58	-8302	-122	-4
9	-3003	-490	28	59	-8330	-122	-4
10	-3121	-484	27	60	-8359	-121	-4
11	-3244	-476	26	61	-8384	-121	-4
12	-3374	-469	25	62	-8410	-120	-4
13	-3513	-460	24	63	-8435	-120	-4
14	-3662	-452	23	64	-8459	-120	-4
15	-3817	-442	22	65	-8482	-119	-4
16	-3978	-432	20	66	-8503	-119	-4
17	-4146	-420	19	67	-8524	-119	-4
18	-4322	-406	18	68	-8544	-118	-4
19	-4504	-395	16	69	-8564	-117	-4
20	-4686	-380	14	70	-8581	-117	-5
21	-4870	-363	13	71	-8599	-116	-5
22	-5055	-346	11	72	-8614	-116	-5
23	-5239	-329	10	73	-8629	-115	-5
24	-5419	-313	9	74	-8643	-114	-5
25	-5601	-298	8	75	-8657	-113	-5
26	-5778	-282	7	76	-8669	-112	-5
27	-5951	-268	6	77	-8681	-111	-5
28	-6114	-253	6	78	-8692	-110	-5
29	-6271	-240	5	79	-8703	-109	-5
30	-6419	-227	4	80	-8714	-108	-5
31	-6554	-215	4	81	-8723	-107	-5
32	-6685	-204	3	82	-8731	-106	-5
33	-6810	-194	3	83	-8739	-105	-5
34	-6929	-185	2	84	-8745	-104	-5
35	-7040	-170	2	85	-8751	-102	-5
36	-7144	-168	1	86	-8753	-101	-5
37	-7239	-161	1	87	-8754	-99	-5
38	-7330	-155	1	88	-8753	-97	-5
39	-7413	-150	0	89	-8751	-95	-5
40	-7489	-145	0	90	-8746	-93	-5
41	-7561	-141	0	91	-8738	-90	-5
42	-7627	-138	-1	92	-8728	-88	-5
43	-7689	-136	-1	93	-8715	-84	-5
44	-7748	-134	-1	94	-8700	-81	-5
45	-7802	-133	-1	95	-8685	-78	-5
46	-7855	-132	-2	96	-8668	-74	-5
47	-7903	-131	-2	97	-8650	-70	-5
48	-7950	-130	-2	98	-8632	-66	-5
49	-7993	-129	-2	99	-8613	-61	-5
50	-8035	-128	-2	100	-8591	-56	-5

The method to operate the Least Squares Reduction using the Table 37a values:

Start Density D 25/4	0.92162	Destination		Table 49 Lookup
T = Observed Temperature ° C.	10.00	40.000		Alpha, Beta, Gamma Lookups
Formula Start Temp 25° C.	25.0	25.0	Fixed #	50.000	Mass % Input
Temp - Formula Start Temp t-25 =	-15.00	15.00		50.00	Round % Mass Whole
Alpha	-803.5	-803.50		51	Exact Match % Row
D 25/4 + Alpha * t-25 Result	12,052.5	-12,052.5		-803.50	Alpha/10
Beta	-1.28	-1.28		-1.2800	Beta/100
t-25 Squared	225.00	225.00		-0.0020	Gamma/1,000
Beta * t-25 Squared	-288.00	-288.00			
Gamma	-0.00200	-0.0020			
t-25 Cubed	-3,375.00	3,375.00			
Gamma * t-25 Cubed	6.75000	-6.75			
Alpha + Beta + Gamma Results =	11,771.3	-12,347.3			
The value of 1 to the minus 6	1.0E-06	1.0E-06			
Multiply Result by 1 to minus 6 = Den Change	0.011771	-0.01235			
Find Density Values at Temp	Vac				
Observed Temp to 25° C. Value Restated	0.011771				
Density Kg. per Liter Restated (Vac)	0.921620				
Density - Formula Change = 25° Density	0.90985				
25° to Destination ° C. Value	-0.01235				
25° Den + Formula Change = Destination Den.	0.89750	0.0015	Tolerence		
Kg. per m3 Vac	897.502	1.35	Kg per m3 Tolerance		
Table 49 Value	897.50	1.50	Grams per m3 Dif.		
Kg. per m3 Difference from Table 49 Value	0.00150	0.11%	% of Tolerance		

Table 49 Values

Temperature ° C.	10	15	15.556	20	25	30	35	40
BOS Alc. Density D t/4	0.92162	0.91776	0.91733	0.91384	0.90985	0.90580	0.90168	0.89750
Kg m3 Vac	921.62	917.76	917.33	913.84	909.85	905.80	901.68	897.50

Table 37a Density

Table 37b Volume

Bulletin of the Bureau of Standards Volume 9 Page 403

% by Wt.	A	B	C	% by Wt.	A	B	C
0	257.2	49.2	-2.9	50	883.1	21.9	0.5
1	259.0	50.0	-3.0	51	890.0	22.0	1.0
2	261.0	51.0	-3.0	52	896.0	22.0	1.0
3	264.0	51.0	-3.0	53	902.0	22.0	1.0
4	267.0	51.0	-3.0	54	906.0	22.0	1.0
5	272.1	51.5	-2.7	55	914.0	22.0	1.0
6	278.0	52.0	-3.0	56	920.0	22.0	1.0
7	286.0	51.0	-3.0	57	926.0	22.0	1.0
8	295.0	51.0	-3.0	58	931.0	22.0	1.0
9	306.0	51.0	-2.0	59	937.0	22.0	1.0
10	318.3	50.4	-2.4	60	942.4	22.5	0.7
11	331.0	50.0	-2.0	61	948.0	23.0	1.0
12	345.0	49.0	-2.0	62	953.0	23.0	1.0
13	360.0	48.0	-2.0	63	958.0	23.0	1.0
14	376.0	48.0	-2.0	64	964.0	23.0	1.0
15	392.0	47.0	-2.0	65	969.0	23.0	1.0
16	409.0	46.0	-2.0	66	974.0	23.0	1.0
17	427.0	45.0	-2.0	67	979.0	23.0	1.0
18	446.0	44.0	-1.0	68	964.0	23.0	1.0
19	465.4	43.0	-1.2	69	969.0	23.0	1.0
20	485.0	42.0	-1.0	70	993.9	23.4	0.9
21	505.0	40.0	-1.0	71	999.0	24.0	1.0
22	525.0	39.0	-1.0	72	1,003.0	24.0	1.0
23	544.6	37.2	-0.6	73	1,006.0	24.0	1.0
24	564.0	36.0	0.0	74	1,012.0	24.0	1.0
25	584.0	34.0	0.0	75	1,017.0	24.0	1.0
26	604.0	33.0	0.0	76	1,021.0	24.0	1.0
27	623.0	32.0	0.0	77	1,026.0	24.0	1.0
28	641.0	31.0	0.0	78	1,030.0	24.0	1.0
29	658.0	30.0	0.0	79	1,034.0	24.0	1.0
30	675.2	28.3	-0.1	80	1,038.5	23.7	0.9
31	691.0	27.0	0.0	81	1,043.0	24.0	1.0
32	706.0	26.0	0.0	82	1,047.0	24.0	1.0
33	720.0	26.0	0.0	83	1,051.0	24.0	1.0
34	734.0	25.0	0.0	84	1,055.0	24.0	1.0
35	748.0	24.0	0.0	85	1,059.0	24.0	1.0
36	760.0	24.0	0.0	86	1,062.0	23.0	1.0
37	772.0	23.0	0.0	87	1,066.0	23.0	1.0
38	784.0	23.0	0.0	88	1,069.0	23.0	1.0
39	794.0	22.0	0.0	89	1,072.0	23.0	1.0
40	804.0	22.0	0.3	90	1,074.9	23.0	0.9
41	814.0	22.0	0.0	91	1,078.0	23.0	1.0
42	822.0	22.0	0.0	92	1,080.0	23.0	1.0
43	831.0	22.0	0.0	93	1,082.0	22.0	1.0
44	839.0	22.0	0.0	94	1,064.0	22.0	1.0
45	847.0	22.0	0.0	95	1,066.0	22.0	1.0
46	855.0	22.0	0.0	96	1,088.0	21.0	1.0
47	862.0	22.0	0.0	97	1,069.0	21.0	1.0
48	870.0	22.0	0.0	98	1,091.0	20.0	1.0
49	876.0	22.0	0.0	99	1,093.0	20.0	1.0
50	883.1	21.9	0.5	100	1,094.3	19.1	0.8

The method to operate the Least Squares Reduction using the Table 37b values:

Present Volume (Any Unit)	1,000.00	125.70	True Proof
Temperature of Liquid ° C.	30.000	1258	Match Proof Row
Destination Temperature ° C.	30.000	55.004	Mass %
37b Two Column Vol to Vol Result	999.98		

Table 7 Vol. Result	999.83	0.0015	Tolerance
Liters Dif. From 37b	-0.15	1.50	Liter Tol.
% of Tol.	-9.8%		

Table 37b Vol. Cor. Two Col. Vol to Vol	Start °		Destination °	
Temperature of Liquid ° C.	30.0000		30.000	Destination °
Calibration Temp of BOS Work	25.000		25.000	BOS Temp
Start ° C. - BOS Calib ° = Temp Dif.	5.000		5.00	Destination - BOS
A	91.40		91.40	
A * Temp Dif. = Result	457.00		457.00	
Temp Dif. Squared	25.00		25.00	
B	0.2200		0.2200	
B * Temp Dif. Squared	5.500		5.500	
Temp Dif. Cubed	125.00		125.00	
C	0.00100		0.0010	
C * Temp Dif. Cubed	0.12500		0.12500	
A + B + C Results =	462.6250		462.63	
1 to the minus 5	1.0E-05		1.0E-05	
Reduce by 1 to the -5 to Output Value	0.004626		0.004626	
The Value of One	1.0		1.0	
1 Minus Correction Value = Net Factor	0.99537		1.00463	One Plus Factor
Present Liters Restated	1,000.00		995.37	25° C Vol.
Net Factor * Present Liters = 25° C. Vol.	995.37		999.98	Vol. @ Target
Lookup Table 37b Values				
Round % Mass Whole	55.00			
Match % Wt. Column	56			
Value A/10	91.40			
Value B/100	0.220			
Value C/1000	0.00100			

Table 37 B Volume

The formula is designed to go to or from 25°C. By chaining the formula to go to that temperature and then to a destination temperature, it operates in any predictive mode desired. It suffers from a limitation, as shown above: when the start temperature and ending temperature are the same, the result is not equivalent to the volume input, in this case by 0.02 liter (20 milliliters) out of 1,000 liters since the input is 1,000 liters but the output above is only 999.98 liters. However, the Table 7 value is on the order of 150 milliliters off in the same translation.

BOS Table 49

Alcoholometric Density Table XLIX (Table 49) Density of mixtures of Ethyl Alcohol and Water D t/4. From the Bulletin of the Bureau of Standards Volume 9 Page 424. The original published values, have been treated as SGs (vacuum) and converted to densities by multiplying by the factor of 0.999972. See Volume 2 for a full discussion of the procedures and rationale. The values at 15.556°C have been added to the original using the IAT General Formula.

Mass %	10° C	15° C	15.556	20° C	25° C	30° C	35° C	40° C	Closest Proof Mass %	Proof at Closest Mass %
0	999.702	999.102	999.040	998.202	997.052	995.652	994.032	992.222	0.000	0.00
1	997.822	997.222	997.170	996.332	995.172	993.762	992.142	990.312	0.995	2.50
2	995.992	995.392	995.340	994.502	993.332	991.912	990.282	988.432	1.993	5.00
3	994.232	993.622	993.570	992.722	991.542	990.112	988.462	986.602	2.994	7.50
4	990.282	991.922	991.870	991.002	989.812	988.362	986.692	984.822	3.996	10.00
5	992.552	990.292	990.240	989.352	988.142	986.672	984.982	983.082	5.004	12.50
6	989.432	988.742	988.680	987.772	986.532	985.042	983.322	981.393	6.015	15.00
7	987.982	987.262	987.190	986.242	984.972	983.442	981.693	979.723	7.030	17.50
8	986.572	985.812	985.730	984.752	983.432	981.863	980.063	978.053	8.004	19.90
9	985.212	984.392	984.310	983.282	981.903	980.283	978.433	976.383	9.022	22.40
10	983.902	983.012	982.920	981.843	980.403	978.723	976.823	974.723	10.004	24.80
11	982.642	981.683	981.590	980.443	978.943	977.203	975.243	973.093	11.028	27.30
12	981.423	980.383	980.280	979.073	977.503	975.703	973.683	971.473	12.013	29.70
13	980.233	979.113	979.000	977.723	976.083	974.213	972.133	969.863	13.002	32.10
14	979.083	977.873	977.760	976.403	974.693	972.753	970.603	968.263	13.990	34.50
15	977.973	976.663	976.530	975.113	973.313	971.303	969.083	966.673	15.021	37.00
16	976.893	975.493	975.340	973.843	971.963	969.873	967.573	965.093	16.016	39.40
17	975.803	974.303	974.140	972.563	970.593	968.413	966.043	963.493	17.014	41.80
18	974.703	973.103	972.930	971.263	969.203	966.943	964.493	961.863	18.012	44.20
19	973.603	971.883	971.710	969.943	967.793	965.443	962.913	960.203	19.013	46.60
20	972.493	970.653	970.460	968.613	966.363	963.923	961.313	958.533	20.018	49.00
21	971.363	969.413	969.210	967.263	964.923	962.393	959.703	956.843	21.026	51.40
22	970.213	968.153	967.940	965.893	963.453	960.843	958.063	955.133	21.997	53.70
23	969.043	966.863	966.640	964.503	961.963	959.263	956.403	953.403	23.012	56.10
24	967.843	965.553	965.310	963.093	960.453	957.663	954.733	951.653	24.027	58.50
25	966.623	964.213	963.970	961.653	958.923	956.043	953.033	949.883	25.008	60.80
26	965.363	962.843	962.580	960.173	957.353	954.393	951.303	948.073	26.033	63.20
27	964.033	961.413	961.140	958.643	955.733	952.693	949.523	946.224	27.021	65.50
28	962.653	959.933	959.640	957.073	954.073	950.953	947.713	944.354	28.012	67.80
29	961.223	958.413	958.110	955.453	952.383	949.193	945.874	942.454	29.007	70.10
30	959.743	956.833	956.530	953.793	950.643	947.383	944.004	940.524	30.009	72.40
31	958.203	955.213	954.890	952.093	948.873	945.544	942.114	938.574	31.017	74.70
32	956.623	953.540	953.220	950.353	947.063	943.674	940.184	936.594	31.985	76.90
33	954.993	951.833	951.510	948.573	945.224	941.774	938.224	934.584	33.003	79.20
34	953.313	950.083	949.750	946.763	943.344	939.834	936.234	932.544	34.024	81.50
35	951.593	948.293	947.950	944.914	941.434	937.874	934.224	930.484	35.009	83.70
36	949.833	946.473	946.120	943.034	939.494	935.884	932.184	928.404	35.997	85.90
37	948.023	944.614	944.250	941.114	937.534	933.874	930.134	926.314	36.992	88.10
38	946.174	942.704	942.340	939.164	935.534	931.834	928.054	924.194	38.040	90.40
39	944.284	940.764	940.400	937.174	933.504	929.764	925.944	922.054	38.998	92.50
40	942.354	938.794	938.420	935.154	931.454	927.674	923.824	919.894	40.009	94.70
41	940.394	936.794	936.410	933.114	929.374	925.554	921.674	917.714	40.984	96.80
42	938.394	934.754	934.370	931.044	927.264	923.414	919.494	915.514	41.966	98.90
43	936.364	932.684	932.300	928.944	925.134	921.254	917.304	913.294	42.997	101.10
44	934.304	930.594	930.210	926.824	922.984	919.074	915.104	911.054	43.988	103.20
45	932.234	928.494	928.100	924.694	920.824	916.894	912.884	908.815	44.983	105.30
46	930.144	926.374	925.980	922.544	918.654	914.694	910.665	906.575	45.988	107.40
47	928.034	924.234	923.840	920.384	916.464	912.474	908.425	904.315	46.999	109.50
48	925.904	922.084	921.680	918.204	914.264	910.255	906.185	902.045	48.010	111.60
49	923.764	919.924	919.520	916.014	912.054	908.025	903.935	899.765	48.982	113.60
50	921.594	917.734	917.330	913.814	909.825	905.775	901.655	897.475	50.010	115.70

Bulletin of the Bureau of Standards Volume 9 Page 424 Table 49 D t/4° C. Revised Values   At 15.556

Mass %	10° C	15° C	15.556	20° C	25° C	30° C	35° C	40° C	Closest Proof Mass %	Proof at Closest Mass %
50	921.594	917.734	917.330	913.814	909.825	905.775	901.655	897.475	50.010	115.70
51	919.404	915.524	915.110	911.574	907.575	903.505	899.375	895.165	50.995	117.70
52	917.204	913.304	912.890	909.335	905.315	901.225	897.075	892.855	51.990	119.70
53	914.994	911.074	910.650	907.085	903.045	898.935	894.765	890.475	52.988	121.70
54	912.764	908.825	908.410	904.825	900.765	896.645	892.455	888.205	53.992	123.70
55	910.525	906.565	906.150	902.555	898.475	894.345	890.135	885.865	55.004	**125.70**
56	908.285	904.305	903.890	900.285	896.185	892.035	887.815	883.535	56.022	127.70
57	906.045	902.045	901.630	898.005	893.895	889.725	885.495	881.195	56.993	129.60
58	903.785	899.775	899.350	895.715	891.595	887.415	883.165	878.855	57.975	131.50
59	901.515	897.495	897.070	893.415	889.285	885.095	880.825	876.475	59.012	133.50
60	899.245	895.205	894.780	891.105	886.965	882.755	878.485	874.146	60.004	135.40
61	896.955	892.905	892.470	888.795	884.635	880.415	876.125	871.776	60.999	137.30
62	894.655	890.595	890.170	886.475	882.305	878.065	873.766	869.406	62.010	139.20
63	892.345	888.275	887.840	884.145	879.955	875.715	871.396	867.026	63.018	141.10
64	890.035	885.945	885.520	881.805	877.605	873.346	869.026	864.636	63.983	142.90
65	887.715	883.615	883.180	879.455	875.245	870.976	866.646	862.246	65.011	144.80
66	885.385	881.275	880.840	877.105	872.886	868.606	864.266	859.846	65.992	146.60
67	883.055	878.925	878.490	874.746	870.516	866.226	861.876	857.446	66.981	148.40
68	880.715	876.575	876.140	872.386	868.146	863.846	859.476	855.046	68.028	150.30
69	878.365	874.216	873.780	870.016	865.766	861.456	857.076	852.636	68.977	152.00
70	875.995	871.846	871.400	867.636	863.376	859.056	854.676	850.226	69.981	153.80
71	873.626	869.466	869.020	865.246	860.976	856.646	852.256	847.806	71.003	155.60
72	871.246	867.076	866.630	862.846	858.566	854.236	849.836	845.376	72.027	157.40
73	868.856	864.676	864.230	860.446	856.156	851.816	847.406	842.946	72.999	159.10
74	866.456	862.266	861.820	858.036	853.736	849.386	844.976	840.506	73.980	160.80
75	864.056	859.856	859.420	855.616	851.316	846.956	842.546	838.067	75.028	162.60
76	861.656	857.446	857.000	853.196	848.886	844.526	840.106	835.617	76.026	164.30
77	859.246	855.026	854.570	850.766	846.446	842.086	837.657	833.167	76.975	165.90
78	856.826	852.596	852.140	848.326	844.006	839.636	835.207	830.717	78.049	167.70
79	854.396	850.156	849.710	845.876	841.556	837.177	832.747	828.247	79.011	169.30
80	851.946	847.696	847.250	843.416	839.087	834.707	830.267	825.757	79.982	170.90
81	849.476	845.226	844.790	840.936	836.617	832.217	827.777	823.267	81.030	172.60
82	846.996	842.746	842.310	838.457	834.127	829.717	825.277	820.767	82.018	174.20
83	844.506	840.256	839.800	835.967	831.617	827.217	822.767	818.257	83.013	175.80
84	842.006	837.747	837.390	833.457	829.107	824.707	820.247	815.737	84.021	177.40
85	839.486	835.227	834.780	830.927	826.577	822.177	817.717	813.197	85.045	179.00
86	836.947	832.687	832.230	828.377	824.027	819.627	815.167	810.647	86.014	180.50
87	834.387	830.117	829.660	825.807	821.457	817.057	812.597	808.087	86.989	182.00
88	831.787	827.517	827.060	823.207	818.857	814.457	810.007	805.497	87.977	183.50
89	829.167	824.897	824.440	820.597	816.237	811.837	807.397	802.888	88.978	185.00
90	826.517	822.247	821.800	817.947	813.597	809.197	804.757	800.258	89.996	186.50
91	823.837	819.567	819.120	815.267	810.917	806.527	802.088	797.588	91.024	188.00
92	821.117	816.857	816.400	812.547	808.207	803.817	799.388	794.888	91.996	189.40
93	818.367	814.107	813.650	809.807	805.467	801.088	796.668	792.178	92.990	190.80
94	815.587	811.317	810.870	807.027	802.698	798.328	793.908	789.448	93.998	192.20
95	812.757	808.497	808.050	804.217	799.888	795.528	791.118	786.678	95.023	193.60
96	809.887	805.637	805.180	801.358	797.038	792.688	788.288	783.858	95.988	194.90
97	806.957	802.718	802.260	798.438	794.128	789.788	785.398	780.978	96.979	196.20
98	803.967	799.728	799.270	795.448	791.148	786.818	782.448	778.038	97.987	197.50
99	800.918	796.678	796.220	792.408	788.118	783.798	779.438	775.048	99.025	198.80
100	797.818	793.578	793.030	789.318	785.038	780.728	776.388	772.008	100.000	200.00

Calibration Temp Table 1

IAT General Formula-based tables created by the iterative calculation process; published at 200 places at selected calibration temperatures. Additional calibration temperature tables are published in Volume 2.

| T4 Vol % | 10° C | 10° C | 10° C | 10° C | 15° C | 15° C | 15° C | 15° C | 15.556° C | 15.556° C | 15.556° C | 15.556° C |
T3 Mass %	3a Den	3b Vol %	4a Den	4b Mass %	3a Den	3b Vol %	4a Den	4b Mass %	3a Den	3b Vol %	4a Den	4b Mass %
0.00	999.695	0.000	999.695	0.000	999.096	0.000	999.096	0.000	999.010	0.000	999.010	0.000
0.50	998.732	0.626	998.925	0.399	998.136	0.629	998.332	0.397	998.051	0.629	998.247	0.397
1.00	997.792	1.251	998.167	0.799	997.198	1.257	997.580	0.795	997.113	1.257	997.495	0.795
1.50	996.874	1.874	997.422	1.200	996.279	1.883	996.839	1.194	996.194	1.884	996.756	1.193
2.00	995.976	2.497	996.691	1.601	995.379	2.509	996.110	1.593	995.294	2.510	996.027	1.592
2.50	995.099	3.118	995.973	2.002	994.497	3.133	995.392	1.993	994.411	3.135	995.308	1.992
3.00	994.241	3.739	995.264	2.405	993.632	3.757	994.684	2.393	993.545	3.759	994.600	2.392
3.50	993.400	4.358	994.570	2.807	992.783	4.379	993.986	2.794	992.695	4.381	993.902	2.793
4.00	992.578	4.977	993.884	3.211	991.950	5.000	993.299	3.195	991.861	5.003	993.213	3.194
4.50	991.773	5.594	993.211	3.614	991.132	5.621	992.620	3.597	991.042	5.624	992.536	3.595
5.00	990.984	6.211	992.547	4.019	990.329	6.240	991.950	4.000	990.238	6.243	991.864	3.998
5.50	990.212	6.827	991.894	4.424	989.541	6.859	991.290	4.403	989.448	6.862	991.205	4.400
6.00	989.456	7.442	991.252	4.829	988.767	7.476	990.639	4.806	988.672	7.480	990.551	4.804
6.50	988.716	8.056	990.620	5.235	988.007	8.093	989.997	5.210	987.909	8.097	989.909	5.207
7.00	987.990	8.669	989.998	5.641	987.260	8.709	989.363	5.614	987.160	8.714	989.274	5.611
7.50	987.279	9.282	989.386	6.047	986.525	9.324	988.738	6.019	986.423	9.329	988.647	6.016
8.00	986.583	9.894	988.783	6.454	985.803	9.939	988.121	6.424	985.698	9.944	988.029	6.421
8.50	985.900	10.505	988.189	6.862	985.094	10.552	987.512	6.830	984.985	10.557	987.417	6.827
9.00	985.230	11.115	987.604	7.270	984.395	11.165	986.911	7.236	984.284	11.170	986.815	7.233
9.50	984.574	11.725	987.030	7.678	983.708	11.777	986.318	7.643	983.593	11.783	986.220	7.639
10.00	983.930	12.334	986.463	8.087	983.031	12.388	985.732	8.050	982.913	12.394	985.632	8.046
10.50	983.297	12.942	985.905	8.496	982.364	12.999	985.154	8.457	982.242	13.005	985.052	8.453
11.00	982.676	13.550	985.355	8.906	981.706	13.609	984.583	8.865	981.581	13.615	984.478	8.861
11.50	982.065	14.157	984.814	9.316	981.057	14.218	984.017	9.274	980.928	14.225	983.911	9.269
12.00	981.463	14.763	984.281	9.726	980.416	14.827	983.460	9.682	980.283	14.833	983.350	9.678
12.50	980.871	15.369	983.755	10.137	979.783	15.434	982.909	10.091	979.645	15.441	982.797	10.086
13.00	980.287	15.974	983.237	10.548	979.156	16.042	982.363	10.501	979.013	16.049	982.248	10.496
13.50	979.710	16.579	982.726	10.959	978.535	16.648	981.823	10.911	978.388	16.655	981.706	10.905
14.00	979.141	17.183	982.221	11.371	977.920	17.254	981.289	11.321	977.768	17.261	981.167	11.316
14.50	978.577	17.787	981.723	11.783	977.309	17.859	980.759	11.732	977.152	17.866	980.635	11.726
15.00	978.018	18.389	981.231	12.195	976.701	18.463	980.235	12.143	976.539	18.471	980.107	12.137
15.50	977.463	18.992	980.744	12.608	976.097	19.067	979.715	12.554	975.929	19.075	979.584	12.548
16.00	976.912	19.593	980.263	13.021	975.495	19.670	979.199	12.966	975.322	19.678	979.064	12.960
16.50	976.363	20.194	979.785	13.435	974.894	20.272	978.686	13.378	974.716	20.280	978.548	13.372
17.00	975.816	20.794	979.313	13.848	974.294	20.873	978.177	13.791	974.110	20.882	978.035	13.784
17.50	975.269	21.394	978.844	14.262	973.694	21.474	977.670	14.204	973.505	21.483	977.524	14.197
18.00	974.723	21.993	978.378	14.677	973.093	22.074	977.166	14.617	972.898	22.083	977.016	14.610
18.50	974.176	22.591	977.916	15.092	972.491	22.673	976.664	15.031	972.289	22.682	976.510	15.024
19.00	973.626	23.189	977.455	15.507	971.886	23.271	976.163	15.445	971.679	23.280	976.005	15.438
19.50	973.074	23.785	976.998	15.922	971.278	23.869	975.665	15.859	971.065	23.878	975.502	15.852
20.00	972.519	24.381	976.541	16.338	970.666	24.465	975.166	16.274	970.447	24.474	974.998	16.267
20.50	971.960	24.977	976.084	16.755	970.050	25.061	974.666	16.690	969.825	25.070	974.494	16.683
21.00	971.395	25.571	975.629	17.171	969.429	25.656	974.168	17.105	969.198	25.665	973.992	17.098
21.50	970.825	26.164	975.173	17.588	968.802	26.250	973.668	17.522	968.565	26.259	973.486	17.515
22.00	970.249	26.757	974.716	18.006	968.169	26.843	973.168	17.938	967.926	26.852	972.982	17.931
22.50	969.666	27.349	974.259	18.424	967.529	27.434	972.664	18.356	967.280	27.444	972.475	18.348
23.00	969.075	27.939	973.800	18.842	966.881	28.025	972.161	18.773	966.626	28.035	971.965	18.766
23.50	968.475	28.529	973.339	19.261	966.226	28.615	971.654	19.191	965.965	28.625	971.453	19.184
24.00	967.867	29.118	972.875	19.680	965.562	29.204	971.144	19.610	965.295	29.213	970.939	19.602
24.50	967.250	29.705	972.408	20.100	964.889	29.792	970.631	20.029	964.617	29.801	970.421	20.021
25.00	966.623	30.292	971.937	20.520	964.207	30.378	970.113	20.449	963.929	30.387	969.899	20.441

T4 Vol % T3 Mass %	10° C 3a Den	10° C 3b Vol %	10° C 4a Den	10° C 4b Mass %	15° C 3a Den	15° C 3b Vol %	15° C 4a Den	15° C 4b Mass %	15.556° C 3a Den	15.556° C 3b Vol %	15.556° C 4a Den	15.556° C 4b Mass %
25.50	965.985	30.877	971.463	20.940	963.516	30.963	969.592	20.869	963.232	30.973	969.373	20.861
26.00	965.337	31.462	970.984	21.361	962.814	31.548	969.066	21.290	962.525	31.557	968.842	21.282
26.50	964.678	32.045	970.500	21.783	962.103	32.130	968.536	21.711	961.808	32.140	968.306	21.703
27.00	964.008	32.627	970.011	22.205	961.381	32.712	967.999	22.133	961.080	32.721	967.765	22.125
27.50	963.326	33.208	969.515	22.628	960.648	33.293	967.458	22.555	960.342	33.302	967.217	22.548
28.00	962.632	33.787	969.014	23.051	959.904	33.872	966.909	22.979	959.593	33.881	966.664	22.971
28.50	961.926	34.365	968.506	23.475	959.149	34.449	966.355	23.402	958.833	34.459	966.106	23.394
29.00	961.208	34.942	967.990	23.900	958.383	35.026	965.793	23.827	958.061	35.035	965.539	23.819
29.50	960.477	35.517	967.467	24.325	957.606	35.601	965.224	24.252	957.279	35.610	964.965	24.244
30.00	959.734	36.091	966.936	24.751	956.817	36.174	964.648	24.678	956.486	36.183	964.384	24.670
30.50	958.978	36.664	966.397	25.178	956.017	36.746	964.064	25.104	955.681	36.755	963.796	25.096
31.00	958.210	37.235	965.850	25.605	955.206	37.317	963.473	25.531	954.865	37.326	963.200	25.523
31.50	957.430	37.805	965.294	26.033	954.383	37.886	962.872	25.959	954.037	37.895	962.595	25.951
32.00	956.637	38.373	964.730	26.461	953.549	38.454	962.263	26.388	953.199	38.463	961.981	26.380
32.50	955.831	38.940	964.155	26.891	952.703	39.020	961.645	26.818	952.350	39.029	961.358	26.810
33.00	955.013	39.505	963.571	27.321	951.847	39.585	961.019	27.248	951.489	39.594	960.727	27.240
33.50	954.183	40.069	962.978	27.752	950.980	40.148	960.383	27.679	950.618	40.157	960.087	27.671
34.00	953.341	40.631	962.374	28.184	950.102	40.710	959.737	28.111	949.736	40.718	959.437	28.103
34.50	952.487	41.192	961.759	28.617	949.213	41.270	959.082	28.544	948.843	41.278	958.777	28.536
35.00	951.621	41.751	961.134	29.051	948.314	41.828	958.417	28.978	947.940	41.837	958.108	28.970
35.50	950.744	42.308	960.499	29.485	947.404	42.385	957.743	29.412	947.027	42.394	957.430	29.404
36.00	949.856	42.864	959.854	29.920	946.484	42.940	957.058	29.848	946.105	42.949	956.741	29.840
36.50	948.957	43.418	959.196	30.357	945.555	43.494	956.364	30.284	945.172	43.502	956.041	30.277
37.00	948.047	43.971	958.528	30.794	944.616	44.046	955.658	30.722	944.230	44.054	955.333	30.714
37.50	947.126	44.521	957.850	31.232	943.668	44.596	954.943	31.160	943.279	44.605	954.613	31.153
38.00	946.196	45.071	957.158	31.672	942.711	45.145	954.217	31.600	942.319	45.153	953.884	31.592
38.50	945.255	45.618	956.457	32.112	941.745	45.692	953.481	32.040	941.350	45.701	953.143	32.033
39.00	944.305	46.164	955.745	32.553	940.770	46.238	952.734	32.482	940.372	46.246	952.394	32.474
39.50	943.346	46.709	955.021	32.995	939.787	46.782	951.976	32.925	939.387	46.790	951.633	32.917
40.00	942.377	47.251	954.285	33.439	938.796	47.324	951.210	33.368	938.394	47.332	950.861	33.361
40.50	941.401	47.793	953.539	33.883	937.798	47.865	950.431	33.813	937.393	47.872	950.081	33.805
41.00	940.415	48.332	952.780	34.329	936.792	48.404	949.643	34.259	936.385	48.411	949.289	34.251
41.50	939.422	48.870	952.011	34.776	935.779	48.941	948.844	34.706	935.370	48.949	948.487	34.698
42.00	938.421	49.406	951.230	35.224	934.759	49.476	948.034	35.154	934.348	49.484	947.675	35.146
42.50	937.413	49.940	950.438	35.673	933.733	50.011	947.215	35.603	933.319	50.018	946.851	35.596
43.00	936.398	50.473	949.636	36.123	932.700	50.543	946.385	36.054	932.285	50.551	946.019	36.046
43.50	935.375	51.004	948.823	36.574	931.661	51.074	945.546	36.505	931.244	51.081	945.176	36.498
44.00	934.347	51.534	947.997	37.027	930.617	51.603	944.695	36.958	930.198	51.610	944.323	36.951
44.50	933.312	52.062	947.163	37.480	929.566	52.130	943.836	37.412	929.146	52.138	943.460	37.405
45.00	932.271	52.588	946.315	37.936	928.511	52.656	942.964	37.868	928.089	52.664	942.588	37.860
45.50	931.224	53.112	945.459	38.392	927.450	53.180	942.086	38.324	927.027	53.188	941.705	38.317
46.00	930.172	53.635	944.593	38.849	926.385	53.703	941.196	38.782	925.960	53.711	940.815	38.774
46.50	929.115	54.157	943.715	39.308	925.315	54.224	940.297	39.241	924.888	54.231	939.914	39.233
47.00	928.053	54.677	942.828	39.768	924.240	54.743	939.390	39.701	923.813	54.751	939.003	39.694
47.50	926.986	55.195	941.931	40.229	923.162	55.261	938.472	40.163	922.733	55.269	938.084	40.155
48.00	925.915	55.711	941.023	40.692	922.079	55.778	937.545	40.626	921.649	55.785	937.156	40.618
48.50	924.840	56.226	940.106	41.156	920.992	56.292	936.610	41.090	920.561	56.299	936.219	41.082
49.00	923.760	56.740	939.181	41.621	919.902	56.805	935.667	41.555	919.469	56.812	935.272	41.548
49.50	922.676	57.251	938.244	42.088	918.808	57.317	934.714	42.022	918.374	57.324	934.317	42.015
50.00	921.589	57.761	937.300	42.556	917.711	57.826	933.753	42.490	917.276	57.833	933.354	42.483

T4 Vol %	10° C	10° C	10° C	10° C	15° C	15° C	15° C	15° C	15.556° C	15.556° C	15.556° C	15.556° C
T3 Mass %	3a Den	3b Vol %	4a Den	4b Mass %	3a Den	3b Vol %	4a Den	4b Mass %	3a Den	3b Vol %	4a Den	4b Mass %
50.50	920.498	58.270	936.347	43.025	916.610	58.335	932.783	42.960	916.174	58.342	932.384	42.952
51.00	919.404	58.777	935.384	43.496	915.506	58.841	931.805	43.431	915.069	58.848	931.405	43.423
51.50	918.306	59.282	934.413	43.968	914.400	59.346	930.820	43.903	913.961	59.353	930.416	43.896
52.00	917.206	59.786	933.432	44.442	913.290	59.850	929.827	44.376	912.851	59.857	929.422	44.369
52.50	916.102	60.288	932.446	44.916	912.177	60.352	928.826	44.851	911.737	60.358	928.419	44.844
53.00	914.995	60.789	931.449	45.393	911.062	60.852	927.816	45.328	910.621	60.859	927.408	45.321
53.50	913.886	61.288	930.446	45.870	909.944	61.351	926.799	45.806	909.502	61.357	926.391	45.798
54.00	912.773	61.786	929.435	46.349	908.824	61.848	925.776	46.285	908.381	61.854	925.365	46.278
54.50	911.658	62.281	928.415	46.830	907.701	62.343	924.746	46.765	907.257	62.350	924.334	46.758
55.00	910.541	62.776	927.388	47.312	906.575	62.837	923.708	47.247	906.130	62.844	923.295	47.240
55.50	909.421	63.269	926.355	47.795	905.447	63.329	922.662	47.731	905.002	63.336	922.247	47.724
56.00	908.298	63.760	925.313	48.280	904.317	63.820	921.610	48.216	903.871	63.827	921.194	48.209
56.50	907.173	64.249	924.266	48.766	903.185	64.309	920.552	48.702	902.737	64.316	920.135	48.695
57.00	906.046	64.737	923.210	49.254	902.050	64.797	919.487	49.190	901.602	64.804	919.069	49.183
57.50	904.916	65.224	922.146	49.744	900.913	65.283	918.413	49.680	900.464	65.290	917.994	49.673
58.00	903.785	65.709	921.079	50.234	899.774	65.767	917.335	50.171	899.324	65.774	916.915	50.164
58.50	902.650	66.192	920.002	50.727	898.632	66.250	916.251	50.663	898.182	66.257	915.830	50.656
59.00	901.514	66.674	918.919	51.221	897.489	66.732	915.159	51.157	897.038	66.738	914.737	51.150
59.50	900.376	67.154	917.831	51.716	896.343	67.211	914.060	51.653	895.891	67.217	913.637	51.646
60.00	899.235	67.632	916.736	52.213	895.196	67.689	912.956	52.150	894.743	67.695	912.532	52.143
60.50	898.092	68.109	915.635	52.711	894.046	68.166	911.847	52.648	893.592	68.172	911.423	52.641
61.00	896.947	68.585	914.527	53.211	892.894	68.640	910.731	53.148	892.440	68.646	910.303	53.142
61.50	895.800	69.058	913.412	53.713	891.740	69.114	909.608	53.650	891.285	69.120	909.182	53.643
62.00	894.651	69.531	912.292	54.216	890.584	69.585	908.478	54.154	890.129	69.591	908.050	54.147
62.50	893.500	70.001	911.165	54.721	889.427	70.055	907.343	54.659	888.970	70.061	906.915	54.652
63.00	892.346	70.470	910.033	55.227	888.267	70.524	906.203	55.165	887.810	70.529	905.774	55.158
63.50	891.191	70.937	908.893	55.735	887.105	70.990	905.054	55.674	886.647	70.996	904.624	55.667
64.00	890.034	71.403	907.747	56.245	885.941	71.455	903.901	56.184	885.483	71.461	903.470	56.177
64.50	888.874	71.867	906.594	56.757	884.775	71.919	902.742	56.695	884.316	71.924	902.308	56.689
65.00	887.713	72.330	905.436	57.270	883.608	72.381	901.575	57.209	883.148	72.386	901.142	57.202
65.50	886.550	72.791	904.274	57.784	882.438	72.841	900.403	57.724	881.977	72.846	899.969	57.717
66.00	885.384	73.250	903.102	58.301	881.267	73.300	899.224	58.241	880.805	73.305	898.790	58.234
66.50	884.217	73.707	901.926	58.819	880.093	73.756	898.040	58.759	879.631	73.762	897.603	58.753
67.00	883.048	74.163	900.740	59.340	878.918	74.212	896.848	59.280	878.455	74.217	896.412	59.273
67.50	881.877	74.618	899.552	59.861	877.741	74.665	895.650	59.802	877.278	74.671	895.214	59.795
68.00	880.703	75.070	898.355	60.385	876.562	75.118	894.446	60.326	876.098	75.123	894.007	60.320
68.50	879.528	75.522	897.151	60.911	875.381	75.568	893.235	60.852	874.916	75.573	892.795	60.846
69.00	878.351	75.971	895.942	61.438	874.198	76.017	892.017	61.380	873.733	76.022	891.579	61.373
69.50	877.172	76.419	894.725	61.968	873.014	76.464	890.793	61.910	872.548	76.469	890.353	61.903
70.00	875.991	76.865	893.502	62.499	871.827	76.909	889.563	62.441	871.361	76.914	889.121	62.435
70.50	874.809	77.309	892.273	63.032	870.639	77.353	888.325	62.975	870.172	77.358	887.882	62.969
71.00	873.624	77.752	891.036	63.567	869.449	77.795	887.079	63.511	868.982	77.800	886.638	63.504
71.50	872.437	78.193	889.793	64.104	868.257	78.236	885.829	64.048	867.789	78.240	885.385	64.042
72.00	871.249	78.633	888.542	64.643	867.064	78.674	884.570	64.588	866.595	78.679	884.125	64.582
72.50	870.058	79.071	887.283	65.185	865.868	79.111	883.306	65.129	865.399	79.116	882.860	65.123
73.00	868.865	79.507	886.019	65.728	864.671	79.547	882.033	65.673	864.201	79.551	881.586	65.667
73.50	867.671	79.942	884.747	66.273	863.472	79.981	880.753	66.219	863.001	79.985	880.305	66.213
74.00	866.474	80.374	883.467	66.821	862.270	80.413	879.466	66.767	861.800	80.417	879.018	66.761
74.50	865.276	80.806	882.181	67.370	861.067	80.843	878.172	67.317	860.596	80.847	877.721	67.312
75.00	864.075	81.235	880.886	67.922	859.862	81.272	876.869	67.870	859.390	81.276	876.419	67.864

T4 Vol % / T3 Mass %	10° C 3a Den	10° C 3b Vol %	10° C 4a Den	10° C 4b Mass %	15° C 3a Den	15° C 3b Vol %	15° C 4a Den	15° C 4b Mass %	15.556° C 3a Den	15.556° C 3b Vol %	15.556° C 4a Den	15.556° C 4b Mass %
75.50	862.872	81.663	879.585	68.476	858.655	81.699	875.558	68.425	858.183	81.703	875.108	68.419
75.75	862.269	81.876	878.931	68.754	858.050	81.912	874.901	68.703	857.578	81.915	874.450	68.697
76.00	861.666	82.089	878.276	69.032	857.445	82.124	874.243	68.981	856.973	82.128	873.790	68.976
76.50	860.459	82.513	876.957	69.591	856.234	82.547	872.917	69.541	855.761	82.551	872.465	69.535
77.00	859.249	82.935	875.632	70.152	855.020	82.969	871.585	70.102	854.546	82.973	871.131	70.097
77.50	858.036	83.356	874.299	70.715	853.803	83.389	870.244	70.666	853.330	83.393	869.789	70.661
78.00	856.821	83.775	872.957	71.281	852.585	83.807	868.897	71.232	852.111	83.811	868.441	71.227
78.50	855.603	84.192	871.610	71.848	851.364	84.224	867.539	71.801	850.889	84.227	867.083	71.796
79.00	854.383	84.608	870.251	72.419	850.140	84.638	866.175	72.372	849.665	84.642	865.718	72.367
79.50	853.159	85.021	868.885	72.992	848.913	85.051	864.800	72.946	848.438	85.054	864.343	72.941
80.00	851.933	85.433	867.511	73.567	847.683	85.462	863.419	73.522	847.208	85.465	862.961	73.517
80.50	850.703	85.843	866.127	74.145	846.451	85.871	862.027	74.101	845.975	85.874	861.569	74.096
81.00	849.470	86.251	864.733	74.726	845.215	86.278	860.626	74.683	844.739	86.281	860.167	74.678
81.50	848.234	86.657	863.329	75.310	843.976	86.684	859.217	75.267	843.499	86.687	858.758	75.262
82.00	846.994	87.061	861.917	75.896	842.733	87.087	857.799	75.854	842.256	87.090	857.336	75.850
82.50	845.750	87.463	860.495	76.485	841.486	87.488	856.369	76.444	841.010	87.491	855.906	76.440
83.00	844.502	87.864	859.062	77.077	840.236	87.888	854.930	77.037	839.759	87.891	854.469	77.032
83.50	843.249	88.262	857.621	77.671	838.982	88.285	853.482	77.632	838.505	88.288	853.018	77.628
84.00	841.993	88.658	856.166	78.269	837.723	88.681	852.021	78.231	837.246	88.683	851.556	78.227
84.50	840.731	89.052	854.701	78.870	836.460	89.074	850.549	78.833	835.982	89.076	850.084	78.829
85.00	839.465	89.444	853.223	79.474	835.192	89.465	849.065	79.438	834.714	89.468	848.600	79.434
85.50	838.194	89.834	851.734	80.081	833.919	89.854	847.570	80.046	833.441	89.857	847.104	80.042
86.00	836.918	90.222	850.230	80.692	832.641	90.241	846.061	80.658	832.163	90.244	845.595	80.654
86.50	835.636	90.607	848.714	81.306	831.358	90.626	844.539	81.273	830.880	90.628	844.070	81.270
87.00	834.348	90.991	847.182	81.924	830.069	91.009	843.002	81.892	829.591	91.011	842.535	81.888
87.50	833.055	91.372	845.635	82.546	828.774	91.389	841.452	82.514	828.296	91.391	840.982	82.511
88.00	831.755	91.751	844.074	83.171	827.474	91.767	839.883	83.141	826.995	91.769	839.413	83.138
88.50	830.448	92.127	842.496	83.800	826.166	92.143	838.300	83.771	825.688	92.144	837.830	83.768
89.00	829.135	92.501	840.898	84.434	824.852	92.516	836.698	84.406	824.374	92.517	836.228	84.403
89.50	827.815	92.872	839.285	85.071	823.531	92.887	835.080	85.044	823.053	92.888	834.610	85.041
90.00	826.488	93.242	837.649	85.714	822.203	93.255	833.440	85.688	821.725	93.256	832.969	85.685
90.50	825.153	93.608	835.995	86.360	820.868	93.621	831.780	86.336	820.389	93.622	831.309	86.333
91.00	823.810	93.972	834.317	87.012	819.524	93.984	830.100	86.988	819.046	93.985	829.627	86.986
91.50	822.458	94.333	832.616	87.669	818.173	94.344	828.395	87.646	817.694	94.346	827.922	87.644
92.00	821.099	94.692	830.891	88.331	816.813	94.702	826.664	88.310	816.334	94.703	826.193	88.307
92.50	819.730	95.048	829.138	88.999	815.444	95.057	824.908	88.979	814.966	95.058	824.434	88.977
93.00	818.352	95.401	827.359	89.672	814.066	95.410	823.123	89.654	813.588	95.411	822.650	89.652
93.50	816.965	95.751	825.549	90.352	812.679	95.759	821.309	90.335	812.200	95.760	820.836	90.333
94.00	815.567	96.099	823.705	91.039	811.281	96.106	819.465	91.022	810.803	96.106	818.989	91.021
94.50	814.158	96.443	821.829	91.732	809.873	96.449	817.584	91.717	809.395	96.450	817.111	91.715
95.00	812.739	96.784	819.917	92.432	808.455	96.790	815.667	92.419	807.976	96.790	815.191	92.418
95.50	811.308	97.122	817.962	93.141	807.024	97.127	813.709	93.129	806.546	97.127	813.233	93.128
96.00	809.864	97.457	815.968	93.857	805.582	97.461	811.710	93.847	805.104	97.461	811.234	93.846
96.50	808.407	97.789	813.924	94.583	804.127	97.792	809.664	94.574	803.649	97.792	809.189	94.573
97.00	806.936	98.116	811.827	95.319	802.658	98.119	807.566	95.311	802.181	98.119	807.088	95.311
97.50	805.450	98.440	809.678	96.064	801.175	98.442	805.411	96.059	800.698	98.443	804.936	96.058
98.00	803.949	98.761	807.461	96.822	799.676	98.762	803.194	96.818	799.200	98.762	802.720	96.817
98.50	802.430	99.077	805.175	97.592	798.161	99.078	800.906	97.590	797.685	99.078	800.429	97.590
99.00	800.892	99.389	802.805	98.377	796.629	99.390	798.539	98.376	796.154	99.390	798.063	98.376
99.50	799.335	99.697	800.337	99.179	795.078	99.697	796.076	99.179	794.603	99.697	795.601	99.179
100.00	797.755	100.000	797.755	100.000	793.506	100.000	793.506	100.000	793.032	100.000	793.032	100.000

Table 7

Table 7 is very handy and quite accurate for only being three decimal places. It also shows the freeze points for alcohol and water mixtures.

Table 7 Volume Correction

Proof|° F.	18.0	20.0	22.0	24.0	26.0	28.0	30.0	32.0	34.0	36.0	38.0	40.0	42.0	44.0
0	...	...	...	...	...	...	...	...	...	...	...	1.001	1.001	1.001
5	...	...	...	...	...	...	...	...	...	...	1.001	1.001	1.001	1.001
10	...	...	...	...	...	...	...	...	...	...	1.001	1.001	1.001	1.001
15	...	...	...	...	...	...	...	...	...	...	1.001	1.001	1.001	1.001
20	...	...	...	...	...	...	...	1.002	1.002	1.002	1.002	1.001	1.001	1.001
25	...	...	...	...	...	...	...	1.002	1.002	1.002	1.002	1.002	1.002	1.001
30	...	...	1.003	1.003	1.003	1.002	1.002	1.002	1.002	1.002	1.002	1.002	1.002	1.002
35	...	...	1.003	1.003	1.003	1.003	1.003	1.003	1.003	1.003	1.003	1.003	1.002	1.002
40	1.005	1.004	1.004	1.004	1.004	1.004	1.004	1.004	1.004	1.003	1.003	1.003	1.003	1.002
45	1.006	1.006	1.006	1.005	1.005	1.005	1.005	1.005	1.004	1.004	1.004	1.004	1.003	1.003
50	1.007	1.007	1.007	1.006	1.006	1.006	1.006	1.005	1.005	1.005	1.004	1.004	1.004	1.003
55	1.009	1.008	1.008	1.008	1.007	1.007	1.007	1.006	1.006	1.006	1.005	1.005	1.004	1.004
60	1.010	1.010	1.009	1.009	1.008	1.008	1.008	1.007	1.007	1.006	1.006	1.005	1.005	1.004
65	1.011	1.011	1.010	1.010	1.009	1.009	1.008	1.008	1.007	1.007	1.006	1.006	1.005	1.005
70	1.013	1.012	1.012	1.011	1.010	1.010	1.009	1.009	1.008	1.008	1.007	1.006	1.006	1.005
75	1.014	1.013	1.013	1.012	1.011	1.011	1.010	1.010	1.009	1.008	1.008	1.007	1.006	1.006
80	1.015	1.014	1.013	1.013	1.012	1.011	1.011	1.010	1.009	1.009	1.008	1.007	1.007	1.006
85	1.016	1.015	1.014	1.014	1.013	1.012	1.011	1.011	1.010	1.009	1.008	1.008	1.007	1.006
90	1.016	1.016	1.015	1.014	1.013	1.013	1.012	1.011	1.010	1.010	1.009	1.008	1.007	1.006
95	1.017	1.016	1.016	1.015	1.014	1.013	1.012	1.012	1.011	1.010	1.009	1.008	1.008	1.007
100	1.018	1.017	1.016	1.015	1.014	1.014	1.013	1.012	1.011	1.010	1.009	1.009	1.008	1.007
105	1.018	1.017	1.017	1.016	1.015	1.014	1.013	1.012	1.011	1.011	1.010	1.009	1.008	1.007
110	1.019	1.018	1.017	1.016	1.015	1.014	1.013	1.013	1.012	1.011	1.010	1.009	1.008	1.007
115	1.019	1.018	1.017	1.016	1.016	1.015	1.014	1.013	1.012	1.011	1.010	1.009	1.008	1.007
120	1.019	1.019	1.018	1.017	1.016	1.015	1.014	1.013	1.012	1.011	1.010	1.009	1.009	1.008
125	1.020	1.019	1.018	1.017	1.016	1.015	1.014	1.013	1.012	1.012	1.011	1.010	1.009	1.008
130	1.020	1.019	1.018	1.017	1.016	1.016	1.015	1.014	1.013	1.012	1.011	1.010	1.009	1.008
135	1.021	1.020	1.019	1.018	1.017	1.016	1.015	1.014	1.013	1.012	1.011	1.010	1.009	1.008
140	1.021	1.020	1.019	1.018	1.017	1.016	1.015	1.014	1.013	1.012	1.011	1.010	1.009	1.008
145	1.021	1.020	1.019	1.018	1.017	1.016	1.015	1.014	1.013	1.012	1.011	1.010	1.009	1.008
150	1.022	1.021	1.020	1.019	1.018	1.017	1.015	1.014	1.013	1.012	1.011	1.010	1.009	1.008
155	1.022	1.021	1.020	1.019	1.018	1.017	1.016	1.015	1.014	1.013	1.012	1.011	1.010	1.009
160	1.022	1.021	1.020	1.019	1.018	1.017	1.016	1.015	1.014	1.013	1.012	1.011	1.010	1.009
165	1.023	1.022	1.020	1.019	1.018	1.017	1.016	1.015	1.014	1.013	1.012	1.011	1.010	1.009
170	1.023	1.022	1.021	1.020	1.019	1.018	1.016	1.015	1.014	1.013	1.012	1.011	1.010	1.009
175	1.023	1.022	1.021	1.020	1.019	1.018	1.017	1.016	1.015	1.013	1.012	1.011	1.010	1.009
180	1.024	1.022	1.021	1.020	1.019	1.018	1.017	1.016	1.015	1.014	1.012	1.011	1.010	1.009
185	1.024	1.023	1.022	1.021	1.019	1.018	1.017	1.016	1.015	1.014	1.013	1.011	1.010	1.009
190	1.024	1.023	1.022	1.021	1.020	1.019	1.017	1.016	1.015	1.014	1.013	1.012	1.010	1.009
195	1.024	1.023	1.022	1.021	1.020	1.019	1.018	1.016	1.015	1.014	1.013	1.012	1.011	1.009
200	1.025	1.024	1.022	1.021	1.020	1.019	1.018	1.017	1.015	1.014	1.013	1.012	1.011	1.010

Table 7 Continued

Proof|° F.	46.0	48.0	50.0	52.0	54.0	56.0	58.0	60.0	62.0	64.0	66.0	68.0	70.0	72.0
0	1.001	1.001	1.001	1.001	1.000	1.000	1.000	1.000	1.000	1.000	0.999	0.999	0.999	0.999
5	1.001	1.001	1.001	1.001	1.000	1.000	1.000	1.000	1.000	1.000	0.999	0.999	0.999	0.999
10	1.001	1.001	1.001	1.001	1.000	1.000	1.000	1.000	1.000	1.000	0.999	0.999	0.999	0.999
15	1.001	1.001	1.001	1.001	1.000	1.000	1.000	1.000	1.000	1.000	0.999	0.999	0.999	0.999
20	1.001	1.001	1.001	1.001	1.001	1.000	1.000	1.000	1.000	1.000	0.999	0.999	0.999	0.998
25	1.001	1.001	1.001	1.001	1.001	1.000	1.000	1.000	1.000	1.000	0.999	0.999	0.999	0.998
30	1.002	1.001	1.001	1.001	1.001	1.001	1.000	1.000	1.000	0.999	0.999	0.999	0.998	0.998
35	1.002	1.002	1.001	1.001	1.001	1.001	1.000	1.000	1.000	0.999	0.999	0.999	0.998	0.998
40	1.002	1.002	1.002	1.001	1.001	1.001	1.000	1.000	1.000	0.999	0.999	0.998	0.998	0.998
45	1.003	1.002	1.002	1.002	1.001	1.001	1.000	1.000	1.000	0.999	0.999	0.998	0.998	0.997
50	1.003	1.003	1.002	1.002	1.001	1.001	1.000	1.000	1.000	0.999	0.999	0.998	0.998	0.997
55	1.003	1.003	1.002	1.002	1.002	1.001	1.000	1.000	0.999	0.999	0.998	0.998	0.997	0.997
60	1.004	1.003	1.003	1.002	1.002	1.001	1.001	1.000	0.999	0.999	0.998	0.998	0.997	0.996
65	1.004	1.004	1.003	1.002	1.002	1.001	1.001	1.000	0.999	0.999	0.998	0.997	0.997	0.996
70	1.005	1.004	1.003	1.003	1.002	1.001	1.001	1.000	0.999	0.999	0.998	0.997	0.997	0.996
75	1.005	1.004	1.003	1.003	1.002	1.001	1.001	1.000	0.999	0.999	0.998	0.997	0.996	0.996
80	1.005	1.004	1.004	1.003	1.002	1.001	1.001	1.000	0.999	0.998	0.998	0.997	0.996	0.995
85	1.005	1.005	1.004	1.003	1.002	1.002	1.001	1.000	0.999	0.998	0.998	0.997	0.996	0.995
90	1.006	1.005	1.004	1.003	1.002	1.002	1.001	1.000	0.999	0.998	0.998	0.997	0.996	0.995
95	1.006	1.005	1.004	1.003	1.003	1.002	1.001	1.000	0.999	0.998	0.997	0.997	0.996	0.995
100	1.006	1.005	1.004	1.004	1.003	1.002	1.001	1.000	0.999	0.998	0.997	0.996	0.996	0.995
105	1.006	1.005	1.004	1.004	1.003	1.002	1.001	1.000	0.999	0.998	0.997	0.996	0.995	0.995
110	1.006	1.005	1.005	1.004	1.003	1.002	1.001	1.000	0.999	0.998	0.997	0.996	0.995	0.994
115	1.007	1.006	1.005	1.004	1.003	1.002	1.001	1.000	0.999	0.998	0.997	0.996	0.995	0.994
120	1.007	1.006	1.005	1.004	1.003	1.002	1.001	1.000	0.999	0.998	0.997	0.996	0.995	0.994
125	1.007	1.006	1.005	1.004	1.003	1.002	1.001	1.000	0.999	0.998	0.997	0.996	0.995	0.994
130	1.007	1.006	1.005	1.004	1.003	1.002	1.001	1.000	0.999	0.998	0.997	0.996	0.995	0.994
135	1.007	1.006	1.005	1.004	1.003	1.002	1.001	1.000	0.999	0.998	0.997	0.996	0.995	0.994
140	1.007	1.006	1.005	1.004	1.003	1.002	1.001	1.000	0.999	0.998	0.997	0.996	0.995	0.994
145	1.007	1.006	1.005	1.004	1.003	1.002	1.001	1.000	0.999	0.998	0.997	0.996	0.995	0.994
150	1.007	1.006	1.005	1.004	1.003	1.002	1.001	1.000	0.999	0.998	0.997	0.996	0.995	0.994
155	1.007	1.006	1.005	1.004	1.003	1.002	1.001	1.000	0.999	0.998	0.997	0.996	0.995	0.994
160	1.008	1.006	1.005	1.004	1.003	1.002	1.001	1.000	0.999	0.998	0.997	0.996	0.995	0.993
165	1.008	1.007	1.005	1.004	1.003	1.002	1.001	1.000	0.999	0.998	0.997	0.996	0.994	0.993
170	1.008	1.007	1.006	1.004	1.003	1.002	1.001	1.000	0.999	0.998	0.997	0.995	0.994	0.993
175	1.008	1.007	1.006	1.004	1.003	1.002	1.001	1.000	0.999	0.998	0.997	0.995	0.994	0.993
180	1.008	1.007	1.006	1.005	1.003	1.002	1.001	1.000	0.999	0.998	0.997	0.995	0.994	0.993
185	1.008	1.007	1.006	1.005	1.003	1.002	1.001	1.000	0.999	0.998	0.997	0.995	0.994	0.993
190	1.008	1.007	1.006	1.005	1.004	1.002	1.001	1.000	0.999	0.998	0.996	0.995	0.994	0.993
195	1.008	1.007	1.006	1.005	1.004	1.002	1.001	1.000	0.999	0.998	0.996	0.995	0.994	0.993
200	1.008	1.007	1.006	1.005	1.004	1.002	1.001	1.000	0.999	0.998	0.996	0.995	0.994	0.993

Table 7 Continued

Proof|° F.	74.0	76.0	78.0	80.0	82.0	84.0	86.0	88.0	90.0	92.0	94.0	96.0	98.0	100.0
0	0.998	0.998	0.998	0.998	0.997	0.997	0.997	0.996	0.996	0.996	0.995	0.995	0.994	0.994
5	0.998	0.998	0.998	0.998	0.997	0.997	0.997	0.996	0.996	0.996	0.995	0.995	0.994	0.994
10	0.998	0.998	0.998	0.997	0.997	0.997	0.996	0.996	0.996	0.995	0.995	0.995	0.994	0.994
15	0.998	0.998	0.998	0.997	0.997	0.997	0.950	0.996	0.996	0.995	0.995	0.994	0.994	0.993
20	0.998	0.998	0.997	0.997	0.997	0.996	0.996	0.996	0.995	0.995	0.994	0.994	0.994	0.993
25	0.998	0.998	0.997	0.997	0.996	0.996	0.996	0.995	0.995	0.994	0.994	0.994	0.993	0.993
30	0.998	0.997	0.997	0.997	0.996	0.996	0.995	0.995	0.994	0.994	0.994	0.993	0.993	0.992
35	0.998	0.997	0.997	0.996	0.996	0.995	0.995	0.994	0.994	0.993	0.993	0.992	0.992	0.991
40	0.997	0.997	0.996	0.996	0.995	0.995	0.994	0.994	0.993	0.993	0.992	0.992	0.991	0.991
45	0.997	0.996	0.996	0.995	0.995	0.994	0.994	0.993	0.993	0.992	0.991	0.991	0.990	0.990
50	0.997	0.996	0.995	0.995	0.994	0.994	0.993	0.993	0.992	0.991	0.991	0.990	0.990	0.989
55	0.996	0.996	0.995	0.994	0.994	0.993	0.993	0.992	0.991	0.991	0.990	0.989	0.989	0.988
60	0.996	0.995	0.995	0.994	0.993	0.993	0.992	0.991	0.991	0.990	0.989	0.988	0.988	0.987
65	0.995	0.995	0.994	0.993	0.993	0.992	0.991	0.991	0.990	0.989	0.988	0.988	0.987	0.986
70	0.995	0.994	0.994	0.993	0.992	0.991	0.991	0.990	0.989	0.988	0.988	0.987	0.986	0.985
75	0.995	0.994	0.993	0.993	0.992	0.991	0.990	0.989	0.989	0.988	0.987	0.986	0.985	0.985
80	0.995	0.994	0.993	0.992	0.991	0.991	0.990	0.989	0.988	0.987	0.986	0.986	0.985	0.984
85	0.994	0.994	0.993	0.992	0.991	0.990	0.989	0.988	0.988	0.987	0.986	0.985	0.984	0.983
90	0.994	0.993	0.992	0.992	0.991	0.990	0.989	0.988	0.987	0.986	0.985	0.984	0.984	0.983
95	0.994	0.993	0.992	0.991	0.990	0.989	0.989	0.988	0.987	0.986	0.985	0.984	0.983	0.982
100	0.994	0.993	0.992	0.991	0.990	0.989	0.988	0.987	0.986	0.985	0.984	0.984	0.983	0.982
105	0.994	0.993	0.992	0.991	0.990	0.989	0.988	0.987	0.986	0.985	0.984	0.983	0.982	0.981
110	0.993	0.992	0.992	0.991	0.990	0.989	0.988	0.987	0.986	0.985	0.984	0.983	0.982	0.981
115	0.993	0.992	0.991	0.990	0.989	0.988	0.987	0.986	0.985	0.984	0.983	0.982	0.981	0.980
120	0.993	0.992	0.991	0.990	0.989	0.988	0.987	0.986	0.985	0.984	0.983	0.982	0.981	0.980
125	0.993	0.992	0.991	0.990	0.989	0.988	0.987	0.986	0.985	0.984	0.983	0.982	0.981	0.980
130	0.993	0.992	0.991	0.990	0.989	0.988	0.987	0.986	0.985	0.984	0.983	0.982	0.981	0.979
135	0.993	0.992	0.991	0.990	0.989	0.988	0.987	0.986	0.985	0.983	0.982	0.981	0.980	0.979
140	0.993	0.992	0.991	0.990	0.989	0.987	0.986	0.985	0.984	0.983	0.982	0.981	0.980	0.979
145	0.993	0.992	0.990	0.989	0.988	0.987	0.986	0.985	0.984	0.983	0.982	0.981	0.980	0.979
150	0.993	0.991	0.990	0.989	0.988	0.987	0.986	0.985	0.984	0.983	0.982	0.980	0.979	0.978
155	0.992	0.991	0.990	0.989	0.988	0.987	0.986	0.985	0.984	0.982	0.981	0.980	0.979	0.978
160	0.992	0.991	0.990	0.989	0.988	0.987	0.986	0.984	0.983	0.982	0.981	0.980	0.979	0.978
165	0.992	0.991	0.990	0.989	0.988	0.987	0.985	0.984	0.983	0.982	0.981	0.980	0.979	0.977
170	0.992	0.991	0.990	0.989	0.988	0.986	0.985	0.984	0.983	0.982	0.981	0.979	0.978	0.977
175	0.992	0.991	0.990	0.989	0.987	0.986	0.985	0.984	0.983	0.982	0.980	0.979	0.978	0.977
180	0.992	0.991	0.990	0.988	0.987	0.986	0.985	0.984	0.982	0.981	0.000	0.979	0.078	0.077
185	0.992	0.991	0.989	0.988	0.987	0.986	0.985	0.984	0.982	0.981	0.980	0.979	0.977	0.976
190	0.992	0.991	0.989	0.988	0.987	0.986	0.985	0.983	0.982	0.981	0.980	0.979	0.977	0.976
195	0.992	0.990	0.989	0.988	0.987	0.986	0.985	0.983	0.982	0.981	0.980	0.978	0.977	0.976
200	0.992	0.990	0.989	0.988	0.987	0.986	0.984	0.983	0.982	0.981	0.980	0.978	0.977	0.976

15.556°C Table Values

The following table is published at 2,000 places. The values have been generated using the general formula iterative process which is described in volume 2 of this work. It is published here since this book is mostly about 15.556°C blending operations. By employing the vacuum mass percents in conjunction with the general formula one can reproduce the density results listed in the table. From there one can derive air densities using the Kaye & Laby equation and the mass percent in air using the methods described in this book. The resulting values can be translated into any desired density type such as SG Vac, or SG Air or many others through simple mathematical operations. I also thought it was necessary to publish expanded parts alcohol and parts water values for use by the blending procedures outlined in this book. The last values in the table are placed here to save a page at the end of the table/book.

Values at 15.556°C (60° F)

Proof	Mass % Vac	Kg m3 Vac	Kg m3 Air	Lbs per Gal. Air	Lbs per Proof Gal Air	Alcohol Parts	Water Parts	Parts Total
196.0	96.8174	802.718	801.612	6.689776	3.413185	98.00	2.56	100.56
196.1	96.8940	802.492	801.386	6.687890	3.410507	98.05	2.50	100.55
196.2	96.9708	802.267	801.161	6.686013	3.407799	98.10	2.43	100.53
196.3	97.0476	802.040	800.934	6.684118	3.405095	98.15	2.37	100.52
196.4	97.1247	801.812	800.706	6.682215	3.402396	98.20	2.31	100.51
196.5	97.2018	801.584	800.478	6.680313	3.399700	98.25	2.25	100.50
196.6	97.2790	801.355	800.249	6.678402	3.397009	98.30	2.18	100.48
196.7	97.3565	801.125	800.019	6.676482	3.394253	98.35	2.12	100.47
196.8	97.4339	800.894	799.788	6.674554	3.391570	98.40	2.06	100.46
196.9	97.5117	800.663	799.557	6.672627	3.388857	98.45	1.99	100.44
197.0	97.5895	800.430	799.324	6.670682	3.386149	98.50	1.93	100.43
197.1	97.6675	800.198	799.092	6.668746	3.383445	98.55	1.87	100.42
197.2	97.7456	799.964	798.858	6.666793	3.380745	98.60	1.81	100.41
197.3	97.8239	799.729	798.623	6.664832	3.378049	98.65	1.74	100.39
197.4	97.9022	799.493	798.387	6.662862	3.375358	98.70	1.68	100.38
197.5	97.9809	799.258	798.152	6.660901	3.372636	98.75	1.62	100.37
197.6	98.0595	799.020	797.914	6.658915	3.369919	98.80	1.55	100.35
197.7	98.1383	798.782	797.675	6.656920	3.367241	98.85	1.49	100.34
197.8	98.2173	798.543	797.436	6.654926	3.364498	98.90	1.42	100.32
197.9	98.2965	798.303	797.196	6.652923	3.361794	98.95	1.36	100.31
198.0	98.3759	798.063	796.956	6.650920	3.359094	99.00	1.30	100.30
198.1	98.4554	797.821	796.714	6.648901	3.356399	99.05	1.23	100.28
198.2	98.5350	797.579	796.472	6.646881	3.353674	99.10	1.17	100.27
198.3	98.6149	797.335	796.228	6.644845	3.350953	99.15	1.11	100.26
198.4	98.6950	797.090	795.983	6.642800	3.348237	99.20	1.04	100.24
198.5	98.7751	796.845	795.738	6.640755	3.345525	99.25	0.98	100.23
198.6	98.8554	796.599	795.492	6.638702	3.342818	99.30	0.91	100.21
198.7	98.9359	796.350	795.243	6.636624	3.340081	99.35	0.85	100.20
198.8	99.0166	796.102	794.995	6.634555	3.337348	99.40	0.78	100.18
198.9	99.0975	795.852	794.745	6.632468	3.334587	99.45	0.72	100.17
199.0	99.1786	795.602	794.495	6.630382	3.331863	99.50	0.65	100.15
199.1	99.2599	795.350	794.243	6.628279	3.329144	99.55	0.59	100.14
199.2	99.3413	795.098	793.991	6.626176	3.326430	99.60	0.52	100.12
199.3	99.4230	794.843	793.736	6.624048	3.323686	99.65	0.46	100.11
199.4	99.5048	794.589	793.482	6.621928	3.320948	99.70	0.39	100.09
199.5	99.5867	794.332	793.225	6.619783	3.318213	99.75	0.33	100.08
199.6	99.6690	794.075	792.968	6.617639	3.315483	99.80	0.26	100.06
199.7	99.7514	793.816	792.709	6.615477	3.312758	99.85	0.20	100.05
199.8	99.8340	793.556	792.449	6.613307	3.310003	99.90	0.13	100.03
199.9	99.9170	793.295	792.188	6.611129	3.307254	99.95	0.07	100.02
200.0	100.000	793.032	791.925	6.608934	3.304509	100.00	0.00	100.00

Find Mass % Traditional Method	Vac	Air
Density of Sample (Vac)	906.12	905.03
Density of AA @ Calib Temp	793.03	791.92
Density of AA/Den of Sample = Ratio	0.87520	0.87503
(Proof/2)/100 = Vol %	62.85%	62.85%
Vol % * Ratio = Mass %	55.006%	54.996%

Methods of finding Mass % in Air and Vacuum. Also see Volume 2 for more methods.

Source Alcohol Find Mass %				
Lbs. per Wine Gal. @ True	7.1926	3.26250	Kg.	
Lbs. per Proof Gal. at True Proof	4.4954			
Lbs. per WG/Lbs. per PG = Proof Gal. per WG.	1.6000			
Divide by 2 = Absolute Alcohol Gal. per WG	0.8000		AA Lbs. per PG	
200 Proof Alc. Lbs. per WG = Absolute Alc.	6.60982	Kg.	3.30491	
AA Lb/WG * AA Gal per WG = Lbs. AA per WG	5.2878	2.39850	2.6439	
Lbs. AA per WG/Lbs. per WG = Fraction AA	0.7352		0.3676	
Result as a Percentage Alcohol % Mass (Air)	73.52%	Kg.	73.52%	Times 2
Lbs. per WG True Proof - Lbs AA = Lbs. H2O	1.9048	0.86401		

Values at 15.556°C (60° F)

Proof	Mass % Vac	Kg m3 Vac	Kg m3 Air	Lbs per Gal. Air	Lbs per Proof Gal Air	Alcohol Parts	Water Parts	Parts Total	Proof	Mass % Vac	Kg m3 Vac	Kg m3 Air	Lbs per Gal. Air	Lbs per Proof Gal Air	Alcohol Parts	Water Parts	Parts Total
0.0	0.0000	999.0160	997.9386	8.328201	N/A	0.00	100.00	100.00	7.0	2.7926	993.902	992.824	8.285518	118.3563	3.50	96.71	100.21
0.1	0.03970	998.933	997.856	8.327512	8261.2720	0.05	99.95	100.00	7.1	2.8327	993.833	992.755	8.284942	116.6846	3.55	96.66	100.21
0.2	0.0794	998.856	997.779	8.326869	4182.9226	0.10	99.91	100.01	7.2	2.8728	993.764	992.686	8.284366	115.0595	3.60	96.62	100.22
0.3	0.1191	998.780	997.703	8.326235	2776.8982	0.15	99.86	100.01	7.3	2.9129	993.695	992.617	8.283790	113.4790	3.65	96.57	100.22
0.4	0.1588	998.703	997.626	8.325592	2078.3074	0.20	99.81	100.01	7.4	2.9530	993.626	992.548	8.283214	111.9414	3.70	96.52	100.22
0.5	0.1985	998.627	997.550	8.324958	1660.5572	0.25	99.76	100.01	7.5	2.9932	993.557	992.479	8.282639	110.4448	3.75	96.48	100.23
0.6	0.2383	998.551	997.474	8.324324	1388.4491	0.30	99.72	100.02	7.6	3.0333	993.488	992.410	8.282063	108.9878	3.80	96.43	100.23
0.7	0.2780	998.474	997.397	8.323681	1188.6722	0.35	99.67	100.02	7.7	3.0734	993.419	992.341	8.281487	107.5686	3.85	96.38	100.23
0.8	0.3177	998.398	997.321	8.323047	1039.1537	0.40	99.62	100.02	7.8	3.1135	993.351	992.273	8.280920	106.1519	3.90	96.34	100.24
0.9	0.3575	998.322	997.245	8.322413	925.6327	0.45	99.57	100.02	7.9	3.1537	993.282	992.204	8.280344	104.8052	3.95	96.29	100.24
1.0	0.3972	998.247	997.169	8.321779	832.3700	0.50	99.53	100.03	8.0	3.1938	993.214	992.136	8.279776	103.4923	4.00	96.24	100.24
1.1	0.4370	998.171	997.093	8.321144	756.1805	0.55	99.48	100.03	8.1	3.2339	993.146	992.068	8.279209	102.2118	4.05	96.20	100.25
1.2	0.4767	998.096	997.018	8.320518	692.7691	0.60	99.43	100.03	8.2	3.2741	993.077	991.999	8.278633	100.9627	4.10	96.15	100.25
1.3	0.5165	998.020	996.942	8.319884	640.4087	0.65	99.38	100.03	8.3	3.3143	993.009	991.931	8.278065	99.743701	4.15	96.10	100.26
1.4	0.5563	997.945	996.867	8.319258	594.3361	0.70	99.34	100.04	8.4	3.3544	992.941	991.863	8.277498	98.553797	4.20	96.06	100.26
1.5	0.5960	997.870	996.792	8.318632	554.4478	0.75	99.29	100.04	8.5	3.3946	992.873	991.795	8.276930	97.363253	4.25	96.01	100.26
1.6	0.6358	997.795	996.717	8.318006	519.5769	0.80	99.24	100.04	8.6	3.4348	992.805	991.727	8.276363	96.229144	4.30	95.96	100.26
1.7	0.6756	997.720	996.642	8.317381	488.8327	0.85	99.19	100.04	8.7	3.4749	992.737	991.659	8.275795	95.121152	4.35	95.92	100.27
1.8	0.7154	997.645	996.567	8.316755	462.1691	0.90	99.15	100.05	8.8	3.5151	992.670	991.592	8.275236	94.038384	4.40	95.87	100.27
1.9	0.7552	997.570	996.492	8.316129	437.6833	0.95	99.10	100.05	8.9	3.5553	992.602	991.524	8.274669	92.979989	4.45	95.83	100.28
2.0	0.7950	997.495	996.417	8.315503	415.6615	1.00	99.05	100.05	9.0	3.5955	992.535	991.457	8.274110	91.945153	4.50	95.78	100.28
2.1	0.8348	997.421	996.343	8.314885	395.7496	1.05	99.01	100.06	9.1	3.6357	992.467	991.389	8.273542	90.908083	4.55	95.73	100.28
2.2	0.8747	997.346	996.268	8.314259	377.6582	1.10	98.96	100.06	9.2	3.6759	992.400	991.322	8.272983	89.918607	4.60	95.69	100.29
2.3	0.9145	997.272	996.194	8.313642	361.5436	1.15	98.91	100.06	9.3	3.7161	992.333	991.255	8.272424	88.950439	4.65	95.64	100.29
2.4	0.9543	997.198	996.120	8.313024	346.3846	1.20	98.87	100.07	9.4	3.7563	992.266	991.188	8.271865	88.002898	4.70	95.59	100.29
2.5	0.9941	997.124	996.046	8.312407	332.4456	1.25	98.82	100.07	9.5	3.7965	992.199	991.121	8.271306	87.052392	4.75	95.55	100.30
2.6	1.0340	997.050	995.972	8.311789	319.5850	1.30	98.77	100.07	9.6	3.8367	992.132	991.054	8.270746	86.144651	4.80	95.50	100.30
2.7	1.0738	996.976	995.898	8.311172	307.6824	1.35	98.72	100.07	9.7	3.8770	992.065	990.987	8.270187	85.255645	4.85	95.45	100.30
2.8	1.1137	996.903	995.825	8.310562	296.6345	1.40	98.68	100.08	9.8	3.9172	991.998	990.920	8.269628	84.384801	4.90	95.41	100.31
2.9	1.1536	996.829	995.751	8.309945	286.3526	1.45	98.63	100.08	9.9	3.9574	991.932	990.854	8.269077	83.531568	4.95	95.36	100.31
3.0	1.1934	996.756	995.678	8.309336	276.9915	1.50	98.58	100.08	10.0	3.9977	991.865	990.787	8.268518	82.674726	5.00	95.32	100.32
3.1	1.2333	996.682	995.604	8.308718	268.0056	1.55	98.54	100.09	10.1	4.0379	991.798	990.720	8.267959	81.855556	5.05	95.27	100.32
3.2	1.2732	996.609	995.531	8.308109	259.5844	1.60	98.49	100.09	10.2	4.0782	991.732	990.654	8.267408	81.052461	5.10	95.22	100.32
3.3	1.3131	996.536	995.458	8.307500	251.6762	1.65	98.44	100.09	10.3	4.1184	991.666	990.588	8.266857	80.264970	5.15	95.18	100.33
3.4	1.3529	996.463	995.385	8.306890	244.2357	1.70	98.39	100.09	10.4	4.1587	991.599	990.521	8.266298	79.473516	5.20	95.13	100.33
3.5	1.3928	996.390	995.312	8.306281	237.2225	1.75	98.35	100.10	10.5	4.1990	991.533	990.455	8.265748	78.716265	5.25	95.08	100.33
3.6	1.4327	996.317	995.239	8.305672	230.6008	1.80	98.30	100.10	10.6	4.2392	991.467	990.389	8.265197	77.973309	5.30	95.04	100.34
3.7	1.4726	996.244	995.166	8.305063	224.3387	1.85	98.25	100.10	10.7	4.2795	991.401	990.322	8.264638	77.226193	5.35	94.99	100.34
3.8	1.5126	996.171	995.093	8.304454	218.4077	1.90	98.21	100.11	10.8	4.3198	991.335	990.256	8.264087	76.510970	5.40	94.94	100.34
3.9	1.5525	996.099	995.021	8.303853	212.9194	1.95	98.16	100.11	10.9	4.3601	991.270	990.191	8.263544	75.808874	5.45	94.90	100.35
4.0	1.5924	996.027	994.949	8.303252	207.5696	2.00	98.11	100.12	11.0	4.4004	991.204	990.125	8.262994	75.119546	5.50	94.85	100.35
4.1	1.6323	995.954	994.876	8.302643	202.6063	2.05	98.07	100.12	11.1	4.4407	991.138	990.059	8.262443	74.425874	5.55	94.81	100.36
4.2	1.6723	995.882	994.804	8.302042	197.7564	2.10	98.02	100.12	11.2	4.4810	991.073	989.994	8.261900	73.761357	5.60	94.76	100.36
4.3	1.7122	995.810	994.732	8.301441	193.1332	2.15	97.97	100.12	11.3	4.5213	991.007	989.928	8.261350	73.108602	5.65	94.71	100.36
4.4	1.7521	995.738	994.660	8.300840	188.7212	2.20	97.93	100.13	11.4	4.5616	990.942	989.863	8.260807	72.451410	5.70	94.67	100.37
4.5	1.7921	995.666	994.588	8.300239	184.5064	2.25	97.88	100.13	11.5	4.6019	990.877	989.798	8.260265	71.821535	5.75	94.62	100.37
4.6	1.8320	995.594	994.516	8.299638	180.4756	2.30	97.83	100.13	11.6	4.6422	990.812	989.733	8.259722	71.202517	5.80	94.57	100.37
4.7	1.8720	995.523	994.445	8.299046	176.6173	2.35	97.79	100.14	11.7	4.6826	990.747	989.668	8.259180	70.579001	5.85	94.53	100.38
4.8	1.9120	995.451	994.373	8.298446	172.9204	2.40	97.74	100.14	11.8	4.7229	990.682	989.603	8.258637	69.981127	5.90	94.48	100.38
4.9	1.9519	995.380	994.302	8.297852	169.3751	2.45	97.69	100.14	11.9	4.7632	990.617	989.538	8.258095	69.393297	5.95	94.44	100.39
5.0	1.9919	995.308	994.230	8.297251	165.9723	2.50	97.64	100.14	12.0	4.8036	990.552	989.473	8.257552	68.800933	6.00	94.39	100.39
5.1	2.0319	995.237	994.159	8.296659	162.7035	2.55	97.60	100.15	12.1	4.8439	990.487	989.408	8.257010	68.232683	6.05	94.34	100.39
5.2	2.0719	995.166	994.088	8.296066	159.5610	2.60	97.55	100.15	12.2	4.8843	990.423	989.344	8.256476	67.673742	6.10	94.30	100.40
5.3	2.1119	995.095	994.017	8.295474	156.5376	2.65	97.50	100.15	12.3	4.9246	990.358	989.279	8.255933	67.123884	6.15	94.25	100.40
5.4	2.1519	995.024	993.946	8.294881	153.6266	2.70	97.46	100.16	12.4	4.9650	990.294	989.215	8.255399	66.582890	6.20	94.21	100.41
5.5	2.1919	994.953	993.875	8.294289	150.8219	2.75	97.41	100.16	12.5	5.0054	990.229	989.150	8.254857	66.050546	6.25	94.16	100.41
5.6	2.2319	994.882	993.804	8.293696	148.1178	2.80	97.36	100.16	12.6	5.0457	990.165	989.086	8.254323	65.513656	6.30	94.11	100.41
5.7	2.2719	994.812	993.734	8.293112	145.5090	2.85	97.32	100.17	12.7	5.0861	990.101	989.022	8.253789	64.998206	6.35	94.07	100.42
5.8	2.3120	994.741	993.663	8.292520	142.9904	2.90	97.27	100.17	12.8	5.1265	990.037	988.958	8.253254	64.490804	6.40	94.02	100.42
5.9	2.3520	994.671	993.593	8.291935	140.5576	2.95	97.22	100.17	12.9	5.1669	989.973	988.894	8.252720	63.978873	6.45	93.98	100.43
6.0	2.3920	994.600	993.522	8.291343	138.2061	3.00	97.18	100.18	13.0	5.2073	989.909	988.830	8.252186	63.487201	6.50	93.93	100.43
6.1	2.4321	994.530	993.452	8.290759	135.9321	3.05	97.13	100.18	13.1	5.2476	989.845	988.766	8.251652	62.991018	6.55	93.88	100.43
6.2	2.4721	994.460	993.382	8.290175	133.7316	3.10	97.08	100.18	13.2	5.2880	989.781	988.702	8.251118	62.514355	6.60	93.84	100.44
6.3	2.5121	994.390	993.312	8.289590	131.6013	3.15	97.04	100.19	13.3	5.3285	989.717	988.638	8.250584	62.044852	6.65	93.79	100.44
6.4	2.5522	994.320	993.242	8.289006	129.5378	3.20	96.99	100.19	13.4	5.3689	989.654	988.575	8.250058	61.570874	6.70	93.75	100.45
6.5	2.5923	994.250	993.172	8.288422	127.5380	3.25	96.94	100.19	13.5	5.4093	989.590	988.511	8.249524	61.115384	6.75	93.70	100.45
6.6	2.6323	994.180	993.102	8.287838	125.5990	3.30	96.90	100.20	13.6	5.4497	989.527	988.448	8.248998	60.655446	6.80	93.65	100.45
6.7	2.6724	994.111	993.033	8.287262	123.7180	3.35	96.85	100.20	13.7	5.4901	989.463	988.384	8.248464	60.213353	6.85	93.61	100.46
6.8	2.7125	994.041	992.963	8.286678	121.8926	3.40	96.80	100.20	13.8	5.5305	989.400	988.321	8.247938	59.766844	6.90	93.56	100.46
6.9	2.7526	993.971	992.893	8.286094	120.0766	3.45	96.76	100.21	13.9	5.5710	989.337	988.258	8.247413	59.337562	6.95	93.51	100.46
7.0	2.7926	993.902	992.824	8.285518	118.3563	3.50	96.71	100.21	14.0	5.6114	989.274	988.195	8.246887	58.914402	7.00	93.47	100.47

Values at 15.556°C (60° F)

Proof	Mass % Vac	Kg m3 Vac	Kg m3 Air	Lbs per Gal. Air	Lbs per Proof Gal Air	Alcohol Parts	Water Parts	Parts Total	Proof	Mass % Vac	Kg m3 Vac	Kg m3 Air	Lbs per Gal. Air	Lbs per Proof Gal Air	Alcohol Parts	Water Parts	Parts Total
14.0	5.6114	989.274	988.195	8.246887	58.914402	7.00	93.47	100.47	21.0	8.4532	985.052	983.973	8.211653	39.106613	10.50	90.27	100.77
14.1	5.6519	989.211	988.132	8.246361	58.486882	7.05	93.42	100.47	21.1	8.4939	984.994	983.915	8.211169	38.917781	10.55	90.22	100.77
14.2	5.6923	989.148	988.069	8.245835	58.075726	7.10	93.38	100.48	21.2	8.5347	984.936	983.857	8.210685	38.730764	10.60	90.18	100.78
14.3	5.7328	989.085	988.006	8.245310	57.660248	7.15	93.33	100.48	21.3	8.5755	984.879	983.800	8.210209	38.550033	10.65	90.13	100.78
14.4	5.7732	989.022	987.943	8.244784	57.260593	7.20	93.29	100.49	21.4	8.6162	984.821	983.742	8.209725	38.366525	10.70	90.09	100.79
14.5	5.8137	988.959	987.880	8.244258	56.856655	7.25	93.24	100.49	21.5	8.6570	984.764	983.685	8.209249	38.184756	10.75	90.04	100.79
14.6	5.8541	988.897	987.818	8.243741	56.468025	7.30	93.19	100.49	21.6	8.6978	984.707	983.628	8.208773	38.004702	10.80	90.00	100.80
14.7	5.8946	988.834	987.755	8.243215	56.075154	7.35	93.15	100.50	21.7	8.7385	984.649	983.570	8.208289	37.826337	10.85	89.95	100.80
14.8	5.9351	988.772	987.693	8.242698	55.697098	7.40	93.10	100.50	21.8	8.7793	984.592	983.513	8.207814	37.653929	10.90	89.91	100.81
14.9	5.9756	988.709	987.630	8.242172	55.314845	7.45	93.05	100.50	21.9	8.8201	984.535	983.455	8.207330	37.478834	10.95	89.86	100.81
15.0	6.0160	988.647	987.568	8.241654	54.946937	7.50	93.01	100.51	22.0	8.8609	984.478	983.398	8.206854	37.305360	11.00	89.81	100.81
15.1	6.0565	988.585	987.506	8.241137	54.574877	7.55	92.96	100.51	22.1	8.9017	984.421	983.341	8.206378	37.133485	11.05	89.77	100.82
15.2	6.0970	988.523	987.444	8.240620	54.216716	7.60	92.92	100.52	22.2	8.9425	984.364	983.284	8.205903	36.967321	11.10	89.72	100.82
15.3	6.1375	988.461	987.382	8.240102	53.854446	7.65	92.87	100.52	22.3	8.9833	984.307	983.227	8.205427	36.798539	11.15	89.68	100.83
15.4	6.1780	988.399	987.320	8.239585	53.505648	7.70	92.83	100.53	22.4	9.0241	984.250	983.170	8.204951	36.631292	11.20	89.63	100.83
15.5	6.2185	988.337	987.258	8.239067	53.152788	7.75	92.78	100.53	22.5	9.0649	984.194	983.114	8.204484	36.465557	11.25	89.59	100.84
15.6	6.2590	988.275	987.196	8.238550	52.812990	7.80	92.73	100.53	22.6	9.1057	984.137	983.057	8.204008	36.301316	11.30	89.54	100.84
15.7	6.2996	988.213	987.134	8.238032	52.469178	7.85	92.69	100.54	22.7	9.1465	984.080	983.000	8.203533	36.138548	11.35	89.50	100.85
15.8	6.3401	988.152	987.073	8.237523	52.138037	7.90	92.64	100.54	22.8	9.1873	984.024	982.944	8.203065	35.981150	11.40	89.45	100.85
15.9	6.3806	988.090	987.011	8.237006	51.802929	7.95	92.60	100.55	22.9	9.2282	983.967	982.887	8.202591	35.821234	11.45	89.41	100.86
16.0	6.4211	988.029	986.950	8.236497	51.480119	8.00	92.55	100.55	23.0	9.2690	983.911	982.831	8.202122	35.662733	11.50	89.36	100.86
16.1	6.4617	987.967	986.888	8.235980	51.153387	8.05	92.50	100.55	23.1	9.3098	983.855	982.775	8.201655	35.505628	11.55	89.31	100.86
16.2	6.5022	987.906	986.827	8.235470	50.838597	8.10	92.46	100.56	23.2	9.3507	983.798	982.718	8.201179	35.349902	11.60	89.27	100.87
16.3	6.5427	987.845	986.766	8.234961	50.519933	8.15	92.41	100.56	23.3	9.3915	983.742	982.662	8.200712	35.195535	11.65	89.22	100.87
16.4	6.5833	987.783	986.704	8.234444	50.212868	8.20	92.37	100.57	23.4	9.4324	983.686	982.606	8.200244	35.046228	11.70	89.18	100.88
16.5	6.6238	987.722	986.643	8.233935	49.901975	8.25	92.32	100.57	23.5	9.4732	983.630	982.550	8.199777	34.894496	11.75	89.13	100.88
16.6	6.6644	987.661	986.582	8.233426	49.602354	8.30	92.28	100.58	23.6	9.5141	983.574	982.494	8.199310	34.744073	11.80	89.09	100.89
16.7	6.7050	987.600	986.521	8.232917	49.298953	8.35	92.23	100.58	23.7	9.5549	983.518	982.438	8.198842	34.594942	11.85	89.04	100.89
16.8	6.7455	987.540	986.461	8.232416	48.999241	8.40	92.18	100.58	23.8	9.5958	983.462	982.382	8.198375	34.447084	11.90	89.00	100.90
16.9	6.7861	987.479	986.400	8.231907	48.710330	8.45	92.14	100.59	23.9	9.6366	983.406	982.326	8.197908	34.300486	11.95	88.95	100.90
17.0	6.8267	987.418	986.339	8.231398	48.417712	8.50	92.09	100.59	24.0	9.6775	983.351	982.271	8.197449	34.155130	12.00	88.91	100.91
17.1	6.8672	987.357	986.278	8.230889	48.135598	8.55	92.05	100.60	24.1	9.7184	983.295	982.215	8.196981	34.014501	12.05	88.86	100.91
17.2	6.9078	987.297	986.218	8.230388	47.849824	8.60	92.00	100.60	24.2	9.7593	983.239	982.159	8.196514	33.871554	12.10	88.82	100.92
17.3	6.9484	987.236	986.157	8.229879	47.574270	8.65	91.96	100.61	24.3	9.8001	983.184	982.104	8.196055	33.729803	12.15	88.77	100.92
17.4	6.9890	987.176	986.097	8.229378	47.295103	8.70	91.91	100.61	24.4	9.8410	983.128	982.048	8.195588	33.589234	12.20	88.73	100.93
17.5	7.0296	987.116	986.037	8.228878	47.019192	8.75	91.86	100.61	24.5	9.8819	983.073	981.993	8.195129	33.449831	12.25	88.68	100.93
17.6	7.0702	987.055	985.976	8.228368	46.753096	8.80	91.82	100.62	24.6	9.9228	983.017	981.937	8.194661	33.311581	12.30	88.64	100.94
17.7	7.1108	986.995	985.916	8.227868	46.483455	8.85	91.77	100.62	24.7	9.9637	982.962	981.882	8.194202	33.174469	12.35	88.59	100.94
17.8	7.1514	986.935	985.856	8.227367	46.223371	8.90	91.73	100.63	24.8	10.0046	982.907	981.827	8.193743	33.038480	12.40	88.54	100.94
17.9	7.1920	986.875	985.796	8.226866	45.959789	8.95	91.68	100.63	24.9	10.0455	982.852	981.772	8.193284	32.903603	12.45	88.50	100.95
18.0	7.2327	986.815	985.736	8.226366	45.699195	9.00	91.63	100.63	25.0	10.0864	982.796	981.716	8.192817	32.773072	12.50	88.46	100.96
18.1	7.2733	986.755	985.676	8.225865	45.447790	9.05	91.59	100.64	25.1	10.1273	982.741	981.661	8.192358	32.640348	12.55	88.41	100.96
18.2	7.3139	986.696	985.617	8.225372	45.192954	9.10	91.54	100.64	25.2	10.1683	982.686	981.606	8.191899	32.508695	12.60	88.36	100.96
18.3	7.3545	986.636	985.557	8.224872	44.940960	9.15	91.50	100.65	25.3	10.2092	982.631	981.551	8.191440	32.378099	12.65	88.32	100.97
18.4	7.3952	986.576	985.497	8.224371	44.697806	9.20	91.45	100.65	25.4	10.2501	982.576	981.496	8.190981	32.248549	12.70	88.27	100.97
18.5	7.4358	986.517	985.438	8.223879	44.451289	9.25	91.41	100.66	25.5	10.2910	982.521	981.441	8.190522	32.120031	12.75	88.23	100.98
18.6	7.4765	986.457	985.378	8.223378	44.213391	9.30	91.36	100.66	25.6	10.3320	982.467	981.387	8.190071	31.992534	12.80	88.18	100.98
18.7	7.5171	986.398	985.319	8.222886	43.972173	9.35	91.32	100.67	25.7	10.3729	982.412	981.332	8.189612	31.866045	12.85	88.14	100.99
18.8	7.5578	986.339	985.260	8.222393	43.733574	9.40	91.27	100.67	25.8	10.4138	982.357	981.277	8.189153	31.740551	12.90	88.09	100.99
18.9	7.5984	986.279	985.200	8.221892	43.503276	9.45	91.22	100.67	25.9	10.4548	982.303	981.223	8.188703	31.616043	12.95	88.05	101.00
19.0	7.6391	986.220	985.141	8.221400	43.269724	9.50	91.18	100.68	26.0	10.4957	982.248	981.168	8.188244	31.492508	13.00	88.00	101.00
19.1	7.6797	986.161	985.082	8.220908	43.038667	9.55	91.13	100.68	26.1	10.5367	982.193	981.113	8.187785	31.369934	13.05	87.96	101.01
19.2	7.7204	986.102	985.023	8.220415	42.815610	9.60	91.09	100.69	26.2	10.5776	982.139	981.059	8.187334	31.248310	13.10	87.91	101.01
19.3	7.7611	986.043	984.964	8.219923	42.589365	9.65	91.04	100.69	26.3	10.6186	982.085	981.005	8.186883	31.127626	13.15	87.87	101.02
19.4	7.8018	985.984	984.905	8.219431	42.365498	9.70	91.00	100.70	26.4	10.6596	982.030	980.950	8.186424	31.007871	13.20	87.82	101.02
19.5	7.8424	985.925	984.846	8.218938	42.149347	9.75	90.95	100.70	26.5	10.7005	981.976	980.896	8.185974	30.889034	13.25	87.78	101.03
19.6	7.8831	985.867	984.788	8.218454	41.930070	9.80	90.91	100.71	26.6	10.7415	981.922	980.842	8.185523	30.771104	13.30	87.73	101.03
19.7	7.9238	985.808	984.729	8.217962	41.713063	9.85	90.86	100.71	26.7	10.7825	981.867	980.787	8.185064	30.656915	13.35	87.69	101.04
19.8	7.9645	985.749	984.670	8.217469	41.498290	9.90	90.81	100.71	26.8	10.8235	981.813	980.733	8.184614	30.540747	13.40	87.64	101.04
19.9	8.0052	985.691	984.612	8.216985	41.290876	9.95	90.77	100.72	26.9	10.8645	981.759	980.679	8.184163	30.425456	13.45	87.60	101.05
20.0	8.0459	985.632	984.553	8.216493	41.080418	10.00	90.72	100.72	27.0	10.9054	981.705	980.625	8.183712	30.311033	13.50	87.55	101.05
20.1	8.0866	985.574	984.495	8.216009	40.872094	10.05	90.68	100.73	27.1	10.9464	981.651	980.571	8.183262	30.197467	13.55	87.51	101.06
20.2	8.1273	985.516	984.437	8.215525	40.670878	10.10	90.63	100.73	27.2	10.9874	981.597	980.517	8.182811	30.084749	13.60	87.46	101.06
20.3	8.1681	985.457	984.378	8.215033	40.466677	10.15	90.59	100.74	27.3	11.0284	981.543	980.463	8.182360	29.972869	13.65	87.42	101.07
20.4	8.2088	985.399	984.320	8.214549	40.264516	10.20	90.54	100.74	27.4	11.0694	981.490	980.410	8.181918	29.861818	13.70	87.37	101.07
20.5	8.2495	985.341	984.262	8.214064	40.064365	10.25	90.49	100.75	27.5	11.1104	981.436	980.356	8.181467	29.751587	13.75	87.33	101.08
20.6	8.2902	985.283	984.204	8.213580	39.871004	10.30	90.45	100.75	27.6	11.1515	981.382	980.302	8.181017	29.642167	13.80	87.28	101.08
20.7	8.3310	985.225	984.146	8.213096	39.679501	10.35	90.40	100.75	27.7	11.1925	981.328	980.248	8.180566	29.533549	13.85	87.24	101.09
20.8	8.3717	985.167	984.088	8.212612	39.485110	10.40	90.36	100.76	27.8	11.2335	981.275	980.195	8.180124	29.423104	13.90	87.19	101.09
20.9	8.4125	985.109	984.030	8.212128	39.297286	10.45	90.31	100.76	27.9	11.2745	981.221	980.141	8.179673	29.316082	13.95	87.15	101.10
21.0	8.4532	985.052	983.973	8.211653	39.106613	10.50	90.27	100.77	28.0	11.3155	981.168	980.088	8.179231	29.209837	14.00	87.10	101.10

Values at 15.556°C (60° F)

Proof	Mass % Vac	Kg m3 Vac	Kg m3 Air	Lbs per Gal. Air	Lbs per Proof Gal Air	Alcohol Parts	Water Parts	Parts Total	Proof	Mass % Vac	Kg m3 Vac	Kg m3 Air	Lbs per Gal. Air	Lbs per Proof Gal Air	Alcohol Parts	Water Parts	Parts Total
28.0	11.3155	981.168	980.088	8.179231	29.209837	14.00	87.10	101.10	35.0	14.1972	977.524	976.443	8.148812	23.282666	17.50	83.96	101.46
28.1	11.3566	981.114	980.034	8.178780	29.104358	14.05	87.06	101.11	35.1	14.2385	977.473	976.392	8.148386	23.215602	17.55	83.91	101.46
28.2	11.3976	981.061	979.981	8.178338	28.999639	14.10	87.01	101.11	35.2	14.2798	977.422	976.341	8.147961	23.147302	17.60	83.87	101.47
28.3	11.4387	981.008	979.928	8.177895	28.895670	14.15	86.97	101.12	35.3	14.3211	977.371	976.290	8.147535	23.081014	17.65	83.82	101.47
28.4	11.4797	980.954	979.874	8.177445	28.792444	14.20	86.92	101.12	35.4	14.3624	977.321	976.240	8.147118	23.015105	17.70	83.78	101.48
28.5	11.5207	980.901	979.821	8.177003	28.689953	14.25	86.88	101.13	35.5	14.4037	977.270	976.189	8.146692	22.947978	17.75	83.73	101.48
28.6	11.5618	980.848	979.768	8.176560	28.588189	14.30	86.83	101.13	35.6	14.4450	977.219	976.138	8.146266	22.882825	17.80	83.69	101.49
28.7	11.6029	980.794	979.714	8.176110	28.487145	14.35	86.79	101.14	35.7	14.4864	977.168	976.087	8.145841	22.818042	17.85	83.65	101.50
28.8	11.6439	980.741	979.661	8.175667	28.386812	14.40	86.74	101.14	35.8	14.5277	977.117	976.036	8.145415	22.752057	17.90	83.60	101.50
28.9	11.6850	980.688	979.608	8.175225	28.287184	14.45	86.70	101.15	35.9	14.5690	977.067	975.986	8.144998	22.688011	17.95	83.56	101.51
29.0	11.7260	980.635	979.555	8.174783	28.188252	14.50	86.65	101.15	36.0	14.6104	977.016	975.935	8.144572	22.624324	18.00	83.51	101.51
29.1	11.7671	980.582	979.502	8.174340	28.090010	14.55	86.61	101.16	36.1	14.6517	976.965	975.884	8.144147	22.559454	18.05	83.47	101.52
29.2	11.8082	980.529	979.449	8.173898	27.992451	14.60	86.56	101.16	36.2	14.6931	976.915	975.834	8.143729	22.496486	18.10	83.42	101.52
29.3	11.8493	980.476	979.396	8.173456	27.895567	14.65	86.52	101.17	36.3	14.7344	976.864	975.783	8.143304	22.433868	18.15	83.38	101.53
29.4	11.8903	980.424	979.344	8.173022	27.801690	14.70	86.47	101.17	36.4	14.7758	976.813	975.732	8.142878	22.370084	18.20	83.33	101.53
29.5	11.9314	980.371	979.291	8.172579	27.706119	14.75	86.43	101.18	36.5	14.8171	976.763	975.682	8.142461	22.308167	18.25	83.29	101.54
29.6	11.9725	980.318	979.238	8.172137	27.608896	14.80	86.38	101.18	36.6	14.8585	976.712	975.631	8.142035	22.245095	18.30	83.24	101.54
29.7	12.0136	980.265	979.185	8.171695	27.514645	14.85	86.34	101.19	36.7	14.8999	976.661	975.580	8.141610	22.183867	18.35	83.20	101.55
29.8	12.0547	980.213	979.133	8.171261	27.421034	14.90	86.29	101.19	36.8	14.9412	976.611	975.530	8.141192	22.122975	18.40	83.15	101.55
29.9	12.0958	980.160	979.080	8.170819	27.328058	14.95	86.25	101.20	36.9	14.9826	976.560	975.479	8.140767	22.060944	18.45	83.11	101.56
30.0	12.1369	980.107	979.027	8.170376	27.235711	15.00	86.20	101.20	37.0	15.0240	976.510	975.429	8.140350	22.000724	18.50	83.06	101.56
30.1	12.1780	980.055	978.975	8.169942	27.143986	15.05	86.16	101.21	37.1	15.0654	976.459	975.378	8.139924	21.940833	18.55	83.02	101.57
30.2	12.2191	980.002	978.922	8.169500	27.052876	15.10	86.11	101.21	37.2	15.1068	976.409	975.328	8.139507	21.879817	18.60	82.97	101.57
30.3	12.2603	979.950	978.870	8.169066	26.962376	15.15	86.07	101.22	37.3	15.1482	976.358	975.277	8.139081	21.820581	18.65	82.93	101.58
30.4	12.3014	979.897	978.817	8.168624	26.872480	15.20	86.02	101.22	37.4	15.1896	976.308	975.227	8.138664	21.760232	18.70	82.88	101.58
30.5	12.3425	979.845	978.765	8.168190	26.781010	15.25	85.98	101.23	37.5	15.2310	976.257	975.176	8.138238	21.701641	18.75	82.84	101.59
30.6	12.3836	979.793	978.713	8.167756	26.692317	15.30	85.93	101.23	37.6	15.2724	976.207	975.126	8.137821	21.643364	18.80	82.80	101.60
30.7	12.4248	979.740	978.660	8.167314	26.604209	15.35	85.89	101.24	37.7	15.3138	976.156	975.075	8.137395	21.583990	18.85	82.75	101.60
30.8	12.4659	979.688	978.608	8.166880	26.516681	15.40	85.84	101.24	37.8	15.3552	976.106	975.025	8.136978	21.526342	18.90	82.71	101.61
30.9	12.5070	979.636	978.556	8.166446	26.429727	15.45	85.80	101.25	37.9	15.3966	976.055	974.974	8.136552	21.467607	18.95	82.66	101.61
31.0	12.5482	979.584	978.504	8.166012	26.343342	15.50	85.75	101.25	38.0	15.4380	976.005	974.924	8.136135	21.410579	19.00	82.62	101.62
31.1	12.5893	979.532	978.452	8.165578	26.257519	15.55	85.71	101.26	38.1	15.4795	975.954	974.873	8.135709	21.353853	19.05	82.57	101.62
31.2	12.6305	979.479	978.399	8.165135	26.172254	15.60	85.66	101.26	38.2	15.5209	975.904	974.823	8.135292	21.296055	19.10	82.53	101.63
31.3	12.6716	979.427	978.347	8.164701	26.085482	15.65	85.62	101.27	38.3	15.5623	975.854	974.773	8.134875	21.239933	19.15	82.48	101.63
31.4	12.7128	979.375	978.295	8.164267	26.001328	15.70	85.57	101.27	38.4	15.6038	975.803	974.722	8.134449	21.184107	19.20	82.44	101.64
31.5	12.7540	979.323	978.243	8.163833	25.917716	15.75	85.53	101.28	38.5	15.6452	975.753	974.672	8.134032	21.128573	19.25	82.39	101.64
31.6	12.7951	979.271	978.191	8.163400	25.834640	15.80	85.48	101.28	38.6	15.6867	975.702	974.621	8.133606	21.071986	19.30	82.35	101.65
31.7	12.8363	979.220	978.140	8.162974	25.752095	15.85	85.44	101.29	38.7	15.7281	975.652	974.571	8.133189	21.017038	19.35	82.30	101.65
31.8	12.8775	979.168	978.088	8.162540	25.670075	15.90	85.39	101.29	38.8	15.7696	975.602	974.521	8.132772	20.961045	19.40	82.26	101.66
31.9	12.9187	979.116	978.036	8.162106	25.586596	15.95	85.35	101.30	38.9	15.8110	975.551	974.470	8.132346	20.906674	19.45	82.21	101.66
32.0	12.9598	979.064	977.984	8.161672	25.505625	16.00	85.30	101.30	39.0	15.8525	975.501	974.420	8.131929	20.851267	19.50	82.17	101.67
32.1	13.0010	979.012	977.932	8.161238	25.425166	16.05	85.26	101.31	39.1	15.8940	975.451	974.370	8.131512	20.797462	19.55	82.12	101.67
32.2	13.0422	978.960	977.880	8.160804	25.345213	16.10	85.21	101.31	39.2	15.9354	975.400	974.319	8.131086	20.743935	19.60	82.08	101.68
32.3	13.0834	978.909	977.829	8.160378	25.265760	16.15	85.17	101.32	39.3	15.9769	975.350	974.269	8.130669	20.689387	19.65	82.03	101.68
32.4	13.1246	978.857	977.777	8.159945	25.184885	16.20	85.12	101.32	39.4	16.0184	975.300	974.219	8.130252	20.636413	19.70	81.99	101.69
32.5	13.1658	978.805	977.725	8.159511	25.106434	16.25	85.08	101.33	39.5	16.0599	975.249	974.168	8.129826	20.582428	19.75	81.95	101.70
32.6	13.2070	978.754	977.674	8.159085	25.028469	16.30	85.03	101.33	39.6	16.1014	975.199	974.118	8.129409	20.530000	19.80	81.90	101.70
32.7	13.2482	978.702	977.622	8.158651	24.950988	16.35	84.99	101.34	39.7	16.1429	975.149	974.068	8.128991	20.476570	19.85	81.86	101.71
32.8	13.2894	978.651	977.571	8.158225	24.873984	16.40	84.95	101.35	39.8	16.1844	975.098	974.017	8.128566	20.424679	19.90	81.81	101.71
32.9	13.3307	978.599	977.519	8.157791	24.795594	16.45	84.90	101.35	39.9	16.2259	975.048	973.967	8.128149	20.371795	19.95	81.77	101.72
33.0	13.3719	978.548	977.468	8.157366	24.719608	16.50	84.85	101.35	40.0	16.2674	974.998	973.917	8.127731	20.320433	20.00	81.72	101.72
33.1	13.4131	978.496	977.416	8.156932	24.643962	16.55	84.81	101.36	40.1	16.3089	974.947	973.866	8.127306	20.268086	20.05	81.68	101.73
33.2	13.4543	978.445	977.365	8.156506	24.568839	16.60	84.77	101.37	40.2	16.3504	974.897	973.816	8.126888	20.217246	20.10	81.63	101.73
33.3	13.4956	978.393	977.313	8.156072	24.492357	16.65	84.72	101.37	40.3	16.3919	974.847	973.766	8.126471	20.165429	20.15	81.59	101.74
33.4	13.5368	978.342	977.262	8.155647	24.418154	16.70	84.68	101.38	40.4	16.4334	974.797	973.716	8.126054	20.115101	20.20	81.54	101.74
33.5	13.5781	978.291	977.211	8.155221	24.344400	16.75	84.63	101.38	40.5	16.4750	974.746	973.665	8.125628	20.063806	20.25	81.50	101.75
33.6	13.6193	978.240	977.160	8.154795	24.271089	16.80	84.59	101.39	40.6	16.5165	974.696	973.615	8.125211	20.013983	20.30	81.45	101.75
33.7	13.6606	978.188	977.108	8.154361	24.196447	16.85	84.54	101.39	40.7	16.5580	974.646	973.565	8.124794	19.963202	20.35	81.41	101.76
33.8	13.7018	978.137	977.057	8.153936	24.124024	16.90	84.50	101.40	40.8	16.5996	974.595	973.514	8.124368	19.912677	20.40	81.36	101.76
33.9	13.7431	978.086	977.006	8.153510	24.052033	16.95	84.45	101.40	40.9	16.6411	974.545	973.464	8.123951	19.863602	20.45	81.32	101.77
34.0	13.7843	978.035	976.955	8.153085	23.980470	17.00	84.41	101.41	41.0	16.6826	974.495	973.414	8.123534	19.813580	20.50	81.27	101.77
34.1	13.8256	977.983	976.903	8.152651	23.907602	17.05	84.36	101.41	41.1	16.7242	974.444	973.363	8.123108	19.764991	20.55	81.23	101.78
34.2	13.8669	977.932	976.852	8.152225	23.836895	17.10	84.32	101.42	41.2	16.7658	974.394	973.313	8.122691	19.715463	20.60	81.18	101.78
34.3	13.9081	977.881	976.801	8.151799	23.766605	17.15	84.27	101.42	41.3	16.8073	974.344	973.263	8.122273	19.667354	20.65	81.14	101.79
34.4	13.9494	977.830	976.750	8.151374	23.696729	17.20	84.23	101.43	41.4	16.8489	974.293	973.212	8.121848	19.618314	20.70	81.10	101.80
34.5	13.9907	977.779	976.699	8.150948	23.625572	17.25	84.18	101.43	41.5	16.8905	974.243	973.162	8.121430	19.570677	20.75	81.05	101.80
34.6	14.0320	977.728	976.648	8.150523	23.556521	17.30	84.14	101.44	41.6	16.9320	974.193	973.112	8.121013	19.522117	20.80	81.01	101.81
34.7	14.0733	977.677	976.597	8.150097	23.487873	17.35	84.09	101.44	41.7	16.9736	974.142	973.061	8.120588	19.473798	20.85	80.96	101.81
34.8	14.1146	977.626	976.545	8.149663	23.417963	17.40	84.05	101.45	41.8	17.0152	974.092	973.011	8.120170	19.426860	20.90	80.92	101.82
34.9	14.1559	977.575	976.494	8.149237	23.350119	17.45	84.00	101.45	41.9	17.0568	974.042	972.961	8.119753	19.379010	20.95	80.87	101.82
35.0	14.1972	977.524	976.443	8.148812	23.282666	17.50	83.96	101.46	42.0	17.0984	973.991	972.910	8.119327	19.332527	21.00	80.83	101.83

Values at 15.556°C (60° F)

Proof	Mass % Vac	Kg m3 Vac	Kg m3 Air	Lbs per Gal. Air	Lbs per Proof Gal Air	Alcohol Parts	Water Parts	Parts Total	Proof	Mass % Vac	Kg m3 Vac	Kg m3 Air	Lbs per Gal. Air	Lbs per Proof Gal Air	Alcohol Parts	Water Parts	Parts Total
42.0	17.0984	973.991	972.910	8.119327	19.332527	21.00	80.83	101.83	49.0	20.0215	970.421	969.339	8.089526	16.509337	24.50	77.69	102.19
42.1	17.1400	973.941	972.860	8.118910	19.285140	21.05	80.78	101.83	49.1	20.0634	970.369	969.287	8.089092	16.475589	24.55	77.65	102.20
42.2	17.1816	973.890	972.809	8.118485	19.237986	21.10	80.74	101.84	49.2	20.1054	970.316	969.234	8.088650	16.441160	24.60	77.60	102.20
42.3	17.2232	973.840	972.759	8.118067	19.192176	21.15	80.69	101.84	49.3	20.1473	970.264	969.182	8.088216	16.406876	24.65	77.56	102.21
42.4	17.2648	973.790	972.709	8.117650	19.145474	21.20	80.65	101.85	49.4	20.1893	970.212	969.130	8.087782	16.372734	24.70	77.51	102.21
42.5	17.3064	973.739	972.658	8.117224	19.100103	21.25	80.60	101.85	49.5	20.2312	970.160	969.078	8.087348	16.338733	24.75	77.47	102.22
42.6	17.3480	973.689	972.608	8.116807	19.053848	21.30	80.56	101.86	49.6	20.2732	970.108	969.026	8.086914	16.304874	24.80	77.42	102.22
42.7	17.3897	973.638	972.557	8.116382	19.007816	21.35	80.51	101.86	49.7	20.3152	970.056	968.974	8.086480	16.271155	24.85	77.38	102.23
42.8	17.4313	973.588	972.507	8.115964	18.963094	21.40	80.47	101.87	49.8	20.3572	970.003	968.921	8.086038	16.237575	24.90	77.33	102.23
42.9	17.4729	973.537	972.456	8.115539	18.917500	21.45	80.42	101.87	49.9	20.3991	969.951	968.869	8.085604	16.204133	24.95	77.29	102.24
43.0	17.5146	973.487	972.406	8.115121	18.872123	21.50	80.38	101.88	50.0	20.4411	969.899	968.817	8.085170	16.170829	25.00	77.24	102.24
43.1	17.5562	973.436	972.355	8.114696	18.828037	21.55	80.34	101.89	50.1	20.4831	969.846	968.764	8.084727	16.137661	25.05	77.20	102.25
43.2	17.5978	973.386	972.305	8.114278	18.783089	21.60	80.29	101.89	50.2	20.5251	969.794	968.712	8.084293	16.104629	25.10	77.15	102.25
43.3	17.6395	973.335	972.254	8.113853	18.738354	21.65	80.25	101.90	50.3	20.5671	969.741	968.659	8.083851	16.071732	25.15	77.11	102.26
43.4	17.6811	973.285	972.204	8.113436	18.694890	21.70	80.20	101.90	50.4	20.6091	969.689	968.607	8.083417	16.038969	25.20	77.06	102.26
43.5	17.7228	973.234	972.153	8.113010	18.650575	21.75	80.16	101.91	50.5	20.6511	969.636	968.554	8.082975	16.006340	25.25	77.02	102.27
43.6	17.7645	973.184	972.103	8.112593	18.607516	21.80	80.11	101.91	50.6	20.6931	969.584	968.502	8.082541	15.973843	25.30	76.97	102.27
43.7	17.8061	973.133	972.052	8.112167	18.563613	21.85	80.07	101.92	50.7	20.7351	969.531	968.449	8.082099	15.941477	25.35	76.93	102.28
43.8	17.8478	973.083	972.002	8.111750	18.519917	21.90	80.02	101.92	50.8	20.7772	969.478	968.396	8.081656	15.909243	25.40	76.88	102.28
43.9	17.8895	973.032	971.951	8.111324	18.476426	21.95	79.98	101.93	50.9	20.8192	969.425	968.343	8.081214	15.877138	25.45	76.84	102.29
44.0	17.9312	972.981	971.900	8.110899	18.434167	22.00	79.93	101.93	51.0	20.8612	969.372	968.290	8.080772	15.845163	25.50	76.79	102.29
44.1	17.9729	972.931	971.850	8.110481	18.391078	22.05	79.89	101.94	51.1	20.9033	969.320	968.238	8.080338	15.813317	25.55	76.75	102.30
44.2	18.0146	972.880	971.799	8.110056	18.348189	22.10	79.84	101.94	51.2	20.9453	969.267	968.185	8.079895	15.781598	25.60	76.70	102.30
44.3	18.0562	972.830	971.749	8.109638	18.306514	22.15	79.80	101.95	51.3	20.9874	969.214	968.132	8.079453	15.750006	25.65	76.66	102.31
44.4	18.0980	972.779	971.698	8.109213	18.264018	22.20	79.75	101.95	51.4	21.0294	969.161	968.079	8.079011	15.718541	25.70	76.61	102.31
44.5	18.1397	972.728	971.647	8.108787	18.221719	22.25	79.71	101.96	51.5	21.0715	969.108	968.026	8.078568	15.686456	25.75	76.57	102.32
44.6	18.1814	972.677	971.596	8.108362	18.180616	22.30	79.66	101.96	51.6	21.1136	969.055	967.973	8.078126	15.655244	25.80	76.52	102.32
44.7	18.2231	972.627	971.546	8.107944	18.138702	22.35	79.62	101.97	51.7	21.1557	969.001	967.919	8.077676	15.624155	25.85	76.48	102.33
44.8	18.2648	972.576	971.495	8.107519	18.096981	22.40	79.57	101.97	51.8	21.1978	968.948	967.866	8.077233	15.593190	25.90	76.43	102.33
44.9	18.3065	972.525	971.444	8.107093	18.055452	22.45	79.53	101.98	51.9	21.2398	968.895	967.813	8.076791	15.562347	25.95	76.39	102.34
45.0	18.3483	972.474	971.393	8.106667	18.015095	22.50	79.48	101.98	52.0	21.2819	968.842	967.760	8.076349	15.531626	26.00	76.34	102.34
45.1	18.3900	972.424	971.343	8.106250	17.973940	22.55	79.44	101.99	52.1	21.3240	968.789	967.707	8.075906	15.501026	26.05	76.30	102.35
45.2	18.4317	972.373	971.292	8.105825	17.932972	22.60	79.39	101.99	52.2	21.3661	968.735	967.653	8.075456	15.470547	26.10	76.25	102.35
45.3	18.4735	972.322	971.241	8.105399	17.892191	22.65	79.35	102.00	52.3	21.4083	968.682	967.600	8.075013	15.440187	26.15	76.21	102.36
45.4	18.5152	972.271	971.190	8.104973	17.852560	22.70	79.31	102.01	52.4	21.4504	968.628	967.546	8.074563	15.409946	26.20	76.16	102.36
45.5	18.5570	972.220	971.139	8.104548	17.812143	22.75	79.26	102.01	52.5	21.4925	968.575	967.493	8.074120	15.379107	26.25	76.12	102.37
45.6	18.5988	972.169	971.088	8.104122	17.771909	22.80	79.22	102.02	52.6	21.5346	968.521	967.439	8.073670	15.349105	26.30	76.07	102.37
45.7	18.6405	972.118	971.037	8.103696	17.731857	22.85	79.17	102.02	52.7	21.5768	968.467	967.385	8.073219	15.319219	26.35	76.03	102.38
45.8	18.6823	972.067	970.986	8.103271	17.692931	22.90	79.13	102.03	52.8	21.6189	968.414	967.332	8.072777	15.289450	26.40	75.98	102.38
45.9	18.7241	972.016	970.935	8.102845	17.653234	22.95	79.08	102.03	52.9	21.6611	968.360	967.278	8.072326	15.259796	26.45	75.94	102.39
46.0	18.7658	971.965	970.884	8.102420	17.613714	23.00	79.04	102.04	53.0	21.7032	968.306	967.224	8.071875	15.230257	26.50	75.89	102.39
46.1	18.8076	971.914	970.833	8.101994	17.574370	23.05	78.99	102.04	53.1	21.7454	968.252	967.170	8.071425	15.200832	26.55	75.85	102.40
46.2	18.8494	971.863	970.782	8.101568	17.536133	23.10	78.95	102.05	53.2	21.7875	968.198	967.116	8.070974	15.170824	26.60	75.80	102.40
46.3	18.8912	971.812	970.731	8.101143	17.497134	23.15	78.90	102.05	53.3	21.8297	968.144	967.062	8.070524	15.141628	26.65	75.76	102.41
46.4	18.9330	971.761	970.680	8.100717	17.458309	23.20	78.86	102.06	53.4	21.8719	968.090	967.008	8.070073	15.112544	26.70	75.71	102.41
46.5	18.9748	971.710	970.629	8.100292	17.419656	23.25	78.81	102.06	53.5	21.9141	968.036	966.954	8.069622	15.083571	26.75	75.67	102.42
46.6	19.0166	971.658	970.577	8.099858	17.381174	23.30	78.77	102.07	53.6	21.9563	967.982	966.900	8.069172	15.054710	26.80	75.62	102.42
46.7	19.0584	971.607	970.526	8.099432	17.343772	23.35	78.72	102.07	53.7	21.9984	967.928	966.846	8.068721	15.025959	26.85	75.58	102.43
46.8	19.1002	971.556	970.475	8.099006	17.305624	23.40	78.68	102.08	53.8	22.0406	967.873	966.791	8.068262	14.996636	26.90	75.53	102.43
46.9	19.1421	971.505	970.424	8.098581	17.267643	23.45	78.63	102.08	53.9	22.0829	967.819	966.737	8.067811	14.968106	26.95	75.49	102.44
47.0	19.1839	971.453	970.372	8.098147	17.229829	23.50	78.59	102.09	54.0	22.1251	967.765	966.683	8.067361	14.939685	27.00	75.44	102.44
47.1	19.2257	971.402	970.321	8.097721	17.192179	23.55	78.54	102.09	54.1	22.1673	967.710	966.628	8.066902	14.911370	27.05	75.40	102.45
47.2	19.2676	971.351	970.270	8.097296	17.154695	23.60	78.50	102.10	54.2	22.2095	967.656	966.574	8.066451	14.882493	27.10	75.35	102.45
47.3	19.3094	971.299	970.218	8.096862	17.118260	23.65	78.45	102.10	54.3	22.2518	967.601	966.519	8.065992	14.854395	27.15	75.31	102.46
47.4	19.3512	971.248	970.167	8.096436	17.081096	23.70	78.41	102.11	54.4	22.2940	967.547	966.465	8.065541	14.826404	27.20	75.26	102.46
47.5	19.3931	971.196	970.115	8.096002	17.044093	23.75	78.36	102.11	54.5	22.3362	967.492	966.410	8.065082	14.798517	27.25	75.22	102.47
47.6	19.4350	971.145	970.064	8.095576	17.008126	23.80	78.32	102.12	54.6	22.3785	967.437	966.355	8.064623	14.770735	27.30	75.17	102.47
47.7	19.4768	971.093	970.012	8.095142	16.971439	23.85	78.28	102.13	54.7	22.4207	967.383	966.301	8.064173	14.742399	27.35	75.13	102.48
47.8	19.5187	971.042	969.961	8.094717	16.934909	23.90	78.23	102.13	54.8	22.4630	967.328	966.246	8.063714	14.714828	27.40	75.08	102.48
47.9	19.5606	970.990	969.909	8.094283	16.898537	23.95	78.19	102.14	54.9	22.5053	967.273	966.191	8.063255	14.687359	27.45	75.04	102.49
48.0	19.6024	970.939	969.858	8.093857	16.863180	24.00	78.14	102.14	55.0	22.5475	967.218	966.136	8.062796	14.659342	27.50	74.99	102.49
48.1	19.6443	970.887	969.806	8.093423	16.827115	24.05	78.10	102.15	55.1	22.5898	967.163	966.081	8.062337	14.632079	27.55	74.94	102.49
48.2	19.6862	970.835	969.754	8.092989	16.791203	24.10	78.05	102.15	55.2	22.6321	967.108	966.026	8.061878	14.604918	27.60	74.90	102.50
48.3	19.7281	970.784	969.702	8.092555	16.755445	24.15	78.01	102.16	55.3	22.6744	967.053	965.971	8.061419	14.577858	27.65	74.85	102.50
48.4	19.7700	970.732	969.650	8.092121	16.719838	24.20	77.96	102.16	55.4	22.7167	966.997	965.915	8.060951	14.550257	27.70	74.81	102.51
48.5	19.8119	970.680	969.598	8.091687	16.684383	24.25	77.92	102.17	55.5	22.7590	966.942	965.860	8.060492	14.523398	27.75	74.76	102.51
48.6	19.8538	970.628	969.546	8.091253	16.649077	24.30	77.87	102.17	55.6	22.8013	966.887	965.805	8.060033	14.496639	27.80	74.72	102.52
48.7	19.8957	970.576	969.494	8.090820	16.613921	24.35	77.83	102.18	55.7	22.8436	966.832	965.750	8.059574	14.469344	27.85	74.67	102.52
48.8	19.9377	970.525	969.443	8.090394	16.578912	24.40	77.78	102.18	55.8	22.8860	966.776	965.694	8.059107	14.442783	27.90	74.63	102.53
48.9	19.9796	970.473	969.391	8.089960	16.544051	24.45	77.74	102.19	55.9	22.9283	966.720	965.638	8.058640	14.416320	27.95	74.58	102.53
49.0	20.0215	970.421	969.339	8.089526	16.509337	24.50	77.69	102.19	56.0	22.9706	966.665	965.583	8.058181	14.389326	28.00	74.54	102.54

Values at 15.556°C (60° F)

Proof	Mass % Vac	Kg m3 Vac	Kg m3 Air	Lbs per Gal. Air	Lbs per Proof Gal Air	Alcohol Parts	Water Parts	Parts Total	Proof	Mass % Vac	Kg m3 Vac	Kg m3 Air	Lbs per Gal. Air	Lbs per Proof Gal Air	Alcohol Parts	Water Parts	Parts Total
56.0	22.9706	966.665	965.583	8.058181	14.389326	28.00	74.54	102.54	63.0	25.9512	962.594	961.511	8.024198	12.737083	31.50	71.35	102.85
56.1	23.0130	966.609	965.527	8.057713	14.363058	28.05	74.49	102.54	63.1	25.9941	962.533	961.450	8.023689	12.716007	31.55	71.31	102.86
56.2	23.0553	966.554	965.472	8.057254	14.336886	28.10	74.45	102.55	63.2	26.0369	962.472	961.389	8.023180	12.695001	31.60	71.26	102.86
56.3	23.0977	966.498	965.416	8.056787	14.310189	28.15	74.40	102.55	63.3	26.0798	962.411	961.328	8.022671	12.674064	31.65	71.21	102.86
56.4	23.1400	966.442	965.360	8.056320	14.284209	28.20	74.36	102.56	63.4	26.1226	962.350	961.267	8.022162	12.653197	31.70	71.17	102.87
56.5	23.1824	966.386	965.304	8.055852	14.258322	28.25	74.31	102.56	63.5	26.1655	962.289	961.206	8.021653	12.632397	31.75	71.12	102.87
56.6	23.2248	966.330	965.248	8.055385	14.231917	28.30	74.27	102.57	63.6	26.2084	962.227	961.144	8.021135	12.612148	31.80	71.08	102.88
56.7	23.2672	966.274	965.192	8.054918	14.206220	28.35	74.22	102.57	63.7	26.2513	962.166	961.083	8.020626	12.591483	31.85	71.03	102.88
56.8	23.3096	966.218	965.136	8.054450	14.180007	28.40	74.17	102.57	63.8	26.2942	962.104	961.021	8.020109	12.570886	31.90	70.99	102.89
56.9	23.3519	966.162	965.080	8.053983	14.154497	28.45	74.13	102.58	63.9	26.3371	962.042	960.959	8.019592	12.550356	31.95	70.94	102.89
57.0	23.3944	966.105	965.023	8.053507	14.129078	28.50	74.09	102.59	64.0	26.3800	961.981	960.898	8.019082	12.529894	32.00	70.89	102.89
57.1	23.4368	966.049	964.967	8.053040	14.103149	28.55	74.04	102.59	64.1	26.4229	961.919	960.836	8.018565	12.509497	32.05	70.85	102.90
57.2	23.4792	965.993	964.911	8.052573	14.077914	28.60	73.99	102.59	64.2	26.4658	961.857	960.774	8.018048	12.489167	32.10	70.80	102.90
57.3	23.5216	965.936	964.854	8.052097	14.052172	28.65	73.95	102.60	64.3	26.5087	961.795	960.712	8.017530	12.468904	32.15	70.76	102.91
57.4	23.5640	965.880	964.798	8.051630	14.027715	28.70	73.90	102.60	64.4	26.5517	961.733	960.650	8.017013	12.448705	32.20	70.71	102.91
57.5	23.6065	965.823	964.741	8.051154	14.002749	28.75	73.86	102.61	64.5	26.5946	961.671	960.588	8.016495	12.428572	32.25	70.66	102.91
57.6	23.6489	965.766	964.684	8.050678	13.977281	28.80	73.81	102.61	64.6	26.6376	961.608	960.525	8.015970	12.408504	32.30	70.62	102.92
57.7	23.6913	965.710	964.628	8.050211	13.952495	28.85	73.77	102.62	64.7	26.6806	961.546	960.463	8.015452	12.388501	32.35	70.57	102.92
57.8	23.7338	965.653	964.571	8.049735	13.927209	28.90	73.72	102.62	64.8	26.7235	961.484	960.401	8.014935	12.368562	32.40	70.53	102.93
57.9	23.7763	965.596	964.514	8.049259	13.902599	28.95	73.68	102.63	64.9	26.7665	961.421	960.338	8.014409	12.348688	32.45	70.48	102.93
58.0	23.8187	965.539	964.457	8.048784	13.877494	29.00	73.63	102.63	65.0	26.8095	961.358	960.275	8.013883	12.328877	32.50	70.43	102.93
58.1	23.8612	965.482	964.400	8.048308	13.853060	29.05	73.59	102.64	65.1	26.8525	961.296	960.213	8.013366	12.309129	32.55	70.39	102.94
58.2	23.9037	965.425	964.343	8.047832	13.828132	29.10	73.54	102.64	65.2	26.8955	961.233	960.150	8.012840	12.289445	32.60	70.34	102.94
58.3	23.9462	965.368	964.286	8.047357	13.803872	29.15	73.50	102.65	65.3	26.9385	961.170	960.087	8.012314	12.269823	32.65	70.30	102.95
58.4	23.9887	965.310	964.228	8.046873	13.779121	29.20	73.45	102.65	65.4	26.9815	961.107	960.024	8.011789	12.250264	32.70	70.25	102.95
58.5	24.0312	965.253	964.171	8.046397	13.755032	29.25	73.41	102.66	65.5	27.0246	961.044	959.961	8.011263	12.230768	32.75	70.20	102.96
58.6	24.0737	965.196	964.114	8.045921	13.730456	29.30	73.36	102.66	65.6	27.0676	960.981	959.898	8.010737	12.211333	32.80	70.16	102.96
58.7	24.1162	965.138	964.056	8.045437	13.706536	29.35	73.31	102.66	65.7	27.1106	960.918	959.835	8.010211	12.191960	32.85	70.11	102.96
58.8	24.1587	965.081	963.999	8.044962	13.682133	29.40	73.27	102.67	65.8	27.1537	960.854	959.771	8.009677	12.172648	32.90	70.07	102.97
58.9	24.2013	965.023	963.941	8.044478	13.658381	29.45	73.22	102.67	65.9	27.1968	960.791	959.708	8.009151	12.153398	32.95	70.02	102.97
59.0	24.2438	964.966	963.884	8.044002	13.634150	29.50	73.18	102.68	66.0	27.2398	960.727	959.644	8.008617	12.134208	33.00	69.97	102.97
59.1	24.2864	964.908	963.826	8.043518	13.610564	29.55	73.13	102.68	66.1	27.2829	960.664	959.581	8.008092	12.115079	33.05	69.93	102.98
59.2	24.3289	964.850	963.768	8.043034	13.586501	29.60	73.09	102.69	66.2	27.3260	960.600	959.517	8.007557	12.096009	33.10	69.88	102.98
59.3	24.3715	964.792	963.710	8.042550	13.563080	29.65	73.04	102.69	66.3	27.3691	960.536	959.453	8.007023	12.077000	33.15	69.84	102.99
59.4	24.4140	964.734	963.652	8.042066	13.539185	29.70	73.00	102.70	66.4	27.4122	960.472	959.389	8.006489	12.058051	33.20	69.79	102.99
59.5	24.4566	964.676	963.594	8.041582	13.515373	29.75	72.95	102.70	66.5	27.4553	960.408	959.325	8.005955	12.039161	33.25	69.74	102.99
59.6	24.4992	964.618	963.536	8.041098	13.492197	29.80	72.90	102.70	66.6	27.4984	960.344	959.261	8.005421	12.020330	33.30	69.70	103.00
59.7	24.5418	964.560	963.478	8.040614	13.468550	29.85	72.86	102.71	66.7	27.5416	960.280	959.197	8.004887	12.001122	33.35	69.65	103.00
59.8	24.5844	964.501	963.419	8.040121	13.445534	29.90	72.81	102.71	66.8	27.5847	960.216	959.133	8.004353	11.982409	33.40	69.61	103.01
59.9	24.6270	964.443	963.361	8.039637	13.422050	29.95	72.77	102.72	66.9	27.6279	960.151	959.068	8.003810	11.963755	33.45	69.56	103.01
60.0	24.6696	964.385	963.303	8.039153	13.398649	30.00	72.72	102.72	67.0	27.6710	960.087	959.004	8.003276	11.945159	33.50	69.51	103.01
60.1	24.7122	964.326	963.244	8.038661	13.375871	30.05	72.68	102.73	67.1	27.7142	960.022	958.939	8.002734	11.926621	33.55	69.47	103.02
60.2	24.7548	964.267	963.185	8.038168	13.352630	30.10	72.63	102.73	67.2	27.7573	959.958	958.875	8.002200	11.908140	33.60	69.42	103.02
60.3	24.7974	964.209	963.127	8.037684	13.330007	30.15	72.59	102.74	67.3	27.8005	959.893	958.810	8.001657	11.889288	33.65	69.37	103.02
60.4	24.8401	964.150	963.068	8.037192	13.306926	30.20	72.54	102.74	67.4	27.8437	959.828	958.745	8.001115	11.870923	33.70	69.33	103.03
60.5	24.8827	964.091	963.009	8.036700	13.283924	30.25	72.49	102.74	67.5	27.8869	959.763	958.680	8.000572	11.852614	33.75	69.28	103.03
60.6	24.9254	964.032	962.950	8.036207	13.261533	30.30	72.45	102.75	67.6	27.9301	959.698	958.615	8.000030	11.834362	33.80	69.24	103.04
60.7	24.9680	963.973	962.891	8.035715	13.238688	30.35	72.40	102.75	67.7	27.9733	959.633	958.550	7.999487	11.816165	33.85	69.19	103.04
60.8	25.0107	963.911	962.831	8.035214	13.215921	30.40	72.36	102.76	67.8	28.0165	959.568	958.485	7.998945	11.797604	33.90	69.14	103.04
60.9	25.0534	963.855	962.772	8.034722	13.193759	30.45	72.31	102.76	67.9	28.0598	959.502	958.419	7.998394	11.779940	33.95	69.10	103.05
61.0	25.0961	963.796	962.713	8.034229	13.171146	30.50	72.27	102.77	68.0	28.1030	959.437	958.354	7.997852	11.761911	34.00	69.05	103.05
61.1	25.1387	963.737	962.654	8.033737	13.148611	30.55	72.22	102.77	68.1	28.1463	959.371	958.288	7.997301	11.743936	34.05	69.01	103.06
61.2	25.1814	963.677	962.594	8.033236	13.126674	30.60	72.18	102.78	68.2	28.1895	959.306	958.223	7.996758	11.725601	34.10	68.96	103.06
61.3	25.2241	963.618	962.535	8.032744	13.104290	30.65	72.13	102.78	68.3	28.2328	959.240	958.157	7.996208	11.707737	34.15	68.91	103.06
61.4	25.2669	963.558	962.475	8.032243	13.081983	30.70	72.08	102.78	68.4	28.2761	959.174	958.091	7.995657	11.689928	34.20	68.87	103.07
61.5	25.3096	963.499	962.416	8.031751	13.059751	30.75	72.04	102.79	68.5	28.3194	959.108	958.025	7.995106	11.672173	34.25	68.82	103.07
61.6	25.3523	963.439	962.356	8.031250	13.038109	30.80	71.99	102.79	68.6	28.3627	959.042	957.959	7.994555	11.654060	34.30	68.77	103.07
61.7	25.3950	963.379	962.296	8.030749	13.016027	30.85	71.95	102.80	68.7	28.4060	958.976	957.893	7.994004	11.636414	34.35	68.73	103.08
61.8	25.4378	963.319	962.236	8.030249	12.994018	30.90	71.90	102.80	68.8	28.4493	958.910	957.827	7.993454	11.618821	34.40	68.68	103.08
61.9	25.4805	963.259	962.176	8.029748	12.972085	30.95	71.86	102.81	68.9	28.4926	958.844	957.761	7.992903	11.600873	34.45	68.64	103.09
62.0	25.5233	963.199	962.116	8.029247	12.950732	31.00	71.81	102.81	69.0	28.5359	958.778	957.695	7.992352	11.583388	34.50	68.59	103.09
62.1	25.5660	963.139	962.056	8.028746	12.928944	31.05	71.76	102.81	69.1	28.5793	958.711	957.628	7.991793	11.565954	34.55	68.54	103.09
62.2	25.6088	963.079	961.996	8.028246	12.907229	31.10	71.72	102.82	69.2	28.6226	958.645	957.562	7.991242	11.548170	34.60	68.50	103.10
62.3	25.6516	963.019	961.936	8.027745	12.885587	31.15	71.67	102.82	69.3	28.6660	958.578	957.495	7.990683	11.530842	34.65	68.45	103.10
62.4	25.6944	962.958	961.875	8.027236	12.864518	31.20	71.63	102.83	69.4	28.7093	958.511	957.428	7.990124	11.513567	34.70	68.40	103.10
62.5	25.7371	962.898	961.815	8.026735	12.843019	31.25	71.58	102.83	69.5	28.7527	958.444	957.361	7.989565	11.495943	34.75	68.36	103.11
62.6	25.7800	962.837	961.754	8.026226	12.821592	31.30	71.54	102.84	69.6	28.7961	958.377	957.294	7.989006	11.478772	34.80	68.31	103.11
62.7	25.8228	962.777	961.694	8.025725	12.800236	31.35	71.49	102.84	69.7	28.8395	958.310	957.227	7.988446	11.461254	34.85	68.26	103.11
62.8	25.8656	962.716	961.633	8.025216	12.778951	31.40	71.44	102.84	69.8	28.8829	958.243	957.160	7.987887	11.444186	34.90	68.22	103.12
62.9	25.9084	962.655	961.572	8.024707	12.758229	31.45	71.40	102.85	69.9	28.9263	958.176	957.093	7.987328	11.427169	34.95	68.17	103.12
63.0	25.9512	962.594	961.511	8.024198	12.737083	31.50	71.35	102.85	70.0	28.9697	958.108	957.025	7.986761	11.409809	35.00	68.13	103.13

Values at 15.556°C (60° F)

Proof	Mass % Vac	Kg m3 Vac	Kg m3 Air	Lbs per Gal. Air	Lbs per Proof Gal Air	Alcohol Parts	Water Parts	Parts Total	Proof	Mass % Vac	Kg m3 Vac	Kg m3 Air	Lbs per Gal. Air	Lbs per Proof Gal Air	Alcohol Parts	Water Parts	Parts Total
70.0	28.9697	958.108	957.025	7.986761	11.409809	35.00	68.13	103.13	77.0	32.0327	953.144	952.060	7.945326	10.318529	38.50	64.85	103.35
70.1	29.0131	958.041	956.958	7.986202	11.392894	35.05	68.08	103.13	77.1	32.0768	953.069	951.985	7.944700	10.304371	38.55	64.80	103.35
70.2	29.0566	957.973	956.890	7.985634	11.375637	35.10	68.03	103.13	77.2	32.1209	952.995	951.911	7.944082	10.290253	38.60	64.76	103.36
70.3	29.1000	957.906	956.823	7.985075	11.358823	35.15	67.99	103.14	77.3	32.1650	952.920	951.836	7.943456	10.276173	38.65	64.71	103.36
70.4	29.1435	957.838	956.755	7.984507	11.341669	35.20	67.94	103.14	77.4	32.2092	952.845	951.761	7.942830	10.262131	38.70	64.66	103.36
70.5	29.1869	957.770	956.687	7.983940	11.324956	35.25	67.89	103.14	77.5	32.2533	952.770	951.686	7.942205	10.248128	38.75	64.61	103.36
70.6	29.2304	957.702	956.619	7.983372	11.308291	35.30	67.85	103.15	77.6	32.2975	952.695	951.611	7.941579	10.233846	38.80	64.57	103.37
70.7	29.2739	957.634	956.551	7.982805	11.291290	35.35	67.80	103.15	77.7	32.3416	952.620	951.536	7.940953	10.219920	38.85	64.52	103.37
70.8	29.3174	957.566	956.483	7.982237	11.274724	35.40	67.75	103.15	77.8	32.3858	952.545	951.461	7.940327	10.206031	38.90	64.47	103.37
70.9	29.3609	957.498	956.415	7.981670	11.257823	35.45	67.71	103.16	77.9	32.4300	952.469	951.385	7.939693	10.192181	38.95	64.43	103.38
71.0	29.4044	957.430	956.347	7.981103	11.241355	35.50	67.66	103.16	78.0	32.4742	952.394	951.310	7.939067	10.178368	39.00	64.38	103.38
71.1	29.4479	957.361	956.278	7.980527	11.224554	35.55	67.61	103.16	78.1	32.5184	952.318	951.234	7.938432	10.164592	39.05	64.33	103.38
71.2	29.4914	957.293	956.210	7.979959	11.208184	35.60	67.57	103.17	78.2	32.5627	952.242	951.158	7.937798	10.150542	39.10	64.28	103.38
71.3	29.5350	957.224	956.141	7.979383	11.191482	35.65	67.52	103.17	78.3	32.6069	952.167	951.083	7.937172	10.136841	39.15	64.24	103.39
71.4	29.5785	957.155	956.072	7.978808	11.174829	35.70	67.47	103.17	78.4	32.6511	952.090	951.006	7.936530	10.123177	39.20	64.19	103.39
71.5	29.6221	957.086	956.003	7.978232	11.158603	35.75	67.43	103.18	78.5	32.6954	952.015	950.931	7.935904	10.109551	39.25	64.14	103.39
71.6	29.6657	957.018	955.934	7.977656	11.142049	35.80	67.38	103.18	78.6	32.7397	951.938	950.854	7.935261	10.095652	39.30	64.09	103.39
71.7	29.7092	956.948	955.864	7.977072	11.125918	35.85	67.34	103.19	78.7	32.7840	951.862	950.778	7.934627	10.082099	39.35	64.05	103.40
71.8	29.7528	956.879	955.795	7.976496	11.109460	35.90	67.29	103.19	78.8	32.8282	951.786	950.702	7.933993	10.068583	39.40	64.00	103.40
71.9	29.7964	956.810	955.726	7.975920	11.093423	35.95	67.24	103.19	78.9	32.8726	951.710	950.626	7.933358	10.054796	39.45	63.95	103.40
72.0	29.8400	956.741	955.657	7.975344	11.077061	36.00	67.20	103.20	79.0	32.9169	951.633	950.549	7.932716	10.041353	39.50	63.91	103.41
72.1	29.8836	956.671	955.587	7.974760	11.060747	36.05	67.15	103.20	79.1	32.9612	951.556	950.472	7.932073	10.027945	39.55	63.86	103.41
72.2	29.9272	956.602	955.518	7.974184	11.044850	36.10	67.10	103.20	79.2	33.0055	951.480	950.396	7.931439	10.014270	39.60	63.81	103.41
72.3	29.9709	956.532	955.448	7.973600	11.028631	36.15	67.05	103.20	79.3	33.0498	951.403	950.319	7.930796	10.000935	39.65	63.76	103.41
72.4	30.0145	956.462	955.378	7.973016	11.012460	36.20	67.01	103.21	79.4	33.0942	951.326	950.242	7.930154	9.987635	39.70	63.72	103.42
72.5	30.0582	956.392	955.308	7.972432	10.996702	36.25	66.96	103.21	79.5	33.1386	951.249	950.165	7.929511	9.974370	39.75	63.67	103.42
72.6	30.1018	956.322	955.238	7.971847	10.980623	36.30	66.91	103.21	79.6	33.1829	951.171	950.087	7.928860	9.961141	39.80	63.62	103.42
72.7	30.1455	956.253	955.169	7.971272	10.964592	36.35	66.87	103.22	79.7	33.2273	951.094	950.010	7.928218	9.947946	39.85	63.57	103.42
72.8	30.1892	956.182	955.098	7.970679	10.948971	36.40	66.82	103.22	79.8	33.2717	951.017	949.933	7.927575	9.934488	39.90	63.53	103.43
72.9	30.2329	956.112	955.028	7.970095	10.933032	36.45	66.77	103.22	79.9	33.3162	950.939	949.855	7.926924	9.921364	39.95	63.48	103.43
73.0	30.2766	956.042	954.958	7.969511	10.917139	36.50	66.73	103.23	80.0	33.3606	950.862	949.778	7.926282	9.907978	40.00	63.43	103.43
73.1	30.3203	955.971	954.887	7.968918	10.901652	36.55	66.68	103.23	80.1	33.4050	950.784	949.700	7.925631	9.894924	40.05	63.38	103.43
73.2	30.3640	955.901	954.817	7.968334	10.885851	36.60	66.63	103.23	80.2	33.4494	950.706	949.622	7.924980	9.881904	40.10	63.34	103.44
73.3	30.4077	955.830	954.746	7.967742	10.870095	36.65	66.59	103.24	80.3	33.4939	950.628	949.544	7.924329	9.868624	40.15	63.29	103.44
73.4	30.4515	955.760	954.676	7.967157	10.854741	36.70	66.54	103.24	80.4	33.5383	950.550	949.466	7.923678	9.855674	40.20	63.24	103.44
73.5	30.4952	955.689	954.605	7.966565	10.839075	36.75	66.49	103.24	80.5	33.5828	950.472	949.388	7.923027	9.842464	40.25	63.19	103.44
73.6	30.5390	955.618	954.534	7.965972	10.823454	36.80	66.45	103.25	80.6	33.6273	950.394	949.310	7.922376	9.829582	40.30	63.15	103.45
73.7	30.5827	955.547	954.463	7.965380	10.807878	36.85	66.40	103.25	80.7	33.6718	950.316	949.232	7.921725	9.816442	40.35	63.10	103.45
73.8	30.6265	955.475	954.391	7.964779	10.792347	36.90	66.35	103.25	80.8	33.7163	950.238	949.154	7.921074	9.803628	40.40	63.05	103.45
73.9	30.6703	955.404	954.320	7.964186	10.777212	36.95	66.31	103.26	80.9	33.7608	950.159	949.074	7.920406	9.790557	40.45	63.00	103.45
74.0	30.7141	955.333	954.249	7.963594	10.761769	37.00	66.26	103.26	81.0	33.8054	950.080	948.995	7.919747	9.777810	40.50	62.96	103.46
74.1	30.7579	955.261	954.177	7.962993	10.746370	37.05	66.21	103.26	81.1	33.8499	950.002	948.917	7.919096	9.764808	40.55	62.91	103.46
74.2	30.8017	955.190	954.106	7.962400	10.731015	37.10	66.17	103.27	81.2	33.8944	949.923	948.838	7.918437	9.752129	40.60	62.86	103.46
74.3	30.8455	955.118	954.034	7.961800	10.715704	37.15	66.12	103.27	81.3	33.9390	949.844	948.759	7.917778	9.739195	40.65	62.81	103.46
74.4	30.8894	955.046	953.962	7.961199	10.700783	37.20	66.07	103.27	81.4	33.9836	949.765	948.680	7.917118	9.726295	40.70	62.77	103.47
74.5	30.9332	954.974	953.890	7.960598	10.685558	37.25	66.03	103.28	81.5	34.0281	949.686	948.601	7.916459	9.713715	40.75	62.72	103.47
74.6	30.9771	954.902	953.818	7.959997	10.670376	37.30	65.98	103.28	81.6	34.0727	949.607	948.522	7.915800	9.700883	40.80	62.67	103.47
74.7	31.0210	954.830	953.746	7.959396	10.655238	37.35	65.93	103.28	81.7	34.1174	949.527	948.442	7.915132	9.688369	40.85	62.62	103.47
74.8	31.0648	954.758	953.674	7.958795	10.640142	37.40	65.88	103.28	81.8	34.1620	949.448	948.363	7.914473	9.675603	40.90	62.58	103.48
74.9	31.1087	954.686	953.602	7.958194	10.625089	37.45	65.84	103.29	81.9	34.2066	949.368	948.283	7.913805	9.662872	40.95	62.53	103.48
75.0	31.1526	954.613	953.529	7.957585	10.610078	37.50	65.79	103.29	82.0	34.2512	949.289	948.204	7.913146	9.650455	41.00	62.48	103.48
75.1	31.1965	954.541	953.457	7.956984	10.595110	37.55	65.74	103.29	82.1	34.2959	949.209	948.124	7.912478	9.637789	41.05	62.43	103.48
75.2	31.2404	954.468	953.384	7.956375	10.580523	37.60	65.70	103.30	82.2	34.3406	949.129	948.044	7.911811	9.625157	41.10	62.38	103.49
75.3	31.2843	954.396	953.312	7.955774	10.565638	37.65	65.65	103.30	82.3	34.3852	949.049	947.964	7.911143	9.612837	41.15	62.34	103.49
75.4	31.3283	954.323	953.239	7.955165	10.550794	37.70	65.60	103.30	82.4	34.4299	948.969	947.884	7.910475	9.600270	41.20	62.29	103.49
75.5	31.3722	954.250	953.166	7.954556	10.535993	37.75	65.56	103.31	82.5	34.4746	948.889	947.804	7.909808	9.587735	41.25	62.24	103.49
75.6	31.4162	954.177	953.093	7.953947	10.521233	37.80	65.51	103.31	82.6	34.5193	948.809	947.724	7.909140	9.575511	41.30	62.19	103.49
75.7	31.4602	954.104	953.020	7.953337	10.506514	37.85	65.46	103.31	82.7	34.5640	948.728	947.643	7.908464	9.563041	41.35	62.15	103.50
75.8	31.5041	954.030	952.946	7.952720	10.491836	37.90	65.42	103.32	82.8	34.6088	948.648	947.563	7.907796	9.550604	41.40	62.10	103.50
75.9	31.5481	953.957	952.873	7.952111	10.477200	37.95	65.37	103.32	82.9	34.6535	948.567	947.482	7.907120	9.538198	41.45	62.05	103.50
76.0	31.5921	953.884	952.800	7.951501	10.462604	38.00	65.32	103.32	83.0	34.6982	948.487	947.402	7.906453	9.526100	41.50	62.00	103.50
76.1	31.6361	953.810	952.726	7.950884	10.448049	38.05	65.27	103.32	83.1	34.7430	948.406	947.321	7.905777	9.513758	41.55	61.96	103.51
76.2	31.6801	953.737	952.653	7.950275	10.433534	38.10	65.23	103.33	83.2	34.7878	948.325	947.240	7.905101	9.501449	41.60	61.91	103.51
76.3	31.7242	953.663	952.579	7.949657	10.419059	38.15	65.18	103.33	83.3	34.8326	948.244	947.159	7.904425	9.489171	41.65	61.86	103.51
76.4	31.7682	953.589	952.505	7.949039	10.404625	38.20	65.13	103.33	83.4	34.8774	948.163	947.078	7.903749	9.477196	41.70	61.81	103.51
76.5	31.8123	953.515	952.431	7.948422	10.390230	38.25	65.09	103.34	83.5	34.9222	948.082	946.997	7.903073	9.464981	41.75	61.76	103.51
76.6	31.8563	953.441	952.357	7.947804	10.375875	38.30	65.04	103.34	83.6	34.9670	948.000	946.915	7.902389	9.452797	41.80	61.72	103.52
76.7	31.9004	953.367	952.283	7.947187	10.361560	38.35	64.99	103.34	83.7	35.0118	947.919	946.834	7.901713	9.440645	41.85	61.67	103.52
76.8	31.9445	953.293	952.209	7.946569	10.347285	38.40	64.95	103.35	83.8	35.0567	947.837	946.752	7.901028	9.428523	41.90	61.62	103.52
76.9	31.9886	953.218	952.134	7.945943	10.332725	38.45	64.90	103.35	83.9	35.1015	947.756	946.671	7.900352	9.416433	41.95	61.57	103.52
77.0	32.0327	953.144	952.060	7.945326	10.318529	38.50	64.85	103.35	84.0	35.1464	947.674	946.589	7.899668	9.404641	42.00	61.53	103.53

213

Alcohol Blending and Accounting, Volume 1

Values at 15.556°C (60° F)

Proof	Mass % Vac	Kg m3 Vac	Kg m3 Air	Lbs per Gal. Air	Lbs per Proof Gal Air	Alcohol Parts	Water Parts	Parts Total	Proof	Mass % Vac	Kg m3 Vac	Kg m3 Air	Lbs per Gal. Air	Lbs per Proof Gal Air	Alcohol Parts	Water Parts	Parts Total
84.0	35.1464	947.674	946.589	7.899668	9.404641	42.00	61.53	103.53	91.0	38.3166	941.706	940.620	7.849854	8.626159	45.50	58.15	103.65
84.1	35.1913	947.592	946.507	7.898984	9.392612	42.05	61.48	103.53	91.1	38.3623	941.617	940.531	7.849112	8.616037	45.55	58.10	103.65
84.2	35.2362	947.510	946.425	7.898299	9.380614	42.10	61.43	103.53	91.2	38.4080	941.529	940.443	7.848377	8.605716	45.60	58.05	103.65
84.3	35.2811	947.428	946.343	7.897615	9.368646	42.15	61.38	103.53	91.3	38.4538	941.440	940.354	7.847634	8.595419	45.65	58.00	103.65
84.4	35.3260	947.346	946.261	7.896931	9.356709	42.20	61.33	103.53	91.4	38.4995	941.350	940.264	7.846883	8.585147	45.70	57.95	103.65
84.5	35.3709	947.264	946.179	7.896246	9.344802	42.25	61.29	103.54	91.5	38.5453	941.261	940.175	7.846141	8.575121	45.75	57.91	103.66
84.6	35.4158	947.182	946.097	7.895562	9.332925	42.30	61.24	103.54	91.6	38.5911	941.172	940.086	7.845398	8.564898	45.80	57.86	103.66
84.7	35.4608	947.099	946.014	7.894869	9.321079	42.35	61.19	103.54	91.7	38.6369	941.083	939.997	7.844655	8.554698	45.85	57.81	103.66
84.8	35.5058	947.017	945.932	7.894185	9.309262	42.40	61.14	103.54	91.8	38.6827	940.994	939.908	7.843912	8.544523	45.90	57.76	103.66
84.9	35.5507	946.934	945.849	7.893492	9.297476	42.45	61.09	103.54	91.9	38.7285	940.904	939.818	7.843161	8.534372	45.95	57.71	103.66
85.0	35.5957	946.852	945.767	7.892808	9.285719	42.50	61.05	103.55	92.0	38.7743	940.814	939.728	7.842410	8.524465	46.00	57.66	103.66
85.1	35.6407	946.769	945.684	7.892115	9.273992	42.55	61.00	103.55	92.1	38.8202	940.725	939.639	7.841667	8.514361	46.05	57.61	103.66
85.2	35.6857	946.686	945.601	7.891423	9.262295	42.60	60.95	103.55	92.2	38.8661	940.635	939.549	7.840916	8.504282	46.10	57.57	103.67
85.3	35.7307	946.603	945.518	7.890730	9.250627	42.65	60.90	103.55	92.3	38.9119	940.545	939.459	7.840165	8.494226	46.15	57.52	103.67
85.4	35.7758	946.520	945.435	7.890037	9.238988	42.70	60.85	103.55	92.4	38.9578	940.455	939.369	7.839414	8.484194	46.20	57.47	103.67
85.5	35.8208	946.436	945.351	7.889336	9.227379	42.75	60.81	103.56	92.5	39.0037	940.365	939.279	7.838663	8.474186	46.25	57.42	103.67
85.6	35.8659	946.353	945.268	7.888644	9.215798	42.80	60.76	103.56	92.6	39.0496	940.275	939.189	7.837912	8.464201	46.30	57.37	103.67
85.7	35.9109	946.270	945.185	7.887951	9.204247	42.85	60.71	103.56	92.7	39.0955	940.185	939.099	7.837161	8.454240	46.35	57.32	103.67
85.8	35.9560	946.186	945.101	7.887250	9.192725	42.90	60.66	103.56	92.8	39.1415	940.094	939.008	7.836402	8.444302	46.40	57.27	103.67
85.9	36.0011	946.102	945.017	7.886549	9.181231	42.95	60.61	103.56	92.9	39.1874	940.004	938.918	7.835650	8.434603	46.45	57.22	103.67
86.0	36.0462	946.019	944.934	7.885856	9.169767	43.00	60.57	103.57	93.0	39.2334	939.913	938.827	7.834891	8.424711	46.50	57.18	103.68
86.1	36.0913	945.935	944.850	7.885155	9.158331	43.05	60.52	103.57	93.1	39.2794	939.823	938.737	7.834140	8.414843	46.55	57.13	103.68
86.2	36.1364	945.851	944.766	7.884454	9.146923	43.10	60.47	103.57	93.2	39.3254	939.732	938.646	7.833380	8.404997	46.60	57.08	103.68
86.3	36.1816	945.767	944.682	7.883753	9.135291	43.15	60.42	103.57	93.3	39.3714	939.641	938.555	7.832621	8.395175	46.65	57.03	103.68
86.4	36.2267	945.683	944.598	7.883052	9.123941	43.20	60.37	103.57	93.4	39.4173	939.550	938.464	7.831862	8.385376	46.70	56.98	103.68
86.5	36.2719	945.599	944.514	7.882351	9.112618	43.25	60.32	103.57	93.5	39.4634	939.459	938.373	7.831102	8.375599	46.75	56.93	103.68
86.6	36.3171	945.514	944.429	7.881642	9.101324	43.30	60.28	103.58	93.6	39.5094	939.368	938.282	7.830343	8.365845	46.80	56.88	103.68
86.7	36.3622	945.430	944.345	7.880941	9.090058	43.35	60.23	103.58	93.7	39.5555	939.277	938.191	7.829583	8.355903	46.85	56.83	103.68
86.8	36.4074	945.345	944.260	7.880232	9.078820	43.40	60.18	103.58	93.8	39.6015	939.186	938.100	7.828824	8.346195	46.90	56.78	103.68
86.9	36.4526	945.261	944.176	7.879531	9.067360	43.45	60.13	103.58	93.9	39.6476	939.095	938.009	7.828064	8.336719	46.95	56.74	103.69
87.0	36.4979	945.176	944.091	7.878821	9.056178	43.50	60.08	103.58	94.0	39.6937	939.003	937.917	7.827297	8.327056	47.00	56.69	103.69
87.1	36.5431	945.091	944.006	7.878112	9.045023	43.55	60.04	103.59	94.1	39.7398	938.912	937.826	7.826537	8.317415	47.05	56.64	103.69
87.2	36.5883	945.006	943.921	7.877403	9.033896	43.60	59.99	103.59	94.2	39.7859	938.820	937.734	7.825769	8.307796	47.10	56.59	103.69
87.3	36.6336	944.921	943.836	7.876693	9.022550	43.65	59.94	103.59	94.3	39.8320	938.728	937.642	7.825002	8.298199	47.15	56.54	103.69
87.4	36.6789	944.836	943.751	7.875984	9.011478	43.70	59.89	103.59	94.4	39.8782	938.637	937.551	7.824242	8.288624	47.20	56.49	103.69
87.5	36.7241	944.751	943.666	7.875274	9.000433	43.75	59.84	103.59	94.5	39.9243	938.545	937.459	7.823475	8.279072	47.25	56.44	103.69
87.6	36.7694	944.666	943.581	7.874565	8.989415	43.80	59.80	103.60	94.6	39.9705	938.453	937.367	7.822707	8.269335	47.30	56.39	103.69
87.7	36.8147	944.580	943.495	7.873847	8.978180	43.85	59.75	103.60	94.7	40.0167	938.361	937.275	7.821939	8.259824	47.35	56.35	103.70
87.8	36.8600	944.494	943.409	7.873130	8.967216	43.90	59.70	103.60	94.8	40.0629	938.269	937.183	7.821171	8.250340	47.40	56.30	103.70
87.9	36.9054	944.409	943.324	7.872420	8.956279	43.95	59.65	103.60	94.9	40.1091	938.176	937.090	7.820395	8.240876	47.45	56.25	103.70
88.0	36.9507	944.323	943.238	7.871703	8.945127	44.00	59.60	103.60	95.0	40.1553	938.084	936.998	7.819627	8.231433	47.50	56.20	103.70
88.1	36.9961	944.237	943.152	7.870985	8.934244	44.05	59.55	103.60	95.1	40.2015	937.991	936.905	7.818851	8.221807	47.55	56.15	103.70
88.2	37.0414	944.152	943.067	7.870276	8.923387	44.10	59.51	103.61	95.2	40.2477	937.899	936.813	7.818083	8.212408	47.60	56.10	103.70
88.3	37.0868	944.065	942.980	7.869549	8.912317	44.15	59.46	103.61	95.3	40.2940	937.806	936.720	7.817307	8.203031	47.65	56.05	103.70
88.4	37.1322	943.979	942.894	7.868832	8.901513	44.20	59.41	103.61	95.4	40.3403	937.713	936.627	7.816531	8.193674	47.70	56.00	103.70
88.5	37.1776	943.893	942.808	7.868114	8.890497	44.25	59.36	103.61	95.5	40.3866	937.621	936.535	7.815763	8.184137	47.75	55.95	103.70
88.6	37.2230	943.807	942.722	7.867396	8.879746	44.30	59.31	103.61	95.6	40.4329	937.528	936.442	7.814987	8.174823	47.80	55.91	103.71
88.7	37.2684	943.720	942.635	7.866670	8.869022	44.35	59.26	103.61	95.7	40.4792	937.435	936.349	7.814211	8.165531	47.85	55.86	103.71
00.0	27.2120	043.634	942.549	7.865953	8.858086	44.40	59.22	103.62	95.8	40.5255	937.342	936.256	7.813435	8.156059	47.90	55.81	103.71
88.9	37.3593	943.547	942.462	7.865227	8.847413	44.45	59.17	103.62	95.9	40.5718	937.249	936.163	7.812659	8.146809	47.95	55.76	103.71
89.0	37.4048	943.461	942.376	7.864509	8.836530	44.50	59.12	103.62	96.0	40.6182	937.156	936.070	7.811883	8.137581	48.00	55.71	103.71
89.1	37.4502	943.374	942.289	7.863783	8.825910	44.55	59.07	103.62	96.1	40.6645	937.062	935.976	7.811098	8.128173	48.05	55.66	103.71
89.2	37.4957	943.287	942.202	7.863057	8.815079	44.60	59.02	103.62	96.2	40.7109	936.969	935.883	7.810322	8.118987	48.10	55.61	103.71
89.3	37.5412	943.200	942.114	7.862322	8.804510	44.65	58.97	103.62	96.3	40.7573	936.875	935.789	7.809538	8.109821	48.15	55.56	103.71
89.4	37.5867	943.113	942.027	7.861596	8.793733	44.70	58.92	103.62	96.4	40.8037	936.782	935.696	7.808762	8.100478	48.20	55.51	103.71
89.5	37.6322	943.026	941.940	7.860870	8.783215	44.75	58.88	103.63	96.5	40.8501	936.688	935.602	7.807977	8.091354	48.25	55.46	103.71
89.6	37.6778	942.938	941.852	7.860136	8.772489	44.80	58.83	103.63	96.6	40.8965	936.594	935.508	7.807193	8.082053	48.30	55.41	103.71
89.7	37.7233	942.851	941.765	7.859410	8.762022	44.85	58.78	103.63	96.7	40.9430	936.500	935.414	7.806408	8.072970	48.35	55.37	103.72
89.8	37.7689	942.763	941.677	7.858675	8.751347	44.90	58.73	103.63	96.8	40.9894	936.406	935.320	7.805624	8.063908	48.40	55.32	103.72
89.9	37.8145	942.676	941.590	7.857949	8.740699	44.95	58.68	103.63	96.9	41.0359	936.312	935.225	7.804831	8.054670	48.45	55.27	103.72
90.0	37.8601	942.588	941.502	7.857215	8.730308	45.00	58.63	103.63	97.0	41.0824	936.218	935.131	7.804046	8.045649	48.50	55.22	103.72
90.1	37.9057	942.500	941.414	7.856481	8.719711	45.05	58.59	103.64	97.1	41.1289	936.124	935.037	7.803262	8.036452	48.55	55.17	103.72
90.2	37.9512	942.412	941.326	7.855746	8.709369	45.10	58.54	103.64	97.2	41.1753	936.030	934.943	7.802477	8.027472	48.60	55.12	103.72
90.3	37.9969	942.325	941.239	7.855011	8.698823	45.15	58.49	103.64	97.3	41.2219	935.935	934.848	7.801685	8.018317	48.65	55.07	103.72
90.4	38.0425	942.237	941.151	7.854286	8.688302	45.20	58.44	103.64	97.4	41.2684	935.841	934.754	7.800900	8.009377	48.70	55.02	103.72
90.5	38.0882	942.148	941.062	7.853543	8.678035	45.25	58.39	103.64	97.5	41.3150	935.746	934.659	7.800107	8.000263	48.75	54.97	103.72
90.6	38.1338	942.060	940.974	7.852809	8.667564	45.30	58.34	103.64	97.6	41.3615	935.652	934.565	7.799323	7.991170	48.80	54.92	103.72
90.7	38.1795	941.972	940.886	7.852074	8.657119	45.35	58.29	103.64	97.7	41.4081	935.557	934.470	7.798530	7.982291	48.85	54.87	103.72
90.8	38.2252	941.883	940.797	7.851331	8.646925	45.40	58.25	103.65	97.8	41.4547	935.462	934.375	7.797737	7.973239	48.90	54.83	103.73
90.9	38.2709	941.795	940.709	7.850597	8.636529	45.45	58.20	103.65	97.9	41.5013	935.367	934.280	7.796944	7.964399	48.95	54.78	103.73
91.0	38.3166	941.706	940.620	7.849854	8.626159	45.50	58.15	103.65	98.0	41.5479	935.272	934.185	7.796152	7.955387	49.00	54.73	103.73

Values at 15.556°C (60° F)

Proof	Mass % Vac	Kg m3 Vac	Kg m3 Air	Lbs per Gal. Air	Lbs per Proof Gal Air	Alcohol Parts	Water Parts	Parts Total	Proof	Mass % Vac	Kg m3 Vac	Kg m3 Air	Lbs per Gal. Air	Lbs per Proof Gal Air	Alcohol Parts	Water Parts	Parts Total
98.0	41.5479	935.272	934.185	7.796152	7.955387	49.00	54.73	103.73	105.0	44.8442	928.419	927.331	7.738952	7.370542	52.50	51.26	103.76
98.1	41.5945	935.177	934.090	7.795359	7.946396	49.05	54.68	103.73	105.1	44.8918	928.318	927.230	7.738109	7.362659	52.55	51.21	103.76
98.2	41.6411	935.082	933.995	7.794566	7.937616	49.10	54.63	103.73	105.2	44.9394	928.217	927.129	7.737266	7.354957	52.60	51.16	103.76
98.3	41.6878	934.987	933.900	7.793773	7.928665	49.15	54.58	103.73	105.3	44.9869	928.117	927.029	7.736432	7.347108	52.65	51.11	103.76
98.4	41.7345	934.891	933.804	7.792972	7.919734	49.20	54.53	103.73	105.4	45.0346	928.016	926.928	7.735589	7.339276	52.70	51.06	103.76
98.5	41.7811	934.796	933.709	7.792179	7.911012	49.25	54.48	103.73	105.5	45.0822	927.915	926.827	7.734746	7.331622	52.75	51.01	103.76
98.6	41.8278	934.701	933.614	7.791386	7.902121	49.30	54.43	103.73	105.6	45.1298	927.813	926.725	7.733895	7.323823	52.80	50.96	103.76
98.7	41.8745	934.605	933.518	7.790585	7.893249	49.35	54.38	103.73	105.7	45.1775	927.712	926.624	7.733052	7.316040	52.85	50.91	103.76
98.8	41.9212	934.509	933.422	7.789784	7.884586	49.40	54.33	103.73	105.8	45.2252	927.611	926.523	7.732209	7.308435	52.90	50.86	103.76
98.9	41.9679	934.413	933.326	7.788983	7.875754	49.45	54.28	103.73	105.9	45.2729	927.510	926.422	7.731366	7.300684	52.95	50.81	103.76
99.0	42.0147	934.318	933.231	7.788190	7.866942	49.50	54.23	103.73	106.0	45.3206	927.408	926.320	7.730515	7.292951	53.00	50.76	103.76
99.1	42.0615	934.222	933.135	7.787389	7.858336	49.55	54.19	103.74	106.1	45.3683	927.307	926.219	7.729672	7.285394	53.05	50.72	103.77
99.2	42.1082	934.126	933.039	7.786588	7.849562	49.60	54.14	103.74	106.2	45.4160	927.206	926.118	7.728829	7.277692	53.10	50.66	103.76
99.3	42.1550	934.030	932.943	7.785787	7.840809	49.65	54.09	103.74	106.3	45.4638	927.104	926.016	7.727978	7.270007	53.15	50.61	103.76
99.4	42.2018	933.934	932.847	7.784985	7.832074	49.70	54.04	103.74	106.4	45.5116	927.002	925.914	7.727127	7.262338	53.20	50.56	103.76
99.5	42.2486	933.837	932.750	7.784176	7.823359	49.75	53.99	103.74	106.5	45.5593	926.901	925.813	7.726284	7.254844	53.25	50.52	103.77
99.6	42.2954	933.741	932.654	7.783375	7.814849	49.80	53.94	103.74	106.6	45.6071	926.799	925.711	7.725433	7.247207	53.30	50.47	103.77
99.7	42.3423	933.644	932.557	7.782565	7.806172	49.85	53.89	103.74	106.7	45.6549	926.697	925.609	7.724581	7.239586	53.35	50.42	103.77
99.8	42.3891	933.548	932.461	7.781764	7.797515	49.90	53.84	103.74	106.8	45.7027	926.595	925.507	7.723730	7.231980	53.40	50.37	103.77
99.9	42.4360	933.451	932.364	7.780955	7.788877	49.95	53.79	103.74	106.9	45.7505	926.493	925.405	7.722879	7.224391	53.45	50.32	103.77
100.0	42.4829	933.355	932.268	7.780154	7.780258	50.00	53.74	103.74	107.0	45.7984	926.391	925.303	7.722028	7.216976	53.50	50.27	103.77
100.1	42.5298	933.258	932.171	7.779344	7.771658	50.05	53.69	103.74	107.1	45.8462	926.289	925.201	7.721177	7.209418	53.55	50.22	103.77
100.2	42.5767	933.161	932.074	7.778535	7.763077	50.10	53.64	103.74	107.2	45.8941	926.186	925.098	7.720317	7.201876	53.60	50.17	103.77
100.3	42.6236	933.064	931.977	7.777725	7.754515	50.15	53.59	103.74	107.3	45.9420	926.084	924.996	7.719466	7.194350	53.65	50.12	103.77
100.4	42.6705	932.967	931.880	7.776915	7.745971	50.20	53.54	103.74	107.4	45.9899	925.982	924.894	7.718614	7.186840	53.70	50.07	103.77
100.5	42.7175	932.870	931.783	7.776106	7.737447	50.25	53.49	103.74	107.5	46.0378	925.879	924.791	7.717755	7.179345	53.75	50.02	103.77
100.6	42.7644	932.773	931.686	7.775296	7.729122	50.30	53.45	103.75	107.6	46.0858	925.777	924.689	7.716904	7.171866	53.80	49.97	103.77
100.7	42.8114	932.676	931.589	7.774487	7.720635	50.35	53.40	103.75	107.7	46.1337	925.674	924.586	7.716044	7.164402	53.85	49.92	103.77
100.8	42.8584	932.578	931.491	7.773669	7.712166	50.40	53.35	103.75	107.8	46.1817	925.571	924.483	7.715185	7.156954	53.90	49.87	103.77
100.9	42.9054	932.481	931.394	7.772860	7.703716	50.45	53.30	103.75	107.9	46.2297	925.468	924.380	7.714325	7.149521	53.95	49.82	103.77
101.0	42.9524	932.384	931.297	7.772050	7.695284	50.50	53.25	103.75	108.0	46.2776	925.366	924.278	7.713474	7.142104	54.00	49.77	103.77
101.1	42.9994	932.286	931.199	7.771232	7.686871	50.55	53.20	103.75	108.1	46.3256	925.262	924.174	7.712606	7.134703	54.05	49.72	103.77
101.2	43.0465	932.188	931.101	7.770414	7.678297	50.60	53.15	103.75	108.2	46.3736	925.160	924.072	7.711755	7.127316	54.10	49.67	103.77
101.3	43.0935	932.090	931.003	7.769597	7.669921	50.65	53.10	103.75	108.3	46.4217	925.057	923.969	7.710895	7.119945	54.15	49.62	103.77
101.4	43.1406	931.993	930.906	7.768787	7.661563	50.70	53.05	103.75	108.4	46.4697	924.954	923.866	7.710035	7.112589	54.20	49.57	103.77
101.5	43.1877	931.895	930.808	7.767969	7.653223	50.75	53.00	103.75	108.5	46.5178	924.850	923.762	7.709167	7.105248	54.25	49.52	103.77
101.6	43.2348	931.797	930.710	7.767151	7.644902	50.80	52.95	103.75	108.6	46.5659	924.747	923.659	7.708308	7.097923	54.30	49.47	103.77
101.7	43.2819	931.699	930.612	7.766334	7.636598	50.85	52.90	103.75	108.7	46.6139	924.644	923.556	7.707448	7.090612	54.35	49.42	103.77
101.8	43.3290	931.601	930.514	7.765516	7.628313	50.90	52.85	103.75	108.8	46.6621	924.540	923.452	7.706580	7.083317	54.40	49.37	103.77
101.9	43.3761	931.502	930.415	7.764689	7.620045	50.95	52.80	103.75	108.9	46.7102	924.437	923.349	7.705721	7.076036	54.45	49.32	103.77
102.0	43.4233	931.404	930.317	7.763872	7.611796	51.00	52.75	103.75	109.0	46.7583	924.333	923.245	7.704853	7.068770	54.50	49.27	103.77
102.1	43.4704	931.306	930.219	7.763054	7.603564	51.05	52.70	103.75	109.1	46.8064	924.230	923.142	7.703993	7.061520	54.55	49.22	103.77
102.2	43.5176	931.207	930.120	7.762228	7.595175	51.10	52.65	103.75	109.2	46.8546	924.126	923.038	7.703125	7.054133	54.60	49.17	103.77
102.3	43.5648	931.109	930.022	7.761410	7.586979	51.15	52.60	103.75	109.3	46.9028	924.022	922.934	7.702258	7.046913	54.65	49.12	103.77
102.4	43.6120	931.010	929.923	7.760584	7.578801	51.20	52.55	103.75	109.4	46.9510	923.918	922.830	7.701390	7.039707	54.70	49.07	103.77
102.5	43.6592	930.911	929.824	7.759757	7.570640	51.25	52.50	103.75	109.5	46.9992	923.814	922.726	7.700522	7.032516	54.75	49.02	103.77
102.6	43.7065	930.813	929.726	7.758939	7.562324	51.30	52.45	103.75	109.6	47.0473	923.710	922.622	7.699654	7.025339	54.80	48.97	103.77
102.7	43.7537	930.714	929.627	7.758113	7.554199	51.35	52.41	103.76	109.7	47.0956	923.606	922.518	7.698786	7.018028	54.85	48.91	103.76
102.8	43.8010	930.615	929.528	7.757287	7.546091	51.40	52.36	103.76	109.8	47.1439	923.502	922.414	7.697918	7.010881	54.90	48.87	103.77
102.9	43.8482	930.516	929.429	7.756461	7.538001	51.45	52.31	103.76	109.9	47.1921	923.398	922.310	7.697050	7.003749	54.95	48.82	103.77
103.0	43.8955	930.417	929.330	7.755635	7.529756	51.50	52.26	103.76	110.0	47.2404	923.294	922.206	7.696182	6.996631	55.00	48.77	103.77
103.1	43.9428	930.318	929.231	7.754809	7.521701	51.55	52.21	103.76	110.1	47.2887	923.189	922.101	7.695306	6.989380	55.05	48.71	103.76
103.2	43.9902	930.219	929.132	7.753982	7.513663	51.60	52.16	103.76	110.2	47.3370	923.085	921.997	7.694438	6.982291	55.10	48.66	103.76
103.3	44.0375	930.119	929.032	7.753148	7.505471	51.65	52.11	103.76	110.3	47.3853	922.981	921.893	7.693570	6.975216	55.15	48.61	103.76
103.4	44.0848	930.020	928.933	7.752322	7.497468	51.70	52.06	103.76	110.4	47.4336	922.876	921.788	7.692694	6.968009	55.20	48.56	103.76
103.5	44.1322	929.920	928.833	7.751487	7.489481	51.75	52.01	103.76	110.5	47.4820	922.772	921.684	7.691826	6.960964	55.25	48.51	103.76
103.6	44.1795	929.821	928.734	7.750661	7.481342	51.80	51.96	103.76	110.6	47.5303	922.667	921.579	7.690949	6.953933	55.30	48.46	103.76
103.7	44.2269	929.721	928.634	7.749826	7.473390	51.85	51.91	103.76	110.7	47.5787	922.562	921.473	7.690065	6.946707	55.35	48.41	103.76
103.8	44.2743	929.621	928.534	7.748992	7.465455	51.90	51.86	103.76	110.8	47.6271	922.457	921.368	7.689189	6.939767	55.40	48.36	103.76
103.9	44.3217	929.521	928.433	7.748149	7.457368	51.95	51.81	103.76	110.9	47.6755	922.353	921.264	7.688321	6.932633	55.45	48.31	103.76
104.0	44.3691	929.422	928.334	7.747323	7.449466	52.00	51.76	103.76	111.0	47.7239	922.248	921.159	7.687444	6.925659	55.50	48.26	103.76
104.1	44.4166	929.322	928.234	7.746488	7.441414	52.05	51.71	103.76	111.1	47.7723	922.142	921.053	7.686560	6.918699	55.55	48.21	103.76
104.2	44.4640	929.222	928.134	7.745654	7.433547	52.10	51.66	103.76	111.2	47.8208	922.037	920.948	7.685684	6.911608	55.60	48.16	103.76
104.3	44.5115	929.121	928.033	7.744811	7.425529	52.15	51.61	103.76	111.3	47.8693	921.932	920.843	7.684807	6.904676	55.65	48.11	103.76
104.4	44.5590	929.021	927.933	7.743976	7.417696	52.20	51.56	103.76	111.4	47.9178	921.827	920.738	7.683931	6.897614	55.70	48.06	103.76
104.5	44.6065	928.921	927.833	7.743142	7.409711	52.25	51.51	103.76	111.5	47.9662	921.722	920.633	7.683055	6.890710	55.75	48.01	103.76
104.6	44.6540	928.821	927.733	7.742307	7.401910	52.30	51.46	103.76	111.6	48.0147	921.617	920.528	7.682178	6.883676	55.80	47.96	103.76
104.7	44.7015	928.720	927.632	7.741464	7.393960	52.35	51.41	103.76	111.7	48.0633	921.511	920.422	7.681294	6.876800	55.85	47.91	103.76
104.8	44.7491	928.620	927.532	7.740630	7.386193	52.40	51.36	103.76	111.8	48.1118	921.406	920.317	7.680418	6.869795	55.90	47.86	103.76
104.9	44.7966	928.520	927.432	7.739795	7.378277	52.45	51.31	103.76	111.9	48.1604	921.300	920.211	7.679533	6.862947	55.95	47.81	103.76
105.0	44.8442	928.419	927.331	7.738952	7.370542	52.50	51.26	103.76	112.0	48.2089	921.194	920.105	7.678648	6.855970	56.00	47.76	103.76

Values at 15.556°C (60° F)

Proof	Mass % Vac	Kg m3 Vac	Kg m3 Air	Lbs per Gal. Air	Lbs per Proof Gal Air	Alcohol Parts	Water Parts	Parts Total	Proof	Mass % Vac	Kg m3 Vac	Kg m3 Air	Lbs per Gal. Air	Lbs per Proof Gal Air	Alcohol Parts	Water Parts	Parts Total
112.0	48.2089	921.194	920.105	7.678648	6.855970	56.00	47.76	103.76	119.0	51.6456	913.638	912.548	7.615582	6.399622	59.50	44.23	103.73
112.1	48.2575	921.089	920.000	7.677772	6.849007	56.05	47.71	103.76	119.1	51.6952	913.528	912.438	7.614664	6.393555	59.55	44.18	103.73
112.2	48.3061	920.983	919.894	7.676887	6.842200	56.10	47.66	103.76	119.2	51.7449	913.418	912.328	7.613746	6.387376	59.60	44.12	103.72
112.3	48.3547	920.877	919.788	7.676003	6.835265	56.15	47.61	103.76	119.3	51.7946	913.307	912.217	7.612820	6.381208	59.65	44.07	103.72
112.4	48.4033	920.771	919.682	7.675118	6.828485	56.20	47.56	103.76	119.4	51.8442	913.197	912.107	7.611902	6.375299	59.70	44.02	103.72
112.5	48.4520	920.665	919.576	7.674234	6.821578	56.25	47.51	103.76	119.5	51.8940	913.087	911.997	7.610984	6.369155	59.75	43.97	103.72
112.6	48.5006	920.559	919.470	7.673349	6.814685	56.30	47.46	103.76	119.6	51.9437	912.976	911.886	7.610057	6.363023	59.80	43.92	103.72
112.7	48.5493	920.453	919.364	7.672464	6.807946	56.35	47.41	103.76	119.7	51.9934	912.865	911.775	7.609131	6.357025	59.85	43.87	103.72
112.8	48.5980	920.347	919.258	7.671580	6.801080	56.40	47.36	103.76	119.8	52.0431	912.755	911.665	7.608213	6.350916	59.90	43.82	103.72
112.9	48.6467	920.241	919.152	7.670695	6.794228	56.45	47.31	103.76	119.9	52.0929	912.644	911.554	7.607287	6.344819	59.95	43.77	103.72
113.0	48.6954	920.135	919.046	7.669811	6.787530	56.50	47.26	103.76	120.0	52.1427	912.533	911.443	7.606360	6.338734	60.00	43.72	103.72
113.1	48.7441	920.028	918.939	7.668918	6.780705	56.55	47.21	103.76	120.1	52.1925	912.422	911.332	7.605434	6.332660	60.05	43.67	103.72
113.2	48.7928	919.922	918.833	7.668033	6.773894	56.60	47.16	103.76	120.2	52.2423	912.311	911.221	7.604508	6.326719	60.10	43.62	103.72
113.3	48.8416	919.815	918.726	7.667140	6.767097	56.65	47.11	103.76	120.3	52.2921	912.200	911.110	7.603581	6.320669	60.15	43.57	103.72
113.4	48.8904	919.709	918.620	7.666255	6.760452	56.70	47.06	103.76	120.4	52.3420	912.089	910.999	7.602655	6.314630	60.20	43.52	103.72
113.5	48.9392	919.602	918.513	7.665362	6.753681	56.75	47.01	103.76	120.5	52.3918	911.978	910.888	7.601729	6.308602	60.25	43.47	103.72
113.6	48.9880	919.496	918.407	7.664478	6.746925	56.80	46.96	103.76	120.6	52.4417	911.867	910.777	7.600802	6.302586	60.30	43.41	103.71
113.7	49.0368	919.389	918.300	7.663585	6.740182	56.85	46.91	103.76	120.7	52.4915	911.756	910.666	7.599876	6.296581	60.35	43.36	103.71
113.8	49.0856	919.282	918.193	7.662692	6.733452	56.90	46.85	103.75	120.8	52.5414	911.645	910.555	7.598950	6.290708	60.40	43.31	103.71
113.9	49.1345	919.175	918.086	7.661799	6.726736	56.95	46.80	103.75	120.9	52.5914	911.533	910.443	7.598015	6.284726	60.45	43.26	103.71
114.0	49.1833	919.068	917.979	7.660906	6.720169	57.00	46.75	103.75	121.0	52.6413	911.422	910.332	7.597089	6.278755	60.50	43.21	103.71
114.1	49.2322	918.961	917.872	7.660013	6.713479	57.05	46.70	103.75	121.1	52.6912	911.310	910.220	7.596154	6.272796	60.55	43.16	103.71
114.2	49.2811	918.854	917.765	7.659120	6.706803	57.10	46.65	103.75	121.2	52.7412	911.199	910.109	7.595228	6.266848	60.60	43.11	103.71
114.3	49.3300	918.747	917.658	7.658227	6.700140	57.15	46.60	103.75	121.3	52.7912	911.087	909.997	7.594293	6.260911	60.65	43.06	103.71
114.4	49.3789	918.640	917.551	7.657334	6.693489	57.20	46.55	103.75	121.4	52.8412	910.976	909.886	7.593367	6.254985	60.70	43.01	103.71
114.5	49.4279	918.532	917.443	7.656433	6.686853	57.25	46.50	103.75	121.5	52.8912	910.864	909.774	7.592432	6.249071	60.75	42.96	103.71
114.6	49.4768	918.425	917.336	7.655540	6.680229	57.30	46.45	103.75	121.6	52.9412	910.752	909.662	7.591497	6.243168	60.80	42.91	103.71
114.7	49.5257	918.318	917.229	7.654647	6.673618	57.35	46.40	103.75	121.7	52.9913	910.640	909.550	7.590563	6.237276	60.85	42.86	103.71
114.8	49.5748	918.210	917.121	7.653746	6.667021	57.40	46.35	103.75	121.8	53.0413	910.529	909.439	7.589636	6.231395	60.90	42.80	103.70
114.9	49.6237	918.102	917.013	7.652844	6.660436	57.45	46.30	103.75	121.9	53.0914	910.417	909.327	7.588702	6.225525	60.95	42.75	103.70
115.0	49.6728	917.995	916.906	7.651951	6.653865	57.50	46.25	103.75	122.0	53.1415	910.304	909.214	7.587759	6.219549	61.00	42.70	103.70
115.1	49.7218	917.887	916.798	7.651050	6.647306	57.55	46.20	103.75	122.1	53.1916	910.192	909.102	7.586824	6.213702	61.05	42.65	103.70
115.2	49.7708	917.780	916.691	7.650157	6.640760	57.60	46.15	103.75	122.2	53.2418	910.080	908.990	7.585889	6.207865	61.10	42.60	103.70
115.3	49.8199	917.672	916.583	7.649256	6.634228	57.65	46.10	103.75	122.3	53.2919	909.968	908.878	7.584955	6.202040	61.15	42.55	103.70
115.4	49.8689	917.564	916.475	7.648355	6.627708	57.70	46.05	103.75	122.4	53.3420	909.856	908.766	7.584020	6.196225	61.20	42.50	103.70
115.5	49.9180	917.456	916.367	7.647453	6.621201	57.75	46.00	103.75	122.5	53.3922	909.743	908.653	7.583077	6.190421	61.25	42.45	103.70
115.6	49.9671	917.348	916.259	7.646552	6.614706	57.80	45.95	103.75	122.6	53.4424	909.631	908.541	7.582142	6.184628	61.30	42.40	103.70
115.7	50.0162	917.240	916.151	7.645651	6.608225	57.85	45.90	103.75	122.7	53.4926	909.519	908.429	7.581207	6.178731	61.35	42.35	103.70
115.8	50.0654	917.132	916.043	7.644749	6.601756	57.90	45.85	103.75	122.8	53.5428	909.406	908.316	7.580264	6.172960	61.40	42.29	103.69
115.9	50.1145	917.023	915.934	7.643840	6.595168	57.95	45.79	103.74	122.9	53.5930	909.293	908.203	7.579321	6.167199	61.45	42.24	103.69
116.0	50.1637	916.915	915.826	7.642938	6.588724	58.00	45.74	103.74	123.0	53.6433	909.181	908.091	7.578387	6.161450	61.50	42.19	103.69
116.1	50.2129	916.807	915.718	7.642037	6.582294	58.05	45.69	103.74	123.1	53.6936	909.068	907.978	7.577444	6.155596	61.55	42.14	103.69
116.2	50.2620	916.699	915.610	7.641136	6.575875	58.10	45.64	103.74	123.2	53.7439	908.956	907.866	7.576509	6.149868	61.60	42.09	103.69
116.3	50.3113	916.590	915.501	7.640226	6.569469	58.15	45.59	103.74	123.3	53.7941	908.842	907.751	7.575549	6.144151	61.65	42.04	103.69
116.4	50.3605	916.482	915.393	7.639325	6.562946	58.20	45.54	103.74	123.4	53.8445	908.730	907.639	7.574615	6.138330	61.70	41.99	103.69
116.5	50.4097	916.373	915.284	7.638415	6.556565	58.25	45.49	103.74	123.5	53.8948	908.617	907.526	7.573672	6.132635	61.75	41.94	103.69
116.6	50.4590	916.264	915.175	7.637506	6.550197	58.30	45.44	103.74	123.6	53.9451	908.504	907.413	7.572728	6.126949	61.80	41.89	103.69
116.7	50.5082	916.156	915.067	7.636604	6.543841	58.35	45.39	103.74	123.7	53.9955	908.390	907.299	7.571777	6.121161	61.85	41.84	103.69
116.8	50.5575	916.047	914.958	7.635695	6.537368	58.40	45.34	103.74	123.8	54.0459	908.278	907.187	7.570842	6.115497	61.90	41.78	103.68
116.9	50.6068	915.938	914.849	7.634785	6.531037	58.45	45.29	103.74	123.9	54.0963	908.164	907.073	7.569901	6.109843	61.95	41.73	103.68
117.0	50.6562	915.829	914.740	7.633875	6.524718	58.50	45.24	103.74	124.0	54.1467	908.051	906.960	7.568948	6.104088	62.00	41.68	103.68
117.1	50.7055	915.720	914.630	7.632957	6.518283	58.55	45.19	103.74	124.1	54.1971	907.938	906.847	7.568005	6.098455	62.05	41.63	103.68
117.2	50.7548	915.611	914.521	7.632048	6.511989	58.60	45.14	103.74	124.2	54.2476	907.824	906.733	7.567054	6.092720	62.10	41.58	103.68
117.3	50.8042	915.502	914.412	7.631138	6.505707	58.65	45.09	103.74	124.3	54.2981	907.711	906.620	7.566111	6.087109	62.15	41.53	103.68
117.4	50.8535	915.393	914.303	7.630228	6.499309	58.70	45.04	103.74	124.4	54.3485	907.598	906.507	7.565168	6.081396	62.20	41.48	103.68
117.5	50.9029	915.284	914.194	7.629319	6.493052	58.75	44.99	103.74	124.5	54.3990	907.484	906.393	7.564216	6.075805	62.25	41.43	103.68
117.6	50.9523	915.175	914.085	7.628409	6.486806	58.80	44.94	103.74	124.6	54.4496	907.370	906.279	7.563265	6.070113	62.30	41.38	103.68
117.7	51.0018	915.065	913.975	7.627491	6.480446	58.85	44.88	103.73	124.7	54.5000	907.257	906.166	7.562322	6.064543	62.35	41.33	103.68
117.8	51.0512	914.956	913.866	7.626581	6.474224	58.90	44.83	103.73	124.8	54.5506	907.143	906.052	7.561370	6.058872	62.40	41.27	103.67
117.9	51.1007	914.846	913.756	7.625663	6.467888	58.95	44.78	103.73	124.9	54.6012	907.029	905.938	7.560419	6.053323	62.45	41.22	103.67
118.0	51.1501	914.737	913.647	7.624754	6.461691	59.00	44.73	103.73	125.0	54.6518	906.915	905.824	7.559468	6.047673	62.50	41.17	103.67
118.1	51.1996	914.627	913.537	7.623836	6.455380	59.05	44.68	103.73	125.1	54.7024	906.801	905.710	7.558516	6.042144	62.55	41.12	103.67
118.2	51.2491	914.518	913.428	7.622926	6.449206	59.10	44.63	103.73	125.2	54.7530	906.687	905.596	7.557565	6.036515	62.60	41.07	103.67
118.3	51.2986	914.408	913.318	7.622008	6.442919	59.15	44.58	103.73	125.3	54.8036	906.573	905.482	7.556614	6.030896	62.65	41.02	103.67
118.4	51.3481	914.298	913.208	7.621090	6.436770	59.20	44.53	103.73	125.4	54.8542	906.459	905.368	7.555662	6.025398	62.70	40.97	103.67
118.5	51.3977	914.188	913.098	7.620172	6.430507	59.25	44.48	103.73	125.5	54.9049	906.345	905.254	7.554711	6.019800	62.75	40.92	103.67
118.6	51.4472	914.078	912.988	7.619254	6.424381	59.30	44.43	103.73	125.6	54.9555	906.230	905.139	7.553751	6.014212	62.80	40.87	103.67
118.7	51.4968	913.968	912.878	7.618336	6.418142	59.35	44.38	103.73	125.7	55.0063	906.116	905.025	7.552800	6.008744	62.85	40.81	103.66
118.8	51.5464	913.858	912.768	7.617418	6.412040	59.40	44.33	103.73	125.8	55.0570	906.002	904.911	7.551848	6.003177	62.90	40.76	103.66
118.9	51.5960	913.748	912.658	7.616500	6.405825	59.45	44.28	103.73	125.9	55.1077	905.887	904.796	7.550889	5.997620	62.95	40.71	103.66
119.0	51.6456	913.638	912.548	7.615582	6.399622	59.50	44.23	103.73	126.0	55.1584	905.773	904.682	7.549937	5.992182	63.00	40.66	103.66

Values at 15.556°C (60° F)

Proof	Mass % Vac	Kg m3 Vac	Kg m3 Air	Lbs per Gal. Air	Lbs per Proof Gal Air	Alcohol Parts	Water Parts	Parts Total	Proof	Mass % Vac	Kg m3 Vac	Kg m3 Air	Lbs per Gal. Air	Lbs per Proof Gal Air	Alcohol Parts	Water Parts	Parts Total
126.0	55.1584	905.773	904.682	7.549937	5.992182	63.00	40.66	103.66	133.0	58.7527	897.604	896.512	7.481755	5.625462	66.50	37.06	103.56
126.1	55.2092	905.658	904.567	7.548977	5.986646	63.05	40.61	103.66	133.1	58.8046	897.485	896.393	7.480762	5.620486	66.55	37.01	103.56
126.2	55.2600	905.544	904.453	7.548026	5.981120	63.10	40.56	103.66	133.2	58.8566	897.366	896.274	7.479769	5.615520	66.60	36.96	103.56
126.3	55.3108	905.429	904.338	7.547066	5.975604	63.15	40.51	103.66	133.3	58.9086	897.247	896.155	7.478776	5.610562	66.65	36.91	103.56
126.4	55.3616	905.314	904.223	7.546107	5.970098	63.20	40.46	103.66	133.4	58.9606	897.128	896.036	7.477783	5.605613	66.70	36.86	103.56
126.5	55.4124	905.200	904.109	7.545155	5.964710	63.25	40.41	103.66	133.5	59.0127	897.009	895.917	7.476790	5.600673	66.75	36.81	103.56
126.6	55.4632	905.085	903.994	7.544196	5.959224	63.30	40.35	103.65	133.6	59.0648	896.889	895.797	7.475788	5.595741	66.80	36.76	103.56
126.7	55.5141	904.970	903.879	7.543236	5.953748	63.35	40.30	103.65	133.7	59.1168	896.770	895.678	7.474795	5.590818	66.85	36.70	103.55
126.8	55.5650	904.855	903.764	7.542276	5.948282	63.40	40.25	103.65	133.8	59.1689	896.650	895.558	7.473794	5.585904	66.90	36.65	103.55
126.9	55.6158	904.740	903.649	7.541316	5.942827	63.45	40.20	103.65	133.9	59.2210	896.531	895.439	7.472801	5.580998	66.95	36.60	103.55
127.0	55.6667	904.625	903.534	7.540357	5.937381	63.50	40.15	103.65	134.0	59.2731	896.412	895.320	7.471808	5.576101	67.00	36.55	103.55
127.1	55.7177	904.509	903.418	7.539389	5.931945	63.55	40.10	103.65	134.1	59.3253	896.292	895.200	7.470806	5.571212	67.05	36.50	103.55
127.2	55.7687	904.394	903.303	7.538429	5.926520	63.60	40.05	103.65	134.2	59.3774	896.172	895.080	7.469805	5.566239	67.10	36.44	103.54
127.3	55.8196	904.279	903.188	7.537469	5.921104	63.65	40.00	103.65	134.3	59.4296	896.053	894.961	7.468812	5.561367	67.15	36.39	103.54
127.4	55.8705	904.164	903.073	7.536509	5.915698	63.70	39.94	103.64	134.4	59.4818	895.933	894.841	7.467810	5.556505	67.20	36.34	103.54
127.5	55.9216	904.048	902.957	7.535541	5.910302	63.75	39.89	103.64	134.5	59.5341	895.813	894.721	7.466809	5.551650	67.25	36.29	103.54
127.6	55.9726	903.933	902.842	7.534582	5.904916	63.80	39.84	103.64	134.6	59.5863	895.693	894.601	7.465807	5.546805	67.30	36.24	103.54
127.7	56.0236	903.817	902.726	7.533614	5.899539	63.85	39.79	103.64	134.7	59.6386	895.573	894.481	7.464806	5.541874	67.35	36.19	103.54
127.8	56.0746	903.701	902.610	7.532646	5.894172	63.90	39.74	103.64	134.8	59.6909	895.453	894.361	7.463804	5.537046	67.40	36.14	103.54
127.9	56.1257	903.586	902.495	7.531686	5.888815	63.95	39.69	103.64	134.9	59.7431	895.333	894.241	7.462803	5.532225	67.45	36.08	103.53
128.0	56.1768	903.470	902.379	7.530718	5.883468	64.00	39.64	103.64	135.0	59.7954	895.213	894.121	7.461801	5.527413	67.50	36.03	103.53
128.1	56.2279	903.354	902.263	7.529750	5.878131	64.05	39.59	103.64	135.1	59.8478	895.093	894.000	7.460792	5.522425	67.55	35.98	103.53
128.2	56.2790	903.239	902.148	7.528790	5.872803	64.10	39.53	103.63	135.2	59.9001	894.972	893.879	7.459782	5.517630	67.60	35.93	103.53
128.3	56.3301	903.122	902.031	7.527814	5.867485	64.15	39.48	103.63	135.3	59.9525	894.852	893.759	7.458780	5.512752	67.65	35.88	103.53
128.4	56.3812	903.007	901.916	7.526854	5.862176	64.20	39.43	103.63	135.4	60.0049	894.732	893.639	7.457779	5.507974	67.70	35.82	103.52
128.5	56.4324	902.890	901.799	7.525877	5.856774	64.25	39.38	103.63	135.5	60.0573	894.611	893.518	7.456769	5.503204	67.75	35.77	103.52
128.6	56.4836	902.775	901.684	7.524918	5.851484	64.30	39.33	103.63	135.6	60.1097	894.491	893.398	7.455768	5.498351	67.80	35.72	103.52
128.7	56.5348	902.658	901.567	7.523941	5.846205	64.35	39.28	103.63	135.7	60.1621	894.370	893.277	7.454758	5.493598	67.85	35.67	103.52
128.8	56.5860	902.542	901.451	7.522973	5.840935	64.40	39.23	103.63	135.8	60.2146	894.249	893.156	7.453748	5.488761	67.90	35.62	103.52
128.9	56.6372	902.426	901.335	7.522005	5.835674	64.45	39.18	103.63	135.9	60.2671	894.129	893.036	7.452747	5.484025	67.95	35.57	103.52
129.0	56.6885	902.309	901.218	7.521029	5.830320	64.50	39.12	103.62	136.0	60.3196	894.008	892.915	7.451737	5.479205	68.00	35.51	103.51
129.1	56.7398	902.193	901.102	7.520061	5.825079	64.55	39.07	103.62	136.1	60.3721	893.887	892.794	7.450727	5.474485	68.05	35.46	103.51
129.2	56.7910	902.077	900.986	7.519093	5.819846	64.60	39.02	103.62	136.2	60.4246	893.766	892.673	7.449717	5.469683	68.10	35.41	103.51
129.3	56.8423	901.960	900.868	7.518108	5.814624	64.65	38.97	103.62	136.3	60.4772	893.645	892.552	7.448707	5.464979	68.15	35.36	103.51
129.4	56.8936	901.843	900.751	7.517131	5.809308	64.70	38.92	103.62	136.4	60.5297	893.524	892.431	7.447698	5.460193	68.20	35.31	103.51
129.5	56.9450	901.727	900.635	7.516163	5.804104	64.75	38.87	103.62	136.5	60.5823	893.403	892.310	7.446688	5.455506	68.25	35.25	103.50
129.6	56.9963	901.610	900.518	7.515187	5.798910	64.80	38.82	103.62	136.6	60.6349	893.281	892.188	7.445670	5.450736	68.30	35.20	103.50
129.7	57.0477	901.493	900.401	7.514211	5.793623	64.85	38.76	103.61	136.7	60.6876	893.160	892.067	7.444660	5.445975	68.35	35.15	103.50
129.8	57.0991	901.376	900.284	7.513234	5.788447	64.90	38.71	103.61	136.8	60.7402	893.039	891.946	7.443650	5.441312	68.40	35.10	103.50
129.9	57.1505	901.260	900.168	7.512266	5.783180	64.95	38.66	103.61	136.9	60.7929	892.917	891.824	7.442632	5.436567	68.45	35.05	103.50
130.0	57.2019	901.142	900.050	7.511281	5.778022	65.00	38.61	103.61	137.0	60.8456	892.796	891.703	7.441622	5.431831	68.50	34.99	103.49
130.1	57.2533	901.026	899.934	7.510313	5.772874	65.05	38.56	103.61	137.1	60.8982	892.674	891.581	7.440604	5.427192	68.55	34.94	103.49
130.2	57.3048	900.908	899.816	7.509328	5.767635	65.10	38.51	103.61	137.2	60.9510	892.553	891.460	7.439594	5.422472	68.60	34.89	103.49
130.3	57.3563	900.791	899.699	7.508352	5.762506	65.15	38.46	103.61	137.3	61.0038	892.431	891.338	7.438576	5.417760	68.65	34.84	103.49
130.4	57.4078	900.674	899.582	7.507376	5.757285	65.20	38.40	103.60	137.4	61.0565	892.309	891.216	7.437558	5.413057	68.70	34.79	103.49
130.5	57.4593	900.557	899.465	7.506399	5.752174	65.25	38.35	103.60	137.5	61.1093	892.188	891.095	7.436548	5.408450	68.75	34.74	103.49
130.6	57.5108	900.439	899.347	7.505414	5.746972	65.30	38.30	103.60	137.6	61.1621	892.066	890.973	7.435530	5.403762	68.80	34.68	103.48
130.7	57.5623	900.322	899.230	7.504438	5.741879	65.35	38.25	103.60	137.7	61.2149	891.944	890.851	7.434512	5.399083	68.85	34.63	103.48
130.8	57.6139	900.204	899.112	7.503453	5.736696	65.40	38.20	103.60	137.8	61.2677	891.822	890.729	7.433494	5.394412	68.90	34.58	103.48
130.9	57.6655	900.087	898.995	7.502477	5.731522	65.45	38.15	103.60	137.9	61.3206	891.700	890.607	7.432476	5.389749	68.95	34.53	103.48
131.0	57.7171	899.969	898.877	7.501492	5.726456	65.50	38.10	103.60	138.0	61.3734	891.578	890.485	7.431457	5.385181	69.00	34.48	103.48
131.1	57.7687	899.852	898.760	7.500516	5.721301	65.55	38.04	103.59	138.1	61.4263	891.456	890.363	7.430439	5.380534	69.05	34.42	103.47
131.2	57.8203	899.734	898.642	7.499531	5.716253	65.60	37.99	103.59	138.2	61.4793	891.333	890.241	7.429413	5.375895	69.10	34.37	103.47
131.3	57.8720	899.616	898.524	7.498546	5.711116	65.65	37.94	103.59	138.3	61.5322	891.211	890.118	7.428395	5.371264	69.15	34.32	103.47
131.4	57.9236	899.498	898.406	7.497561	5.705988	65.70	37.89	103.59	138.4	61.5851	891.089	889.996	7.427377	5.366640	69.20	34.27	103.47
131.5	57.9753	899.380	898.288	7.496577	5.700968	65.75	37.84	103.59	138.5	61.6381	890.966	889.873	7.426350	5.362025	69.25	34.22	103.47
131.6	58.0270	899.262	898.170	7.495592	5.695858	65.80	37.79	103.59	138.6	61.6911	890.843	889.750	7.425324	5.357418	69.30	34.16	103.46
131.7	58.0787	899.144	898.052	7.494607	5.690757	65.85	37.73	103.58	138.7	61.7441	890.721	889.628	7.424305	5.352818	69.35	34.11	103.46
131.8	58.1305	899.026	897.934	7.493622	5.685666	65.90	37.68	103.58	138.8	61.7972	890.598	889.505	7.423279	5.348227	69.40	34.06	103.46
131.9	58.1822	898.908	897.816	7.492638	5.680681	65.95	37.63	103.58	138.9	61.8501	890.476	889.383	7.422261	5.343643	69.45	34.01	103.46
132.0	58.2340	898.790	897.698	7.491653	5.675607	66.00	37.58	103.58	139.0	61.9032	890.353	889.260	7.421234	5.339067	69.50	33.96	103.46
132.1	58.2857	898.671	897.579	7.490660	5.670543	66.05	37.53	103.58	139.1	61.9563	890.230	889.137	7.420208	5.334499	69.55	33.91	103.46
132.2	58.3376	898.553	897.461	7.489675	5.665487	66.10	37.48	103.58	139.2	62.0094	890.107	889.014	7.419181	5.329939	69.60	33.85	103.45
132.3	58.3894	898.435	897.343	7.488690	5.660538	66.15	37.43	103.58	139.3	62.0626	889.984	888.891	7.418155	5.325301	69.65	33.80	103.45
132.4	58.4413	898.316	897.224	7.487697	5.655500	66.20	37.37	103.57	139.4	62.1157	889.861	888.768	7.417128	5.320756	69.70	33.75	103.45
132.5	58.4931	898.198	897.106	7.486712	5.650472	66.25	37.32	103.57	139.5	62.1688	889.738	888.645	7.416102	5.316219	69.75	33.70	103.45
132.6	58.5450	898.079	896.987	7.485719	5.645452	66.30	37.27	103.57	139.6	62.2220	889.614	888.521	7.415067	5.311690	69.80	33.64	103.44
132.7	58.5969	897.960	896.868	7.484726	5.640441	66.35	37.22	103.57	139.7	62.2753	889.491	888.398	7.414041	5.307169	69.85	33.59	103.44
132.8	58.6488	897.842	896.750	7.483741	5.635439	66.40	37.17	103.57	139.8	62.3285	889.368	888.275	7.413014	5.302570	69.90	33.54	103.44
132.9	58.7007	897.723	896.631	7.482748	5.630446	66.45	37.12	103.57	139.9	62.3817	889.244	888.151	7.411979	5.298065	69.95	33.49	103.44
133.0	58.7527	897.604	896.512	7.481755	5.625462	66.50	37.06	103.56	140.0	62.4350	889.121	888.028	7.410953	5.293566	70.00	33.44	103.44

Values at 15.556°C (60° F)

Proof	Mass % Vac	Kg m3 Vac	Kg m3 Air	Lbs per Gal. Air	Lbs per Proof Gal Air	Alcohol Parts	Water Parts	Parts Total	Proof	Mass % Vac	Kg m3 Vac	Kg m3 Air	Lbs per Gal. Air	Lbs per Proof Gal Air	Alcohol Parts	Water Parts	Parts Total
140.0	62.4350	889.121	888.028	7.410953	5.293566	70.00	33.44	103.44	147.0	66.2132	880.305	879.210	7.337363	4.991479	73.50	29.78	103.28
140.1	62.4883	888.997	887.904	7.409918	5.289076	70.05	33.38	103.43	147.1	66.2680	880.176	879.081	7.336286	4.987336	73.55	29.72	103.27
140.2	62.5415	888.874	887.781	7.408892	5.284508	70.10	33.33	103.43	147.2	66.3227	880.048	878.953	7.335218	4.983199	73.60	29.67	103.27
140.3	62.5949	888.750	887.657	7.407857	5.280033	70.15	33.28	103.43	147.3	66.3775	879.919	878.824	7.334142	4.979069	73.65	29.62	103.27
140.4	62.6482	888.626	887.533	7.406822	5.275566	70.20	33.23	103.43	147.4	66.4322	879.790	878.695	7.333065	4.975022	73.70	29.57	103.27
140.5	62.7015	888.502	887.409	7.405787	5.271022	70.25	33.18	103.43	147.5	66.4870	879.662	878.567	7.331997	4.970905	73.75	29.51	103.26
140.6	62.7549	888.379	887.286	7.404761	5.266569	70.30	33.12	103.42	147.6	66.5419	879.533	878.438	7.330920	4.966796	73.80	29.46	103.26
140.7	62.8083	888.255	887.161	7.403717	5.262124	70.35	33.07	103.42	147.7	66.5967	879.404	878.309	7.329844	4.962694	73.85	29.41	103.26
140.8	62.8617	888.131	887.037	7.402683	5.257603	70.40	33.02	103.42	147.8	66.6516	879.275	878.180	7.328767	4.958598	73.90	29.35	103.25
140.9	62.9152	888.007	886.913	7.401648	5.253174	70.45	32.97	103.42	147.9	66.7065	879.146	878.051	7.327691	4.954509	73.95	29.30	103.25
141.0	62.9687	887.882	886.788	7.400605	5.248668	70.50	32.92	103.42	148.0	66.7614	879.017	877.922	7.326614	4.950501	74.00	29.25	103.25
141.1	63.0221	887.758	886.664	7.399570	5.244253	70.55	32.86	103.41	148.1	66.8163	878.888	877.793	7.325538	4.946425	74.05	29.20	103.25
141.2	63.0756	887.634	886.540	7.398535	5.239763	70.60	32.81	103.41	148.2	66.8713	878.758	877.663	7.324453	4.942356	74.10	29.14	103.24
141.3	63.1291	887.510	886.416	7.397500	5.235363	70.65	32.76	103.41	148.3	66.9262	878.629	877.534	7.323376	4.938294	74.15	29.09	103.24
141.4	63.1827	887.385	886.291	7.396457	5.230888	70.70	32.71	103.41	148.4	66.9812	878.500	877.405	7.322300	4.934238	74.20	29.04	103.24
141.5	63.2362	887.260	886.166	7.395414	5.226503	70.75	32.66	103.41	148.5	67.0362	878.370	877.275	7.321215	4.930190	74.25	28.99	103.24
141.6	63.2898	887.136	886.042	7.394379	5.222043	70.80	32.60	103.40	148.6	67.0912	878.240	877.145	7.320130	4.926147	74.30	28.93	103.23
141.7	63.3434	887.011	885.917	7.393336	5.217673	70.85	32.55	103.40	148.7	67.1463	878.111	877.016	7.319053	4.922112	74.35	28.88	103.23
141.8	63.3970	886.887	885.793	7.392301	5.213228	70.90	32.50	103.40	148.8	67.2014	877.981	876.886	7.317968	4.918083	74.40	28.83	103.23
141.9	63.4506	886.762	885.668	7.391258	5.208791	70.95	32.45	103.40	148.9	67.2566	877.851	876.756	7.316883	4.913987	74.45	28.78	103.23
142.0	63.5043	886.637	885.543	7.390214	5.204443	71.00	32.39	103.39	149.0	67.3116	877.721	876.626	7.315799	4.909971	74.50	28.72	103.22
142.1	63.5580	886.512	885.418	7.389171	5.200020	71.05	32.34	103.39	149.1	67.3668	877.591	876.496	7.314714	4.905962	74.55	28.67	103.22
142.2	63.6117	886.387	885.293	7.388128	5.195605	71.10	32.29	103.39	149.2	67.4219	877.461	876.366	7.313629	4.901959	74.60	28.62	103.22
142.3	63.6654	886.262	885.168	7.387085	5.191279	71.15	32.24	103.39	149.3	67.4772	877.331	876.236	7.312544	4.897963	74.65	28.57	103.22
142.4	63.7191	886.137	885.043	7.386042	5.186879	71.20	32.19	103.39	149.4	67.5324	877.201	876.106	7.311459	4.893974	74.70	28.51	103.21
142.5	63.7729	886.012	884.918	7.384999	5.182486	71.25	32.13	103.38	149.5	67.5876	877.071	875.976	7.310374	4.889918	74.75	28.46	103.21
142.6	63.8266	885.887	884.793	7.383955	5.178101	71.30	32.08	103.38	149.6	67.6429	876.941	875.846	7.309289	4.885942	74.80	28.41	103.21
142.7	63.8805	885.761	884.667	7.382904	5.173723	71.35	32.03	103.38	149.7	67.6982	876.810	875.715	7.308196	4.881971	74.85	28.35	103.20
142.8	63.9343	885.636	884.542	7.381861	5.169434	71.40	31.98	103.38	149.8	67.7535	876.680	875.585	7.307111	4.877936	74.90	28.30	103.20
142.9	63.9881	885.510	884.416	7.380809	5.165071	71.45	31.92	103.37	149.9	67.8088	876.549	875.454	7.306018	4.873979	74.95	28.25	103.20
143.0	64.0420	885.385	884.291	7.379766	5.160715	71.50	31.87	103.37	150.0	67.8641	876.419	875.324	7.304933	4.870028	75.00	28.20	103.20
143.1	64.0959	885.259	884.165	7.378715	5.156366	71.55	31.82	103.37	150.1	67.9195	876.288	875.193	7.303840	4.866012	75.05	28.14	103.19
143.2	64.1498	885.133	884.039	7.377663	5.152025	71.60	31.77	103.37	150.2	67.9749	876.157	875.062	7.302746	4.862074	75.10	28.09	103.19
143.3	64.2036	885.008	883.914	7.376620	5.147691	71.65	31.71	103.36	150.3	68.0303	876.026	874.931	7.301653	4.858143	75.15	28.04	103.19
143.4	64.2576	884.882	883.788	7.375568	5.143364	71.70	31.66	103.36	150.4	68.0857	875.896	874.801	7.300568	4.854147	75.20	27.98	103.18
143.5	64.3116	884.756	883.662	7.374517	5.139045	71.75	31.61	103.36	150.5	68.1412	875.764	874.669	7.299467	4.850228	75.25	27.93	103.18
143.6	64.3656	884.630	883.536	7.373465	5.134733	71.80	31.56	103.36	150.6	68.1967	875.633	874.538	7.298373	4.846245	75.30	27.88	103.18
143.7	64.4195	884.504	883.410	7.372414	5.130428	71.85	31.51	103.36	150.7	68.2522	875.502	874.407	7.297280	4.842339	75.35	27.83	103.18
143.8	64.4736	884.378	883.284	7.371362	5.126131	71.90	31.45	103.35	150.8	68.3077	875.371	874.276	7.296187	4.838368	75.40	27.77	103.17
143.9	64.5276	884.251	883.157	7.370302	5.121840	71.95	31.40	103.35	150.9	68.3633	875.240	874.145	7.295094	4.834475	75.45	27.72	103.17
144.0	64.5817	884.126	883.032	7.369259	5.117557	72.00	31.35	103.35	151.0	68.4189	875.108	874.013	7.293992	4.830518	75.50	27.67	103.17
144.1	64.6358	883.999	882.905	7.368199	5.113281	72.05	31.30	103.35	151.1	68.4744	874.977	873.882	7.292899	4.826637	75.55	27.62	103.17
144.2	64.6898	883.873	882.779	7.367148	5.109012	72.10	31.24	103.34	151.2	68.5301	874.846	873.751	7.291805	4.822692	75.60	27.56	103.16
144.3	64.7440	883.746	882.652	7.366088	5.104750	72.15	31.19	103.34	151.3	68.5857	874.714	873.619	7.290704	4.818754	75.65	27.51	103.16
144.4	64.7981	883.620	882.526	7.365036	5.100495	72.20	31.14	103.34	151.4	68.6414	874.582	873.487	7.289602	4.814892	75.70	27.46	103.16
144.5	64.8523	883.493	882.399	7.363977	5.096248	72.25	31.09	103.34	151.5	68.6971	874.450	873.354	7.288492	4.810967	75.75	27.40	103.15
144.6	64.9065	883.367	882.273	7.362925	5.092007	72.30	31.04	103.34	151.6	68.7527	874.319	873.223	7.287399	4.807048	75.80	27.35	103.15
144.7	64.9607	883.240	882.146	7.361865	5.087695	72.35	30.98	103.33	151.7	68.8085	874.187	873.091	7.286298	4.803135	75.85	27.30	103.15
144.0	65.0149	883.113	882.019	7.360805	5.083469	72.40	30.93	103.33	151.8	68.8642	874.055	872.959	7.285196	4.799298	75.90	27.25	103.15
144.9	65.0692	882.986	881.892	7.359745	5.079249	72.45	30.88	103.33	151.9	68.9200	873.922	872.828	7.284000	4.795398	75.95	27.19	103.14
145.0	65.1234	882.859	881.765	7.358686	5.075037	72.50	30.83	103.33	152.0	68.9758	873.790	872.694	7.282984	4.791504	76.00	27.14	103.14
145.1	65.1777	882.732	881.638	7.357626	5.070754	72.55	30.77	103.32	152.1	69.0317	873.526	872.562	7.281883	4.787617	76.05	27.09	103.14
145.2	65.2320	882.605	881.511	7.356566	5.066555	72.60	30.72	103.32	152.2	69.0875	873.526	872.430	7.280781	4.783736	76.10	27.03	103.13
145.3	65.2864	882.478	881.384	7.355506	5.062364	72.65	30.67	103.32	152.3	69.1433	873.393	872.297	7.279671	4.779930	76.15	26.98	103.13
145.4	65.3407	882.350	881.256	7.354438	5.058102	72.70	30.62	103.32	152.4	69.1993	873.261	872.165	7.278570	4.776061	76.20	26.93	103.13
145.5	65.3951	882.223	881.129	7.353378	5.053925	72.75	30.56	103.31	152.5	69.2552	873.129	872.033	7.277468	4.772198	76.25	26.87	103.12
145.6	65.4495	882.096	881.002	7.352318	5.049677	72.80	30.51	103.31	152.6	69.3111	872.996	871.900	7.276358	4.768342	76.30	26.82	103.12
145.7	65.5039	881.968	880.874	7.351250	5.045514	72.85	30.46	103.31	152.7	69.3671	872.863	871.767	7.275248	4.764492	76.35	26.77	103.12
145.8	65.5583	881.841	880.747	7.350190	5.041357	72.90	30.41	103.31	152.8	69.4230	872.730	871.634	7.274138	4.760648	76.40	26.72	103.12
145.9	65.6128	881.713	880.619	7.349122	5.037131	72.95	30.35	103.30	152.9	69.4791	872.598	871.502	7.273037	4.756811	76.45	26.66	103.11
146.0	65.6673	881.585	880.491	7.348054	5.032988	73.00	30.30	103.30	153.0	69.5351	872.465	871.369	7.271927	4.752979	76.50	26.61	103.11
146.1	65.7218	881.458	880.364	7.346994	5.028775	73.05	30.25	103.30	153.1	69.5912	872.332	871.236	7.270817	4.749154	76.55	26.56	103.11
146.2	65.7763	881.330	880.235	7.345917	5.024646	73.10	30.20	103.30	153.2	69.6472	872.199	871.103	7.269707	4.745335	76.60	26.50	103.10
146.3	65.8309	881.202	880.107	7.344849	5.020448	73.15	30.14	103.29	153.3	69.7033	872.066	870.970	7.268597	4.741522	76.65	26.45	103.10
146.4	65.8854	881.074	879.979	7.343781	5.016332	73.20	30.09	103.29	153.4	69.7595	871.932	870.836	7.267479	4.737647	76.70	26.40	103.10
146.5	65.9400	880.946	879.851	7.342712	5.012147	73.25	30.04	103.29	153.5	69.8156	871.799	870.703	7.266369	4.733846	76.75	26.34	103.09
146.6	65.9946	880.818	879.723	7.341644	5.007970	73.30	29.99	103.29	153.6	69.8718	871.666	870.570	7.265259	4.730052	76.80	26.29	103.09
146.7	66.0492	880.690	879.595	7.340576	5.003875	73.35	29.93	103.28	153.7	69.9280	871.532	870.436	7.264140	4.726263	76.85	26.24	103.09
146.8	66.1039	880.561	879.466	7.339499	4.999711	73.40	29.88	103.28	153.8	69.9842	871.399	870.303	7.263031	4.722481	76.90	26.19	103.09
146.9	66.1586	880.433	879.338	7.338431	4.995554	73.45	29.83	103.28	153.9	70.0405	871.265	870.169	7.261912	4.718637	76.95	26.13	103.08
147.0	66.2132	880.305	879.210	7.337363	4.991479	73.50	29.78	103.28	154.0	70.0968	871.131	870.035	7.260794	4.714867	77.00	26.08	103.08

Values at 15.556°C (60° F)

Proof	Mass % Vac	Kg m3 Vac	Kg m3 Air	Lbs per Gal. Air	Lbs per Proof Gal Air	Alcohol Parts	Water Parts	Parts Total	Proof	Mass % Vac	Kg m3 Vac	Kg m3 Air	Lbs per Gal. Air	Lbs per Proof Gal Air	Alcohol Parts	Water Parts	Parts Total
154.0	70.0968	871.131	870.035	7.260794	4.714867	77.00	26.08	103.08	161.0	74.0964	861.568	860.471	7.180979	4.460309	80.50	22.34	102.84
154.1	70.1530	870.998	869.902	7.259684	4.711103	77.05	26.03	103.08	161.1	74.1545	861.428	860.331	7.179810	4.456759	80.55	22.29	102.84
154.2	70.2093	870.864	869.768	7.258566	4.707345	77.10	25.97	103.07	161.2	74.2125	861.288	860.191	7.178642	4.453276	80.60	22.24	102.84
154.3	70.2657	870.730	869.634	7.257447	4.703525	77.15	25.92	103.07	161.3	74.2706	861.149	860.052	7.177482	4.449798	80.65	22.18	102.83
154.4	70.3220	870.596	869.500	7.256329	4.699779	77.20	25.87	103.07	161.4	74.3287	861.009	859.912	7.176313	4.446325	80.70	22.13	102.83
154.5	70.3784	870.461	869.365	7.255203	4.695972	77.25	25.81	103.06	161.5	74.3869	860.869	859.772	7.175145	4.442858	80.75	22.07	102.82
154.6	70.4348	870.328	869.232	7.254093	4.692238	77.30	25.76	103.06	161.6	74.4450	860.729	859.631	7.173968	4.439396	80.80	22.02	102.82
154.7	70.4913	870.193	869.097	7.252966	4.688510	77.35	25.71	103.06	161.7	74.5033	860.588	859.490	7.172792	4.435940	80.85	21.97	102.82
154.8	70.5477	870.059	868.963	7.251848	4.684721	77.40	25.65	103.05	161.8	74.5615	860.448	859.350	7.171623	4.432429	80.90	21.91	102.81
154.9	70.6042	869.924	868.828	7.250721	4.681005	77.45	25.60	103.05	161.9	74.6197	860.308	859.210	7.170455	4.428983	80.95	21.86	102.81
155.0	70.6607	869.790	868.694	7.249603	4.677229	77.50	25.55	103.05	162.0	74.6780	860.167	859.069	7.169278	4.425543	81.00	21.81	102.81
155.1	70.7173	869.655	868.559	7.248476	4.673524	77.55	25.50	103.05	162.1	74.7364	860.026	858.928	7.168102	4.422108	81.05	21.75	102.80
155.2	70.7738	869.520	868.424	7.247350	4.669760	77.60	25.44	103.04	162.2	74.7947	859.886	858.788	7.166933	4.418620	81.10	21.70	102.80
155.3	70.8304	869.386	868.290	7.246231	4.666067	77.65	25.39	103.04	162.3	74.8531	859.745	858.647	7.165756	4.415195	81.15	21.64	102.79
155.4	70.8870	869.251	868.155	7.245105	4.662315	77.70	25.34	103.04	162.4	74.9115	859.604	858.506	7.164580	4.411718	81.20	21.59	102.79
155.5	70.9436	869.116	868.020	7.243978	4.658568	77.75	25.28	103.03	162.5	74.9699	859.463	858.365	7.163403	4.408304	81.25	21.54	102.79
155.6	71.0003	868.981	867.885	7.242851	4.654893	77.80	25.23	103.03	162.6	75.0283	859.322	858.224	7.162226	4.404896	81.30	21.48	102.78
155.7	71.0569	868.846	867.750	7.241725	4.651159	77.85	25.18	103.03	162.7	75.0868	859.181	858.083	7.161050	4.401434	81.35	21.43	102.78
155.8	71.1136	868.711	867.615	7.240598	4.647430	77.90	25.12	103.02	162.8	75.1453	859.040	857.942	7.159873	4.398037	81.40	21.38	102.78
155.9	71.1703	868.576	867.480	7.239471	4.643773	77.95	25.07	103.02	162.9	75.2039	858.898	857.800	7.158688	4.394586	81.45	21.32	102.77
156.0	71.2271	868.441	867.345	7.238345	4.640056	78.00	25.02	103.02	163.0	75.2624	858.757	857.659	7.157511	4.391199	81.50	21.27	102.77
156.1	71.2839	868.305	867.209	7.237210	4.636345	78.05	24.96	103.01	163.1	75.3210	858.615	857.517	7.156326	4.387759	81.55	21.21	102.76
156.2	71.3406	868.170	867.074	7.236083	4.632641	78.10	24.91	103.01	163.2	75.3796	858.473	857.375	7.155141	4.384324	81.60	21.16	102.76
156.3	71.3975	868.034	866.938	7.234948	4.628942	78.15	24.86	103.01	163.3	75.4382	858.332	857.234	7.153964	4.380953	81.65	21.11	102.76
156.4	71.4543	867.899	866.803	7.233822	4.625313	78.20	24.80	103.00	163.4	75.4969	858.190	857.092	7.152779	4.377529	81.70	21.05	102.75
156.5	71.5112	867.763	866.667	7.232687	4.621626	78.25	24.75	103.00	163.5	75.5556	858.048	856.950	7.151594	4.374110	81.75	21.00	102.75
156.6	71.5681	867.627	866.531	7.231552	4.617945	78.30	24.70	103.00	163.6	75.6144	857.906	856.808	7.150409	4.370754	81.80	20.95	102.75
156.7	71.6250	867.491	866.394	7.230408	4.614205	78.35	24.64	102.99	163.7	75.6731	857.764	856.666	7.149224	4.367346	81.85	20.89	102.74
156.8	71.6820	867.355	866.258	7.229273	4.610535	78.40	24.59	102.99	163.8	75.7319	857.621	856.523	7.148031	4.363943	81.90	20.84	102.74
156.9	71.7389	867.219	866.122	7.228138	4.606871	78.45	24.54	102.99	163.9	75.7907	857.479	856.381	7.146846	4.360546	81.95	20.78	102.73
157.0	71.7959	867.083	865.986	7.227003	4.603213	78.50	24.48	102.98	164.0	75.8495	857.337	856.239	7.145661	4.357154	82.00	20.73	102.73
157.1	71.8529	866.947	865.850	7.225868	4.599561	78.55	24.43	102.98	164.1	75.9084	857.194	856.096	7.144467	4.353767	82.05	20.67	102.72
157.2	71.9099	866.810	865.713	7.224725	4.595915	78.60	24.38	102.98	164.2	75.9673	857.052	855.954	7.143282	4.350442	82.10	20.62	102.72
157.3	71.9670	866.674	865.577	7.223590	4.592274	78.65	24.32	102.97	164.3	76.0263	856.909	855.811	7.142089	4.347066	82.15	20.57	102.72
157.4	72.0241	866.538	865.441	7.222455	4.588640	78.70	24.27	102.97	164.4	76.0851	856.767	855.669	7.140904	4.343694	82.20	20.51	102.71
157.5	72.0813	866.401	865.304	7.221312	4.585011	78.75	24.22	102.97	164.5	76.1441	856.623	855.525	7.139702	4.340328	82.25	20.46	102.71
157.6	72.1383	866.264	865.167	7.220169	4.581387	78.80	24.16	102.96	164.6	76.2031	856.480	855.382	7.138509	4.336967	82.30	20.41	102.71
157.7	72.1955	866.128	865.031	7.219034	4.577706	78.85	24.11	102.96	164.7	76.2622	856.337	855.239	7.137315	4.333612	82.35	20.35	102.70
157.8	72.2527	865.991	864.894	7.217890	4.574094	78.90	24.06	102.96	164.8	76.3213	856.194	855.096	7.136122	4.330261	82.40	20.30	102.70
157.9	72.3100	865.854	864.757	7.216747	4.570488	78.95	24.00	102.95	164.9	76.3804	856.051	854.953	7.134929	4.326916	82.45	20.24	102.69
158.0	72.3672	865.717	864.620	7.215604	4.566888	79.00	23.95	102.95	165.0	76.4395	855.907	854.809	7.133727	4.323519	82.50	20.19	102.69
158.1	72.4245	865.580	864.483	7.214460	4.563230	79.05	23.90	102.95	165.1	76.4987	855.764	854.666	7.132533	4.320184	82.55	20.13	102.68
158.2	72.4818	865.443	864.346	7.213317	4.559641	79.10	23.84	102.94	165.2	76.5579	855.620	854.522	7.131332	4.316854	82.60	20.08	102.68
158.3	72.5391	865.306	864.209	7.212174	4.556058	79.15	23.79	102.94	165.3	76.6170	855.477	854.379	7.130138	4.313530	82.65	20.03	102.68
158.4	72.5965	865.168	864.071	7.211022	4.552418	79.20	23.73	102.93	165.4	76.6763	855.333	854.235	7.128937	4.310210	82.70	19.97	102.67
158.5	72.6538	865.031	863.934	7.209879	4.548846	79.25	23.68	102.93	165.5	76.7355	855.189	854.091	7.127735	4.306840	82.75	19.92	102.67
158.6	72.7112	864.893	863.796	7.208727	4.545279	79.30	23.63	102.93	165.6	76.7948	855.045	853.947	7.126533	4.303530	82.80	19.86	102.66
158.7	72.7686	864.756	863.659	7.207584	4.541656	79.35	23.57	102.92	165.7	76.8542	854.900	853.802	7.125323	4.300226	82.85	19.81	102.66
158.8	72.8261	864.618	863.521	7.206432	4.538101	79.40	23.52	102.92	165.8	76.9135	854.757	853.659	7.124130	4.296871	82.90	19.76	102.66
158.9	72.8835	864.480	863.383	7.205280	4.534489	79.45	23.47	102.92	165.9	76.9729	854.612	853.514	7.122920	4.293577	82.95	19.70	102.65
159.0	72.9410	864.343	863.246	7.204137	4.530945	79.50	23.41	102.91	166.0	77.0323	854.468	853.370	7.121718	4.290288	83.00	19.65	102.65
159.1	72.9985	864.205	863.108	7.202985	4.527345	79.55	23.36	102.91	166.1	77.0918	854.323	853.225	7.120508	4.286949	83.05	19.59	102.64
159.2	73.0561	864.067	862.970	7.201834	4.523812	79.60	23.31	102.91	166.2	77.1512	854.179	853.081	7.119306	4.283670	83.10	19.54	102.64
159.3	73.1136	863.929	862.832	7.200682	4.520223	79.65	23.25	102.90	166.3	77.2108	854.034	852.936	7.118096	4.280341	83.15	19.49	102.64
159.4	73.1712	863.790	862.693	7.199522	4.516701	79.70	23.20	102.90	166.4	77.2703	853.889	852.791	7.116886	4.277072	83.20	19.43	102.63
159.5	73.2289	863.652	862.555	7.198370	4.513123	79.75	23.15	102.90	166.5	77.3299	853.744	852.645	7.115667	4.273698	83.25	19.38	102.63
159.6	73.2865	863.514	862.417	7.197219	4.509551	79.80	23.09	102.89	166.6	77.3895	853.599	852.500	7.114457	4.270384	83.30	19.32	102.62
159.7	73.3442	863.376	862.279	7.196067	4.506046	79.85	23.04	102.89	166.7	77.4491	853.453	852.354	7.113239	4.267131	83.35	19.27	102.62
159.8	73.4019	863.237	862.140	7.194907	4.502485	79.90	22.99	102.89	166.8	77.5087	853.309	852.210	7.112037	4.263827	83.40	19.21	102.61
159.9	73.4596	863.099	862.002	7.193755	4.498930	79.95	22.93	102.88	166.9	77.5684	853.163	852.064	7.110819	4.260584	83.45	19.16	102.61
160.0	73.5174	862.960	861.863	7.192595	4.495441	80.00	22.88	102.88	167.0	77.6281	853.018	851.919	7.109609	4.257290	83.50	19.11	102.61
160.1	73.5752	862.821	861.724	7.191435	4.491897	80.05	22.83	102.88	167.1	77.6879	852.872	851.773	7.108390	4.254002	83.55	19.05	102.60
160.2	73.6330	862.682	861.585	7.190275	4.488358	80.10	22.77	102.87	167.2	77.7477	852.726	851.627	7.107172	4.250719	83.60	19.00	102.60
160.3	73.6908	862.543	861.446	7.189115	4.484825	80.15	22.72	102.87	167.3	77.8074	852.580	851.481	7.105953	4.247441	83.65	18.94	102.59
160.4	73.7487	862.404	861.307	7.187955	4.481298	80.20	22.66	102.86	167.4	77.8673	852.434	851.335	7.104735	4.244222	83.70	18.89	102.59
160.5	73.8066	862.265	861.168	7.186795	4.477776	80.25	22.61	102.86	167.5	77.9271	852.289	851.190	7.103525	4.240954	83.75	18.83	102.58
160.6	73.8645	862.126	861.029	7.185635	4.474259	80.30	22.56	102.86	167.6	77.9871	852.142	851.043	7.102298	4.237691	83.80	18.78	102.58
160.7	73.9225	861.986	860.889	7.184467	4.470748	80.35	22.50	102.85	167.7	78.0470	851.996	850.897	7.101080	4.234433	83.85	18.73	102.58
160.8	73.9804	861.847	860.750	7.183307	4.467303	80.40	22.45	102.85	167.8	78.1069	851.850	850.751	7.099861	4.231179	83.90	18.67	102.57
160.9	74.0383	861.708	860.611	7.182147	4.463803	80.45	22.40	102.85	167.9	78.1670	851.703	850.604	7.098634	4.227931	83.95	18.62	102.57
161.0	74.0964	861.568	860.471	7.180979	4.460309	80.50	22.34	102.84	168.0	78.2269	851.557	850.458	7.097416	4.224688	84.00	18.56	102.56

Values at 15.556°C (60° F)

Proof	Mass % Vac	Kg m3 Vac	Kg m3 Air	Lbs per Gal. Air	Lbs per Proof Gal Air	Alcohol Parts	Water Parts	Parts Total	Proof	Mass % Vac	Kg m3 Vac	Kg m3 Air	Lbs per Gal. Air	Lbs per Proof Gal Air	Alcohol Parts	Water Parts	Parts Total
168.0	78.2269	851.557	850.458	7.097416	4.224688	84.00	18.56	102.56	175.0	82.5110	840.982	839.882	7.009155	4.005271	87.50	14.72	102.22
168.1	78.2870	851.410	850.311	7.096189	4.221450	84.05	18.51	102.56	175.1	82.5735	840.826	839.726	7.007853	4.002215	87.55	14.67	102.22
168.2	78.3471	851.263	850.164	7.094962	4.218217	84.10	18.45	102.55	175.2	82.6360	840.670	839.570	7.006551	3.999212	87.60	14.61	102.21
168.3	78.4072	851.116	850.017	7.093736	4.214988	84.15	18.40	102.55	175.3	82.6985	840.513	839.413	7.005241	3.996165	87.65	14.56	102.21
168.4	78.4674	850.969	849.870	7.092509	4.211765	84.20	18.34	102.54	175.4	82.7611	840.357	839.257	7.003939	3.993171	87.70	14.50	102.20
168.5	78.5275	850.822	849.723	7.091282	4.208493	84.25	18.29	102.54	175.5	82.8238	840.200	839.100	7.002629	3.990133	87.75	14.45	102.20
168.6	78.5878	850.674	849.575	7.090047	4.205280	84.30	18.24	102.54	175.6	82.8864	840.044	838.943	7.001319	3.987149	87.80	14.39	102.19
168.7	78.6480	850.527	849.428	7.088820	4.202071	84.35	18.18	102.53	175.7	82.9491	839.887	838.786	7.000008	3.984120	87.85	14.34	102.19
168.8	78.7082	850.380	849.281	7.087593	4.198868	84.40	18.13	102.53	175.8	83.0119	839.730	838.629	6.998698	3.981096	87.90	14.28	102.18
168.9	78.7686	850.231	849.132	7.086350	4.195616	84.45	18.07	102.52	175.9	83.0747	839.572	838.471	6.997380	3.978077	87.95	14.23	102.18
169.0	78.8289	850.084	848.985	7.085123	4.192422	84.50	18.02	102.52	176.0	83.1374	839.415	838.314	6.996069	3.975110	88.00	14.17	102.17
169.1	78.8893	849.936	848.837	7.083888	4.189233	84.55	17.96	102.51	176.1	83.2003	839.257	838.156	6.994751	3.972100	88.05	14.12	102.17
169.2	78.9497	849.788	848.689	7.082653	4.185996	84.60	17.91	102.51	176.2	83.2633	839.099	837.998	6.993432	3.969094	88.10	14.06	102.16
169.3	79.0101	849.640	848.541	7.081418	4.182817	84.65	17.85	102.50	176.3	83.3262	838.941	837.840	6.992114	3.966093	88.15	14.00	102.15
169.4	79.0706	849.491	848.392	7.080174	4.179589	84.70	17.80	102.50	176.4	83.3892	838.783	837.682	6.990795	3.963096	88.20	13.95	102.15
169.5	79.1310	849.343	848.244	7.078939	4.176420	84.75	17.75	102.50	176.5	83.4522	838.624	837.523	6.989468	3.960104	88.25	13.89	102.14
169.6	79.1916	849.195	848.096	7.077704	4.173203	84.80	17.69	102.49	176.6	83.5153	838.466	837.365	6.988150	3.957116	88.30	13.84	102.14
169.7	79.2522	849.046	847.947	7.076461	4.170043	84.85	17.64	102.49	176.7	83.5783	838.308	837.207	6.986831	3.954133	88.35	13.78	102.13
169.8	79.3127	848.898	847.799	7.075226	4.166802	84.90	17.58	102.48	176.8	83.6415	838.149	837.048	6.985504	3.951108	88.40	13.73	102.13
169.9	79.3734	848.749	847.650	7.073982	4.163685	84.95	17.53	102.48	176.9	83.7047	837.990	836.889	6.984177	3.948134	88.45	13.67	102.12
170.0	79.4340	848.600	847.501	7.072739	4.160488	85.00	17.47	102.47	177.0	83.7679	837.830	836.729	6.982842	3.945164	88.50	13.62	102.12
170.1	79.4947	848.451	847.352	7.071495	4.157295	85.05	17.42	102.47	177.1	83.8312	837.671	836.570	6.981515	3.942199	88.55	13.56	102.11
170.2	79.5554	848.301	847.202	7.070243	4.154107	85.10	17.36	102.46	177.2	83.8945	837.511	836.410	6.980180	3.939191	88.60	13.50	102.10
170.3	79.6162	848.152	847.053	7.069000	4.150976	85.15	17.31	102.46	177.3	83.9579	837.352	836.251	6.978853	3.936235	88.65	13.45	102.10
170.4	79.6770	848.002	846.903	7.067748	4.147798	85.20	17.25	102.45	177.4	84.0213	837.192	836.091	6.977518	3.933284	88.70	13.39	102.09
170.5	79.7379	847.853	846.754	7.066505	4.144624	85.25	17.20	102.45	177.5	84.0847	837.032	835.931	6.976182	3.930290	88.75	13.34	102.09
170.6	79.7986	847.703	846.604	7.065253	4.141456	85.30	17.14	102.44	177.6	84.1482	836.871	835.770	6.974839	3.927347	88.80	13.28	102.08
170.7	79.8596	847.554	846.455	7.064009	4.138292	85.35	17.09	102.44	177.7	84.2117	836.711	835.610	6.973503	3.924362	88.85	13.23	102.08
170.8	79.9205	847.403	846.304	7.062749	4.135133	85.40	17.03	102.43	177.8	84.2753	836.550	835.449	6.972160	3.921428	88.90	13.17	102.07
170.9	79.9814	847.253	846.154	7.061497	4.131979	85.45	16.98	102.43	177.9	84.3389	836.390	835.289	6.970825	3.918452	88.95	13.11	102.06
171.0	80.0425	847.103	846.004	7.060246	4.128830	85.50	16.93	102.43	178.0	84.4025	836.229	835.128	6.969481	3.915480	89.00	13.06	102.06
171.1	80.1035	846.953	845.853	7.058985	4.125685	85.55	16.87	102.42	178.1	84.4662	836.068	834.967	6.968137	3.912560	89.05	13.00	102.05
171.2	80.1645	846.802	845.702	7.057725	4.122546	85.60	16.82	102.42	178.2	84.5299	835.907	834.806	6.966794	3.909597	89.10	12.95	102.05
171.3	80.2256	846.652	845.552	7.056473	4.119411	85.65	16.76	102.41	178.3	84.5937	835.745	834.644	6.965442	3.906639	89.15	12.89	102.04
171.4	80.2868	846.501	845.401	7.055213	4.116280	85.70	16.71	102.41	178.4	84.6576	835.583	834.482	6.964090	3.903686	89.20	12.83	102.03
171.5	80.3480	846.350	845.250	7.053953	4.113155	85.75	16.65	102.40	178.5	84.7214	835.421	834.320	6.962738	3.900736	89.25	12.78	102.03
171.6	80.4092	846.199	845.099	7.052693	4.110034	85.80	16.60	102.40	178.6	84.7854	835.259	834.158	6.961386	3.897838	89.30	12.72	102.02
171.7	80.4704	846.048	844.948	7.051433	4.106919	85.85	16.54	102.39	178.7	84.8493	835.097	833.996	6.960034	3.894897	89.35	12.67	102.02
171.8	80.5316	845.897	844.797	7.050173	4.103756	85.90	16.49	102.39	178.8	84.9133	834.934	833.833	6.958674	3.891961	89.40	12.61	102.01
171.9	80.5929	845.745	844.645	7.048904	4.100650	85.95	16.43	102.38	178.9	84.9773	834.772	833.671	6.957322	3.889030	89.45	12.56	102.01
172.0	80.6543	845.594	844.494	7.047644	4.097548	86.00	16.38	102.38	179.0	85.0415	834.609	833.508	6.955961	3.886057	89.50	12.50	102.00
172.1	80.7157	845.442	844.342	7.046375	4.094401	86.05	16.32	102.37	179.1	85.1056	834.446	833.345	6.954601	3.883135	89.55	12.44	101.99
172.2	80.7770	845.290	844.190	7.045107	4.091308	86.10	16.27	102.37	179.2	85.1697	834.283	833.182	6.953241	3.880171	89.60	12.39	101.99
172.3	80.8384	845.139	844.039	7.043847	4.088221	86.15	16.21	102.36	179.3	85.2340	834.120	833.019	6.951880	3.877257	89.65	12.33	101.98
172.4	80.8999	844.987	843.887	7.042578	4.085087	86.20	16.16	102.36	179.4	85.2982	833.956	832.855	6.950512	3.874348	89.70	12.27	101.97
172.5	80.9615	844.834	843.734	7.041301	4.081959	86.25	16.10	102.35	179.5	85.3626	833.791	832.690	6.949135	3.871398	89.75	12.22	101.97
172.6	81.0230	844.682	843.582	7.040033	4.078885	86.30	16.05	102.35	179.6	85.4269	833.628	832.527	6.947775	3.868497	89.80	12.16	101.96
172.7	81.0846	844.530	843.430	7.038764	4.075766	86.35	15.99	102.34	179.7	85.4913	833.463	832.362	6.946398	3.865601	89.85	12.11	101.96
172.8	81.1462	844.377	843.277	7.037488	4.072702	86.40	15.94	102.34	179.8	85.5558	833.299	832.198	6.945029	3.862664	89.90	12.05	101.95
172.9	81.2079	844.224	843.124	7.036211	4.069592	86.45	15.88	102.33	179.9	85.6203	833.134	832.032	6.943644	3.859778	89.95	11.99	101.94
173.0	81.2695	844.071	842.971	7.034934	4.066487	86.50	15.83	102.33	180.0	85.6848	832.970	831.868	6.942275	3.856848	90.00	11.94	101.94
173.1	81.3313	843.918	842.818	7.033657	4.063437	86.55	15.77	102.32	180.1	85.7494	832.804	831.702	6.940890	3.853924	90.05	11.88	101.93
173.2	81.3930	843.765	842.665	7.032380	4.060341	86.60	15.72	102.32	180.2	85.8141	832.639	831.537	6.939513	3.851050	90.10	11.83	101.93
173.3	81.4548	843.611	842.511	7.031095	4.057251	86.65	15.66	102.31	180.3	85.8788	832.473	831.371	6.938127	3.848135	90.15	11.77	101.92
173.4	81.5167	843.458	842.358	7.029818	4.054115	86.70	15.61	102.31	180.4	85.9435	832.308	831.206	6.936750	3.845224	90.20	11.71	101.91
173.5	81.5786	843.304	842.204	7.028533	4.051033	86.75	15.55	102.30	180.5	86.0083	832.142	831.040	6.935365	3.842363	90.25	11.66	101.91
173.6	81.6404	843.150	842.050	7.027248	4.047957	86.80	15.50	102.30	180.6	86.0731	831.976	830.874	6.933980	3.839461	90.30	11.60	101.90
173.7	81.7023	842.997	841.897	7.025971	4.044934	86.85	15.44	102.29	180.7	86.1380	831.810	830.708	6.932594	3.836563	90.35	11.54	101.89
173.8	81.7643	842.842	841.742	7.024677	4.041867	86.90	15.39	102.29	180.8	86.2030	831.643	830.541	6.931201	3.833670	90.40	11.49	101.89
173.9	81.8264	842.689	841.589	7.023401	4.038804	86.95	15.33	102.28	180.9	86.2679	831.476	830.374	6.929807	3.830782	90.45	11.43	101.88
174.0	81.8884	842.534	841.434	7.022107	4.035746	87.00	15.28	102.28	181.0	86.3329	831.310	830.208	6.928422	3.827897	90.50	11.37	101.87
174.1	81.9505	842.380	841.280	7.020822	4.032643	87.05	15.22	102.27	181.1	86.3980	831.142	830.040	6.927020	3.825017	90.55	11.32	101.87
174.2	82.0126	842.225	841.125	7.019528	4.029594	87.10	15.17	102.27	181.2	86.4631	830.975	829.873	6.925626	3.822141	90.60	11.26	101.86
174.3	82.0748	842.070	840.970	7.018235	4.026550	87.15	15.11	102.26	181.3	86.5283	830.807	829.705	6.924224	3.819270	90.65	11.21	101.86
174.4	82.1370	841.915	840.815	7.016941	4.023510	87.20	15.06	102.26	181.4	86.5936	830.639	829.537	6.922822	3.816359	90.70	11.15	101.85
174.5	82.1993	841.760	840.660	7.015648	4.020475	87.25	15.00	102.25	181.5	86.6588	830.471	829.369	6.921420	3.813496	90.75	11.09	101.84
174.6	82.2615	841.605	840.505	7.014354	4.017396	87.30	14.95	102.25	181.6	86.7241	830.303	829.201	6.920018	3.810638	90.80	11.04	101.84
174.7	82.3239	841.449	840.349	7.013052	4.014370	87.35	14.89	102.24	181.7	86.7896	830.134	829.032	6.918607	3.807740	90.85	10.98	101.83
174.8	82.3862	841.293	840.193	7.011750	4.011349	87.40	14.84	102.24	181.8	86.8549	829.966	828.864	6.917205	3.804890	90.90	10.92	101.82
174.9	82.4486	841.138	840.038	7.010457	4.008283	87.45	14.78	102.23	181.9	86.9204	829.797	828.695	6.915795	3.802044	90.95	10.87	101.82
175.0	82.5110	840.982	839.882	7.009155	4.005271	87.50	14.72	102.22	182.0	86.9860	829.627	828.525	6.914376	3.799159	91.00	10.81	101.81

Values at 15.556°C (60° F)

Proof	Mass % Vac	Kg m3 Vac	Kg m3 Air	Lbs per Gal. Air	Lbs per Proof Gal Air	Alcohol Parts	Water Parts	Parts Total	Proof	Mass % Vac	Kg m3 Vac	Kg m3 Air	Lbs per Gal. Air	Lbs per Proof Gal Air	Alcohol Parts	Water Parts	Parts Total
182.0	86.9860	829.627	828.525	6.914376	3.799159	91.00	10.81	101.81	189.0	91.7153	817.110	816.006	6.809900	3.603176	94.50	6.78	101.28
182.1	87.0515	829.458	828.356	6.912966	3.796279	91.05	10.75	101.80	189.1	91.7852	816.920	815.816	6.808314	3.600428	94.55	6.72	101.27
182.2	87.1172	829.288	828.186	6.911547	3.793446	91.10	10.70	101.80	189.2	91.8552	816.729	815.625	6.806720	3.597684	94.60	6.66	101.26
182.3	87.1828	829.119	828.017	6.910137	3.790574	91.15	10.64	101.79	189.3	91.9253	816.538	815.434	6.805127	3.594944	94.65	6.60	101.25
182.4	87.2486	828.948	827.846	6.908710	3.787707	91.20	10.58	101.78	189.4	91.9954	816.347	815.243	6.803533	3.592209	94.70	6.54	101.24
182.5	87.3143	828.778	827.676	6.907291	3.784887	91.25	10.53	101.78	189.5	92.0655	816.156	815.052	6.801939	3.589438	94.75	6.48	101.23
182.6	87.3802	828.607	827.505	6.905864	3.782028	91.30	10.47	101.77	189.6	92.1357	815.963	814.859	6.800328	3.586711	94.80	6.42	101.22
182.7	87.4461	828.436	827.334	6.904437	3.779173	91.35	10.41	101.76	189.7	92.2061	815.771	814.667	6.798726	3.583988	94.85	6.37	101.22
182.8	87.5120	828.265	827.163	6.903010	3.776323	91.40	10.36	101.76	189.8	92.2765	815.579	814.475	6.797123	3.581230	94.90	6.31	101.21
182.9	87.5780	828.093	826.991	6.901574	3.773476	91.45	10.30	101.75	189.9	92.3469	815.386	814.282	6.795513	3.578516	94.95	6.25	101.20
183.0	87.6440	827.922	826.820	6.900147	3.770635	91.50	10.24	101.74	190.0	92.4175	815.191	814.087	6.793885	3.575766	95.00	6.19	101.19
183.1	87.7102	827.750	826.648	6.898712	3.767797	91.55	10.19	101.74	190.1	92.4882	814.998	813.894	6.792275	3.573060	95.05	6.13	101.18
183.2	87.7763	827.578	826.476	6.897276	3.764964	91.60	10.13	101.73	190.2	92.5589	814.803	813.699	6.790647	3.570319	95.10	6.07	101.17
183.3	87.8425	827.406	826.304	6.895841	3.762092	91.65	10.07	101.72	190.3	92.6297	814.609	813.505	6.789028	3.567582	95.15	6.01	101.16
183.4	87.9088	827.233	826.131	6.894397	3.759267	91.70	10.01	101.71	190.4	92.7006	814.414	813.310	6.787401	3.564850	95.20	5.95	101.15
183.5	87.9751	827.060	825.958	6.892954	3.756447	91.75	9.96	101.71	190.5	92.7715	814.219	813.115	6.785774	3.562121	95.25	5.89	101.14
183.6	88.0415	826.887	825.785	6.891510	3.753588	91.80	9.90	101.70	190.6	92.8425	814.022	812.918	6.784129	3.559397	95.30	5.83	101.13
183.7	88.1078	826.713	825.611	6.890058	3.750776	91.85	9.84	101.69	190.7	92.9137	813.826	812.722	6.782494	3.556677	95.35	5.77	101.12
183.8	88.1744	826.540	825.438	6.888614	3.747968	91.90	9.79	101.69	190.8	92.9848	813.630	812.526	6.780858	3.553961	95.40	5.71	101.11
183.9	88.2409	826.366	825.264	6.887162	3.745122	91.95	9.73	101.68	190.9	93.0561	813.432	812.328	6.779206	3.551250	95.45	5.66	101.11
184.0	88.3075	826.192	825.089	6.885701	3.742238	92.00	9.67	101.67	191.0	93.1275	813.235	812.131	6.777562	3.548504	95.50	5.60	101.10
184.1	88.3741	826.018	824.915	6.884249	3.739443	92.05	9.61	101.66	191.1	93.1990	813.037	811.933	6.775909	3.545801	95.55	5.54	101.09
184.2	88.4408	825.843	824.740	6.882789	3.736610	92.10	9.56	101.66	191.2	93.2706	812.839	811.735	6.774257	3.543101	95.60	5.48	101.08
184.3	88.5076	825.668	824.565	6.881328	3.733782	92.15	9.50	101.65	191.3	93.3421	812.639	811.535	6.772588	3.540330	95.65	5.42	101.07
184.4	88.5744	825.492	824.389	6.879860	3.730957	92.20	9.44	101.64	191.4	93.4139	812.440	811.335	6.770919	3.537601	95.70	5.36	101.06
184.5	88.6413	825.317	824.214	6.878399	3.728179	92.25	9.39	101.64	191.5	93.4857	812.240	811.135	6.769250	3.534877	95.75	5.30	101.05
184.6	88.7083	825.141	824.038	6.876930	3.725363	92.30	9.33	101.63	191.6	93.5575	812.040	810.935	6.767581	3.532156	95.80	5.24	101.04
184.7	88.7753	824.965	823.862	6.875462	3.722551	92.35	9.27	101.62	191.7	93.6294	811.840	810.735	6.765911	3.529478	95.85	5.18	101.03
184.8	88.8423	824.789	823.686	6.873993	3.719744	92.40	9.21	101.61	191.8	93.7016	811.638	810.533	6.764226	3.526728	95.90	5.12	101.02
184.9	88.9095	824.612	823.509	6.872516	3.716899	92.45	9.16	101.61	191.9	93.7737	811.437	810.332	6.762548	3.524020	95.95	5.06	101.01
185.0	88.9765	824.436	823.333	6.871047	3.714100	92.50	9.10	101.60	192.0	93.8459	811.235	810.130	6.760863	3.521316	96.00	5.00	101.00
185.1	89.0439	824.259	823.156	6.869570	3.711305	92.55	9.04	101.59	192.1	93.9183	811.032	809.927	6.759168	3.518617	96.05	4.94	100.99
185.2	89.1111	824.081	822.978	6.868084	3.708514	92.60	8.98	101.58	192.2	93.9907	810.829	809.724	6.757474	3.515884	96.10	4.88	100.98
185.3	89.1785	823.903	822.800	6.866599	3.705686	92.65	8.93	101.58	192.3	94.0631	810.626	809.521	6.755780	3.513192	96.15	4.82	100.97
185.4	89.2459	823.725	822.622	6.865113	3.702904	92.70	8.87	101.57	192.4	94.1357	810.421	809.316	6.754069	3.510468	96.20	4.76	100.96
185.5	89.3133	823.547	822.444	6.863628	3.700126	92.75	8.81	101.56	192.5	94.2085	810.217	809.112	6.752367	3.507748	96.25	4.70	100.95
185.6	89.3809	823.368	822.265	6.862134	3.697311	92.80	8.75	101.55	192.6	94.2812	810.012	808.907	6.750656	3.505069	96.30	4.64	100.94
185.7	89.4484	823.190	822.087	6.860649	3.694542	92.85	8.70	101.55	192.7	94.3541	809.807	808.702	6.748945	3.502357	96.35	4.58	100.93
185.8	89.5161	823.010	821.907	6.859146	3.691735	92.90	8.64	101.54	192.8	94.4271	809.601	808.496	6.747226	3.499649	96.40	4.52	100.92
185.9	89.5838	822.831	821.728	6.857653	3.688932	92.95	8.58	101.53	192.9	94.5001	809.395	808.290	6.745507	3.496946	96.45	4.46	100.91
186.0	89.6516	822.651	821.548	6.856150	3.686134	93.00	8.52	101.52	193.0	94.5733	809.188	808.083	6.743779	3.494246	96.50	4.40	100.90
186.1	89.7194	822.471	821.368	6.854648	3.683381	93.05	8.47	101.52	193.1	94.6465	808.980	807.875	6.742044	3.491514	96.55	4.34	100.89
186.2	89.7873	822.290	821.187	6.853138	3.680592	93.10	8.41	101.51	193.2	94.7199	808.772	807.667	6.740308	3.488823	96.60	4.28	100.88
186.3	89.8553	822.110	821.007	6.851635	3.677806	93.15	8.35	101.50	193.3	94.7934	808.564	807.459	6.738572	3.486137	96.65	4.22	100.87
186.4	89.9233	821.929	820.826	6.850125	3.675025	93.20	8.29	101.49	193.4	94.8670	808.355	807.250	6.736828	3.483417	96.70	4.15	100.85
186.5	89.9914	821.748	820.645	6.848614	3.672248	93.25	8.23	101.48	193.5	94.9406	808.146	807.041	6.735084	3.480702	96.75	4.09	100.84
186.6	90.0596	821.566	820.463	6.847096	3.669434	93.30	8.18	101.48	193.6	95.0143	807.936	806.831	6.733331	3.478028	96.80	4.03	100.83
186.7	90.1278	821.384	820.281	6.845577	3.666665	93.35	8.12	101.47	193.7	95.0882	807.725	806.620	6.731570	3.475321	96.85	3.97	100.82
186.8	90.1961	821.202	820.099	6.844058	3.663901	93.40	8.06	101.46	193.8	95.1622	807.513	806.408	6.729801	3.472618	96.90	3.91	100.81
186.9	90.2644	821.019	819.916	6.842531	3.661100	93.45	8.00	101.45	193.9	95.2363	807.302	806.197	6.728040	3.469920	96.95	3.85	100.80
187.0	90.3328	820.837	819.734	6.841012	3.658344	93.50	7.94	101.44	194.0	95.3104	807.090	805.985	6.726271	3.467226	97.00	3.79	100.79
187.1	90.4012	820.653	819.550	6.839476	3.655592	93.55	7.89	101.44	194.1	95.3847	806.877	805.772	6.724493	3.464499	97.05	3.73	100.78
187.2	90.4698	820.470	819.367	6.837949	3.652804	93.60	7.83	101.43	194.2	95.4591	806.663	805.558	6.722707	3.461777	97.10	3.67	100.77
187.3	90.5385	820.286	819.183	6.836413	3.650020	93.65	7.77	101.42	194.3	95.5336	806.450	805.345	6.720930	3.459060	97.15	3.61	100.76
187.4	90.6071	820.102	818.999	6.834878	3.647280	93.70	7.71	101.41	194.4	95.6082	806.235	805.130	6.719135	3.456382	97.20	3.55	100.75
187.5	90.6758	819.918	818.815	6.833342	3.644505	93.75	7.65	101.40	194.5	95.6829	806.020	804.915	6.717341	3.453673	97.25	3.48	100.73
187.6	90.7447	819.732	818.629	6.831790	3.641733	93.80	7.60	101.40	194.6	95.7578	805.804	804.699	6.715539	3.450968	97.30	3.42	100.72
187.7	90.8135	819.548	818.445	6.830255	3.638966	93.85	7.54	101.39	194.7	95.8327	805.588	804.482	6.713728	3.448267	97.35	3.36	100.71
187.8	90.8825	819.362	818.258	6.828694	3.636163	93.90	7.48	101.38	194.8	95.9077	805.371	804.265	6.711917	3.445570	97.40	3.30	100.70
187.9	90.9516	819.176	818.072	6.827142	3.633405	93.95	7.42	101.37	194.9	95.9828	805.153	804.047	6.710097	3.442878	97.45	3.24	100.69
188.0	91.0207	818.990	817.886	6.825589	3.630650	94.00	7.36	101.36	195.0	96.0581	804.936	803.830	6.708286	3.440190	97.50	3.18	100.68
188.1	91.0898	818.803	817.699	6.824029	3.627900	94.05	7.30	101.35	195.1	96.1335	804.717	803.611	6.706459	3.437470	97.55	3.11	100.66
188.2	91.1590	818.617	817.513	6.822477	3.625154	94.10	7.25	101.35	195.2	96.2091	804.497	803.391	6.704623	3.434790	97.60	3.05	100.65
188.3	91.2283	818.430	817.326	6.820916	3.622412	94.15	7.19	101.34	195.3	96.2847	804.277	803.171	6.702787	3.432079	97.65	2.99	100.64
188.4	91.2977	818.242	817.138	6.819347	3.619634	94.20	7.13	101.33	195.4	96.3604	804.057	802.951	6.700951	3.429408	97.70	2.93	100.63
188.5	91.3672	818.054	816.950	6.817778	3.616901	94.25	7.07	101.32	195.5	96.4363	803.835	802.729	6.699098	3.426705	97.75	2.87	100.62
188.6	91.4366	817.866	816.762	6.816209	3.614131	94.30	7.01	101.31	195.6	96.5123	803.613	802.507	6.697245	3.424007	97.80	2.81	100.61
188.7	91.5062	817.677	816.573	6.814632	3.611406	94.35	6.95	101.30	195.7	96.5883	803.390	802.284	6.695384	3.421312	97.85	2.74	100.59
188.8	91.5758	817.489	816.385	6.813063	3.608645	94.40	6.89	101.29	195.8	96.6646	803.167	802.061	6.693523	3.418587	97.90	2.68	100.58
188.9	91.6456	817.299	816.195	6.811477	3.605889	94.45	6.84	101.29	195.9	96.7410	802.943	801.837	6.691654	3.415901	97.95	2.62	100.57
189.0	91.7153	817.110	816.006	6.809900	3.603176	94.50	6.78	101.28	196.0	96.8174	802.718	801.612	6.689776	3.413185	98.00	2.56	100.56

Notes

Text of Gauging Regulations

The following regulations were published in 1978 as part of Title 27, Chapter I, Subchapter M, Part 186 of the Code of Federal Regulations. The gauging manual is known as Publication # 455 and alternately as ATF-P5110.6. While this material is readily available from the TTB I have decided to include an abridged version for convenience since some of the definitions and procedures are so carefully examined in the text.

The main internet address where these regulations can be found is http://www.ttb.gov/spirits/spirits-regs.shtml.

There is some discrepancy in the text as obtained online in 2015 and the regulations published in Publication 455 in 1978. I have used the 2015 text as the primary source. Also, the regulations now appearing online have been assigned different section numbers. I have made an effort to list both the old and the new numbers in the following material.

Subpart A-Scope of Regulations

§30.1 Gauging of distilled spirits. **Old Style §186.1** Gauging of distilled spirits. (a) General. This part relates to the gauging of distilled spirits. The term "gauging" means the determination of the proof and the quantity of distilled spirits. The procedures prescribed in or authorized under the provisions of this part, except as may be otherwise authorized in this chapter, shall be followed in making any determination of quantity or proof of distilled spirits required by or under the authority of regulations in this chapter. The Tables referred to in subpart E of this part appear in the "Gauging Manual Embracing Instructions and Tables for Determining Quantity of Distilled Spirits by Proof and Weight" as incorporated by reference in this part (see paragraph (c) of this section). These Tables, together with their instructions, shall be used, wherever applicable, in making the necessary computations from gauge data.

(b) Tables referred to in subpart E of this part. Table 1 provides a method of correcting hydrometer indications at temperatures between 0 and 100 degrees Fahrenheit to true proof. If distilled spirits contain dissolved solids, temperature correction of the hydrometer reading by the use of this Table would result in apparent proof rather than true proof. Tables 2 and 3 show the gallonage of spirituous liquor according to weight and proof. Table 4 shows the gallons per pound at each one-tenth proof from 1 to 200 proof. Table 5 shows the weight per wine gallon and proof gallon at each proof. Table 6 shows the volumes of alcohol and water and the specific gravity (air and vacuum) of spirituous liquor at each proof. Table 7 provides a means of ascertaining the volume (at 60 degrees Fahrenheit) of spirits at various temperatures ranging from 18 degrees through 100 degrees Fahrenheit.

Subpart B-Definitions §186.11 Meaning of terms. §30.11

When used in this part, where not otherwise distinctly expressed or manifestly incompatible with the intent thereof, terms shall have the meanings ascribed in this section. Words in the plural form shall include the singular, and vice versa, and words importing the masculine gender shall include the feminine. The terms "includes" and "including" do not exclude things not enumerated which are in the same general class.

Administrator. The Administrator, Alcohol and Tobacco Tax and Trade Bureau, Department of the Treasury, Washington, DC. Appropriate TTB officer. An officer or employee of the Alcohol and Tobacco Tax and Trade Bureau (TTB) authorized to perform any functions relating to the administration or enforcement of this part by TTB Order 1135.30, Delegation of the Administrator's Authorities in 27 CFR Part 30, Gauging Manual.

Bulk conveyance. Any tank car, tank truck, tank ship, tank barge, or other similar container approved by the appropriate TTB officer, authorized for the conveyance of spirits (including denatured spirits) in bulk.

CFR. The Code of Federal Regulations.

Container. Any receptacle, vessel, or form of package, bottle, tank, or pipeline used, or capable of use, for holding, storing, transferring or conveying distilled spirits.

Denatured spirits or denatured alcohol. Spirits to which denaturants have been added pursuant to formulas prescribed in 27 CFR Part 21.

Gallon or wine gallon. The liquid measure equivalent to the volume of 231 cubic inches.

I.R.C. The Internal Revenue Code of 1954, as amended.

Package. Any cask, barrel, drum, or similar container approved under the provisions of this chapter.

Proof. The ethyl alcohol content of a liquid at 60 degrees Fahrenheit, stated as twice the percent of ethyl alcohol by volume.

Proof gallon. A United States gallon of proof spirits, or the alcoholic equivalent thereof.

See Also: Proof Gallon: Subpart B Section 5.11 Meaning of Terms - A Proof Gallon = 1 U.S. Gal containing 50% by Volume of Ethyl Alcohol which at 60° F has a specific gravity of .7939 referenced to water at 60°F. This value is in reference to the specific gravity of alcohol in a vacuum rather than in air. Table 6 lists the specific gravity of alcohol at 60°F as .79389 in a vacuum and .79365 in air. This is very odd since the entire TTB regimen is based upon atmospheric densities.

Proof spirits. That liquid which contains one-half its volume of ethyl alcohol of a specific gravity of seven thousand nine hundred and thirty-nine ten-thousandths (0.7939) in vacuum at 60 degrees Fahrenheit referred to water at 60 degrees Fahrenheit as unity.

Spirits, spirituous liquor, or distilled spirits. That substance known as ethyl alcohol, ethanol, or spirits of wine in any form, including all dilutions and mixtures thereof, from whatever source or by whatever process produced, but not denatured spirits unless specifically stated. For the sole purpose of gauging wine and alcoholic flavoring materials on the bonded premises of a distilled spirits plant, such alcoholic ingredients shall have the same meaning described herein to spirits, spirituous liquor, or distilled spirits.

Subpart C-Gauging Instruments

§30.21 Requirements.

(a) General. The proof of distilled spirits shall be determined by the use of gauging instruments as prescribed in this part.

(b) Proprietors. Proprietors shall use only accurate hydrometers and thermometers that show subdivisions or graduations of proof and temperature which are at least as delimitated as the instruments described in §30.22.

(c) Appropriate TTB officers. Appropriate TTB officers shall use only hydrometers and thermometers furnished by the Government. However, where this part requires the use of a specific gravity hydrometer, TTB officers shall use precision grade specific gravity hydrometers conforming to the provisions of §30.24, furnished by the proprietor. However, the appropriate TTB officer may authorize the use of other instruments approved by the appropriate TTB officer as being equally satisfactory for determination of specific gravity and for gauging. From time to time appropriate TTB officers shall verify the accuracy of hydrometers and thermometers used by proprietors.

§186.22 §30.22 Hydrometers and thermometers. §30.22 Hydrometers and thermometers.

The hydrometers used are graduated to read the proof of aqueous alcoholic solutions at 60 degrees Fahrenheit; thus, they read 0 for water, 100 for proof spirits, and 200 for absolute alcohol. Because of temperature-density relationships and the selection of 60 degrees Fahrenheit for reporting proof, the hydrometer readings will be less than the true percent of proof at temperatures below 60 degrees Fahrenheit and greater than the true percent of proof at temperatures above 60 degrees Fahrenheit. Hence, corrections are necessary for hydrometer readings at temperatures other than 60 degrees Fahrenheit. Precision hydrometers shall be used for gauging spirits. Hydrometers and thermometers shall be used and the true percent of proof shall be determined in accordance with §30.31. Hydrometers are designated by letter according to range of proof and are provided in ranges and subdivisions of stems as follows:

US Standard	Range	Subdivision
F	0 to 20	0.2°
G	20 to 40	0.2°
H	40 to 60	0.2°
I	60 to 80	0.2°
K	75 to 95	0.2°
L	90 to 110	0.2°

M................................ 105 to 125 .... 0.2°
N .... 125 to 145 .... 0.2°
P................................ 145 to 165 .... 0.2°
Q .... 165 to 185 .... 0.2°
R................................ 185 to 206 .... 0.2°

Thermometers are designated by type according to range of degrees Fahrenheit and are provided in ranges and subdivisions of degrees as follows:

Type	Range	Subdivision
Pencil type	10° to 100°	1°
V-back	10° to 100°	1°
Glass shell (earlier model)	40° to 100°	1/2°
Glass shell (later model)	40° to 100°	1/4°

§186.23 §30.23 Use of precision hydrometers and thermometers.

Care should be exercised to obtain accurate hydrometer and thermometer readings. In order to accomplish this result, the following precautions should be observed. Bulk spirits should be thoroughly agitated so that the test samples will be representative of the entire quantity. The hydrometers should be kept clean and free of any oily substance. Immediately before readings are taken, the glass cylinder containing the thermometer should be rinsed several times with the spirits which are to be gauged so as to bring both the cylinder and the thermometer to the temperature of the spirits (if time permits, it is desirable to bring both the spirits and the instruments to room temperature). If the outer surface of the cylinder becomes wet, it should be wiped dry to avoid the cooling effect of rapid evaporation. During the readings the cylinder should be protected from drafts or other conditions which might affect its temperature or that of the spirits which it contains. The hands should not be placed on the cylinder in such a manner as to warm the liquid contained therein. The hydrometer should be inserted in the liquid and the hydrometer bulb raised and lowered from top to bottom 5 or 6 times to obtain an even temperature distribution over its surface, and, while the hydrometer bulb remains in the liquid, the stem should be dried and the hydrometer allowed to come to rest without wetting more than a few tenths degrees of the exposed stem. Special care should be taken to ascertain the exact point at which the level of the surface liquid intersects the scale of proof in the stem of the hydrometer. The hydrometer and thermometer should be immediately read, as nearly simultaneously as possible. In reading the hydrometer, a sighting should be made slightly below the plane of the surface of the liquid and the line of sight should then be raised slowly, being kept perpendicular to the hydrometer stem, until the appearance of the surface changes from an ellipse to a straight line. The point where this line intersects the hydrometer scale is the correct reading of the hydrometer. When the correct readings of the hydrometer and the thermometer have been determined, the true percent of proof shall be ascertained from Table 1. Another sample of the spirits should then be taken and be tested in the same manner so as to verify the proof originally ascertained. Hydrometer readings should be made to the nearest 0.05 degree and thermometer readings should be made to the nearest 0.1 degree, and instrument correction factors, if any, should be applied. It is necessary to interpolate in Table 1 for fractional hydrometer and thermometer readings.

Example. A hydrometer reads 192.85° at 72.10°F. The correction factors for the hydrometer and the thermometer, respectively are minus 0.03° and plus 0.05°. The corrected reading, then, is 192.82° at 72.15°F.

193.0° at 72.0°F = 190.2°, 192.0° at 72.0°F = 189.1°, Difference = 1.1°, 192.0° at 72.0°F = 189.1°, 192.0° at 73.0°F = 188.9°, Difference = 0.2°

The hydrometer difference (1.1°) multiplied by the fractional degree of the hydrometer reading (0.82°)=0.902.

The temperature difference (0.2°) multiplied by the fractional degree of the temperature reading (0.15°)=0.03°.

Proof at 60°F=189.1+0.902-0.03=189.972°=190.0°.

As shown, the final proof is rounded to the nearest tenth of a degree of proof. In such cases, if the hundredths decimal is less than five, it will be dropped; if it is five or over, a unit will be added.

§30.24 Specific gravity hydrometers.

(a) The specific gravity hydrometers furnished by proprietors to appropriate TTB officers shall conform to the standard specifications of the American Society for Testing and Materials (ASTM) for such instruments. Such specific gravity hydrometers shall be of a precision grade, standardization temperature 60 °/60°F, and provided in the following ranges and subdivisions:

Range	Subdivision
1.0000 to 1.0500	0.0005
1.0500 to 1.1000	0.0005
1.1000 to 1.1500	0.0005
1.1500 to 1.2000	0.0005
1.2000 to 1.2500	0.0005

No instrument shall be in error by more than 0.0005 specific gravity.

(b) A certificate of accuracy prepared by the instrument manufacturer for the instrument shall be furnished to the appropriate TTB officer.

§30.25 Use of precision specific gravity hydrometers.

The provisions of §30.23 respecting the care, handling, and use of precision instruments shall be followed with respect to the care, handling, and use of precision grade specific gravity hydrometers. Specific gravity hydrometers shall be read to the nearest subdivision. Because of temperature density relationships and the selection of the standardization temperature of 60 °/60°F, the specific gravity readings will be greater at temperatures below 60 degrees Fahrenheit and less at temperatures above 60 degrees Fahrenheit. Hence, correction of the specific gravity readings will be made for temperature other than 60 degrees Fahrenheit. Such correction may be ascertained by dividing the specific gravity hydrometer reading by the applicable correction factor in Table 7.

Example: The specific gravity hydrometer reading is 1.1525, the thermometer reading is 68 degrees Fahrenheit, and the true proof of the spirits is 115 degrees. The correct specific gravity reading will be ascertained as follows:

(a) From Table 7, the correction factor for 115° proof at 68°F is 0.996.

(b) 1.1525 divided by 0.996=1.1571, the corrected specific gravity.

Subpart D -Gauging Procedures

§30.31 Determination of proof.

(a) General. The proof of spirits shall be determined to the nearest tenth degree which shall be the proof used in determining the proof gallons.

(b) Solids content not more than 600 milligrams. Except as otherwise authorized by the appropriate TTB officer, the proof of spirits containing not more than 600 milligrams of solids per 100 milliliters of spirits shall be determined by the use of a hydrometer and thermometer in accordance with the provisions of §30.23 except that if such spirits contain solids in excess of 400 milligrams but not in excess of 600 milligrams per 100 milliliters at gauge proof, there shall be added to the proof so determined the obscuration determined as prescribed in §30.32.

(c) Solids content over 600 milligrams. If such spirits contain solids in excess of 600 milligrams per 100 milliliters at gauge proof, the proof shall be determined on the basis of true proof determined as follows:

(1) By the use of a hydrometer and a thermometer after the spirits have been distilled in a small laboratory still and restored to the original volume and temperature by the addition of pure water to the distillate; or

(2) By a recognized laboratory method which is equal or superior in accuracy to the distillation method.

(d) Initial proof. Except when the proof of spirits is used in making the gauge prescribed in 27 CFR 19.383 or in making a gauge for determination of tax, the initial determination of proof made on the bonded premises of a distilled spirits plant for such spirits may be used whenever a subsequent gauge is required to be made at that same plant provided that no material has been added to change the proof of the spirits.

§30.32 Determination of proof obscuration.

(a) General. Proof obscuration of spirits containing more than 400 but not more than 600 milligrams of solids per 100 milliliters shall be determined by one of the following methods. The evaporation method may be used only for spirits in the range of 80-100 degrees at gauge proof.

(b) Evaporation method. Evaporate the water and alcohol from a carefully measured 25 milliliter sample of spirits, dry the residue at 100 degrees centigrade for 30 minutes and then weigh the residue precisely. Multiply the weight of the residue by 4 to determine the weight of solids in 100 milliliters. The resulting weight per 100 milliliters multiplied by 4 will give the obscuration. Experience has shown that 0.1 gram (100 milligrams) of solids per 100 milliliters of spirits in the range of 80-100 degrees proof will obscure the true proof by 0.4 of one degree of proof. For example, if the weight of solids remaining after evaporation of 25 milliliters 0.125 gram, the amount of solids present in 100 milliliters of the spirits is 0.50 gram (4 times 0.125). The obscuration is 4 times 0.50, which is two degrees of proof. This value added to the temperature corrected hydrometer reading will give the true proof.

(c) Distillation method. Determine the apparent proof and temperature of the sample of spirits and then distill a carefully measured sample in a small laboratory still, and collect a quantity of the distillate, 1 or 2 milliliters less than the original sample. The distillate is adjusted to the original temperature and restored to the original volume by addition of distilled water. The proof of the restored distillate is then determined by use of a precision hydrometer and thermometer in accordance with the provisions of §13.23 to the nearest 0.1 degree of proof. The difference between the proof so determined and the apparent proof of the undistilled sample is the obscuration; or

(d) Pycnometer method. Determine the specific gravity of the undistilled sample, distill and restore the samples as provided in paragraph (c) of this section and determine the specific gravity of the restored distillate by means of a pycnometer. The specific gravities so obtained will be converted to degrees of proof by interpolation of Table 6 to the nearest 0.1 degree of proof. The difference in proof so obtained is the obscuration.

Determination of Quantity

§30.36 General requirements.

The quantity determination of distilled spirits that are withdrawn from bond in bulk upon tax determination or payment shall be by weight. The quantity of other distilled spirits or denatured spirits may be determined by weight or by volume. When the quantity of distilled spirits or denatured distilled spirits is determined by volume, such determination may be by meter as provided in 27 CFR Part 19, or when approved by the appropriate TTB officer, another method or device.

Determination of Quantity by Weight

§30.41 Bulk spirits. **§186.41**

When spirits (including denatured spirits) are to be gauged by weight in bulk quantities, the weight shall be determined by means of weighing tanks, mounted on accurate scales. Before each use, the scales shall be balanced at zero load; thereupon the spirits shall be run into the weighing tank and proofed as prescribed in §30.31. However, if the spirits are to be reduced in proof, the spirits shall be so reduced before final determination of the proof. The scales shall then be brought to a balanced condition and the weight of the spirits determined by reading the beam to the nearest graduation mark. From the weight and the proof thus ascertained, the quantity of the spirits in proof gallons shall be determined by reference to Table 4. However, in the case of spirits which contain solids in excess of 600 milligrams per 100 milliliters, the quantity in proof gallons shall be determined by first ascertaining the wine gallons per pound of the spirits and multiplying the wine gallons per pound by the weight, in pounds, of the spirits being gauged and by the true proof (determined as prescribed in §30.31) and dividing the result by 100. The wine gallons per pound of spirits containing solids in excess of 600 milligrams per 100 milliliters shall be ascertained by:

(a) Use of a precision hydrometer and thermometer, in accordance with the provisions of §30.23, to determine the apparent proof of the spirits (if specific gravity at the temperature of the spirits is not more than 1.0) and reference to Table 4 for the wine gallons per pound, or

(b) Use of a specific gravity hydrometer, in accordance with the provisions of §30.25, to determine the specific gravity of the spirits (if the specific gravity at the temperature of the spirits is more than 1.0) and dividing that specific gravity (corrected to 60 degrees Fahrenheit) into the factor 0.120074 (the wine gallons per pound for water at 60 degrees Fahrenheit). When withdrawing a portion of the contents of a weighing tank, the difference between the quantity (ascertained by proofing and weighing) in the tank immediately before the removal of the spirits and the quantity (ascertained by proofing and weighing) in the tank immediately after the removal of the spirits shall be the quantity considered to be withdrawn.

§30.42 Denatured spirits. §186.42

The quantity, in gallons, of any lot or package of specially denatured spirits may be determined by weighing it and then dividing its weight by the weight per gallon of the formula concerned, as given in the appropriate Tables in subpart H of 27 CFR Part 21. In the case of completely denatured spirits, the gallonage of any lot or package may be ascertained by determining its weight and apparent proof (hydrometer indication, corrected to 60 degrees Fahrenheit) and then multiplying the weight of the wine gallons per pound factor shown in Table 4 for the (apparent) proof.

§30.43 Packaged spirits. §186.43

When the quantity of spirits (including denatured spirits when gauged by weight) in packages, such as barrels, drums, and similar portable containers, is to be determined by gauge of the individual packages, such quantity shall, except as provided in paragraph (b) of this section, be determined by weighing each package on an accurate weighing beam or platform scale having a beam or dial showing weight in pounds and half pounds, where packages having a capacity in excess of 10 wine gallons are to be gauged, or in pounds and ounces, or pounds and hundredths of a pound, where packages designed to hold 10 wine gallons or less are to be gauged. In either case the tare must be determined and subtracted from the gross weight to obtain the net weight. From the proof and weight ascertained, the quantity of the spirits in proof gallons shall be determined by reference to Table 2, 3, or 4. However, if the spirits contain solids in excess of 600 milligrams per 100 milliliters, the proof gallons shall be determined as prescribed for such spirits in §30.41. Notwithstanding the provisions of this section or of §30.44, (a) gross weights and tares of packages being filled need not be taken in any case where the gauge of the spirits is not derived from such weights under the gauging procedure being utilized, and (b) meters, other devices, or other methods may be used for determining the quantity of spirits in individual packages, when such meter is used as provided in 27 CFR Part 19, or, when such other device or method has been approved by the appropriate TTB officer.

§30.44 Weighing containers. §186.44

(a) Weighing containers of more than 10 wine gallons. The weight of containers having a capacity in excess of 10 wine gallons shall be determined and recorded in pounds and half pounds.

(b) Omitted by Author, Section not relevant and taking up too much space in the book.

(c) Containers of other proofs or sizes. Where containers of proofs or sizes not shown above are to be filled, the following rule may be used for ascertaining the weight of the spirits to be placed in the container:

Divide the number of gallons representing the quantity of spirits to be placed in the container by the fractional part of a gallon equivalent to 1 pound, to obtain the weight of the spirits in pounds and fractions of a pound to two decimal places. Reduce the decimal fraction of a pound to ounces by multiplying by 16, calling any fraction of an ounce a whole ounce. The pounds and ounces thus obtained will determine the point to which the spirits must be weighed to produce the results desired. If the weight must be marked on the container in pounds and decimal fractions of a pound, it will be necessary to convert the ounces to hundredths of a pound. The fraction of a gallon equivalent to 1 pound at any given proof shall be ascertained by reference to Table 4. However, if the spirits contain solids in excess of 600 milligrams per 100 milliliters, the fraction of a gallon equivalent to 1 pound shall be determined as prescribed for such spirits in §30.41.

Example. It is desired to fill a 1-gallon can with precisely 1 wine gallon of 194-proof spirits:

1.00 divided by 0.14866=6.73 pounds. 0.73 multiplied by 16=11.68 ounces, rounded to 12 ounces.

Weight of spirits-6 pounds, 12 ounces. Weight, if required, to be marked on can-6.75 pounds.

§30.45 Withdrawal gauge for packages. **§186.45**

When wooden packages are to be individually gauged for withdrawal, actual tare of the packages shall be determined. The actual tare of a package shall be determined by weighing it after its contents (including rinse water, if any) have been temporarily removed to a separate container or vessel. Where the contents of packages have been temporarily removed for determination of tare, the proof, if any rinse water is added to the spirits, shall be determined after a thorough mixing of the rinse water and the spirits and before return of the spirits to the rinsed packages, and the gross weight shall be determined after the spirits and any added rinse water have been returned to the packages. In the case of metal packages the tare established at the time of filling may be used unless it appears to be incorrect. From the proofs and the net weights of the packages, the wine gallons (if desired) and the proof gallons of spirits shall be determined by the use of Table 2. However, if the spirits contain solids in excess of 600 milligrams per 100 milliliters, the wine gallon and proof gallon contents shall be determined as prescribed for such spirits in §30.41. If either the weight or the proof is beyond the limitations of Table 2, either Table 3 or Table 4 may be used.

Determination of Quantity by Volume

§30.51 Procedures for measurement of bulk spirits. **§186.51**

Where the quantity of spirits (including denatured spirits) in bulk is to be determined by volume as authorized by this chapter, the measurement shall be made in tanks, by meters as provided in 27 CFR part 19, or by other devices or methods authorized by the appropriate TTB officer, or as otherwise provided in this chapter, or such measurement may be made in tank cars or tank trucks if calibration charts for such conveyances are provided and such charts have been accurately prepared, and certified as accurate, by engineers or other persons qualified to calibrate such conveyances. Volumetric measurements in tanks shall be made only in accurately calibrated tanks equipped with suitable measuring devices, whereby the actual contents can be correctly ascertained. If the temperature of spirits (including denatured spirits) is other than the standard of 60 degrees Fahrenheit, gallonage determined by volumetric measurements shall be corrected to the standard temperature by means of Table 7. In the case of denatured spirits, the temperature-correction factor for the proof of the spirits used in denaturation will give sufficiently accurate results, except that the temperature-correction factor used for specially denatured spirits, Formula No. 18, should be that given in Table 7 for 100-proof spirits. When the quantity of spirits, in wine gallons, has been determined by volumetric measurement, the number of proof gallons shall be obtained by multiplying the wine gallons by the proof of the spirits as determined under §30.31.

Example Gauge glass reading inches 88. Wine gallons per inch-48.96. Temperature °F 72. Proof of spirits-86.8. Temperature correction factor (Table 7) 0.995. 48.96 W.G.×88=4308.48 wine gallons. 4308.48 W.G.×0.995=4286.94 wine gallons. 4286.94 W.G.×0.868=3721.06392=3721.1 proof gallons.

[T.D. ATF-198, 50 FR 8535, Mar. 1, 1985, as amended by T.D. ATF-381, 61 FR 37004, July 16, 1996]

§30.52 Procedure for measurement of cased spirits. New Section Not in Manual

Where the quantity of spirits in a case is to be determined by volume, such determination shall be made by ascertaining the contents of one bottle in the case and multiplying that figure by the number of bottles in the case. For cases containing bottles filled according to the metric system of measure, the quantity determined shall be converted to wine gallons, as provided in §19.722 of this chapter. The wine gallons of spirits thus determined for one case may then be multiplied by the number of cases containing spirits at the same proof when determining the quantity of spirits for more than one case. The proof gallons of spirits in cases shall be determined by multiplying the wine gallons by the proof (divided by 100).

Subpart E-Prescribed Tables

Note. The Tables referred to in this subpart appear in their entirety in the "Gauging Manual Embracing Instructions and Tables for Determining Quantity of Distilled Spirits by Proof and Weight" which is incorporated by reference in this part (see §30.1).

§30.61 §186.61 Table 1, showing the true percent of proof spirit for any indication of the hydrometer at temperatures between zero and 100 degrees Fahrenheit.

Table 1

This Table shows the true percent of proof of distilled spirits for indications of the hydrometer likely to occur in practice at temperatures between zero and 100 degrees Fahrenheit and shall be used in determining the proof of spirits. The left-hand column contains the reading of the hydrometer and on the same horizontal line, in the body of the Table, in the "Temperature" column corresponding to the reading of the thermometer is the corrected reading or "true percent of proof." The Table is computed for tenths of a percent.

Example.
Temperature, °F..75
Hydrometer reading................................193
True percent of proof............................. 189.5
Where fractional readings are ascertained, the proper interpolations will be made (see §30.23). If the distilled spirits contain dissolved solids, temperature-correction of the hydrometer reading by the use of this Table would result in apparent proof rather than true proof.

§30.62 §186.62 Table 2, showing wine gallons and proof gallons by weight.

Table 2

The wine and proof gallon content by weight and proof of packages of distilled spirits usually found in actual practice will be ascertained from this Table. The left-hand column contains the weights. The true percent of proof is shown on the heading of each page in a range from 90 degrees to 200 degrees. Under the true percent of proof and on the same horizontal line with the weight will be found the wine gallons (at 60 degrees Fahrenheit) and the proof gallons respectively. Where either the weight or the proof of a quantity of spirits is beyond the limitations of this Table, the number of proof gallons may be ascertained by reference to Table 3. This Table may also be used to ascertain the wine gallons (at 60 degrees Fahrenheit) and proof gallons of spirituous liquor containing dissolved solids where the weight, apparent proof (hydrometer indication corrected to 60 degrees Fahrenheit), and obscuration factor have been determined.

Example. 334 lbs. of distilled spirits. Apparent proof 96.0°. Obscuration + 0.8°. True Proof 96.0°+0.8°=96.8°. 334 lbs. at 96.0° apparent proof=42.8 wine gallons. 42.8 wine gallons×96.8°=41.4 proof gallons.

In addition this Table may be used to obtain the wine gallons, at the prevailing temperature, of most liquids within the range of the Table, from the weight of the liquid and the uncorrected reading of the hydrometer stem. An application of this would be in determining the capacity of a package.

Example. It is desired to determine, or to check the rated capacity of a package having a net weight of 395 pounds when completely filled with spirits having an uncorrected hydrometer reading of 113.0°. The full capacity of the package, 51.5 wine gallons, would be found by referring to the Table at 395 pounds and 113° proof (hydrometer reading).

§30.63 §186.63 Table 3, for determining the number of proof gallons from the weight and proof of spirituous liquor.

Table 3 When the weight or proof of a quantity of distilled spirits is not found in Table 2, the proof gallons may be ascertained from Table 3. The wine gallons (at 60 degrees Fahrenheit) may be ascertained by dividing the proof gallons by the proof.

Example. A tank car of spirits of 190 degrees of proof weighed 60,378 pounds net. We find

	Proof gallons
60,000 pounds equal to	16,778.4
300 pounds equal to	83.9
70 pounds equal to	19.6
8 pounds equal to	2.2
Total	16,884.1

That is, the total weight of 60,378 pounds of spirits at 190 proof is equal to 16,884.1 proof gallons. The equivalent gallonage for 70 pounds is found from the column 700 pounds by moving the decimal point one place to the left; that for 8 pounds from the column 800 pounds by moving the decimal point two places to the left.

Example. A package of spirits at 86 proof weighed 321 1/2 pounds net. We find-

	Proof gallons
300 pounds equal to	32.7
20 pounds equal to	2.2
1 pound equal to	0.1
1/2 pound equal to	0.1
Total	35.1

That is, 321 1/2 pounds of spirits at 86 proof is equal to 35.1 proof gallons. The equivalent gallonage for 20 pounds is found from the column 200 pounds by moving the decimal point one place to the left; that for 1 pound from the column 100 pounds by moving the decimal point two places to the left; that for the 1/2 pound from the column 500 pounds by moving the decimal point three places to the left.

Fractional gallons beyond the first decimal ascertained through use of this Table will be dropped if less than 0.05 or will be added as 0.1 if 0.05 or more. The wine gallons (at 60 degrees Fahrenheit) may be determined by dividing the proof gallons by the proof. For example: 35.1 divided by 0.86 equals 40.8 wine gallons.

§30.64 §186.64 Table 4, showing the fractional part of a gallon per pound at each percent and each tenth percent of proof of spirituous liquor.

Table 4 [TTB editorial note: Erratum on page 549 of CFR Book Title 27 , Proof of 173.7 proof should read Wine gallon per pound of 0.14233]

This Table provides a method for use in ascertaining the wine gallon (at 60 degrees Fahrenheit) and/or proof gallon contents of containers of spirits by multiplying the net weight of the spirits by the fractional part of a gallon per pound shown in the Table for spirits of the same proof. Fractional gallons beyond the first decimal will be dropped if less than 0.05 or will be added as 0.1 if 0.05 or more.

Example. It is desired to ascertain the wine gallons and proof gallons of a tank of 190-proof spirits weighing 81,000 pounds. 81,000×0.14718=11,921.58=11,921.6 wine gallons. 81,000×0.27964=22,650.84=22,650.8 proof gallons.

This Table may also be used for ascertaining the quantity of water required to reduce to a given proof. To do this, divide the proof gallons of spirits to be reduced by the fractional part of a proof gallon per pound of spirits at the proof to which the spirits are to be reduced, and subtract from the quotient the net weight of the spirits before reduction. The remainder will be the pounds of water needed to reduce the spirits to the desired proof.

Example. It is desired to ascertain the quantity of water needed to reduce 1,000 pounds of 200-proof spirits, 302.58 proof gallons, to 190 proof: 302.58 divided by 0.27964 equals 1,082.03 pounds, weight of spirits after reduction. 1.082.03 minus 1,000 equals 82.03 pounds, weight of water required to reduce to desired proof.

The slight variation between this Table and Tables 2, 3, and 5 on some calculations is due to the dropping or adding of fractions beyond the first decimal in those Tables. This Table may also be used to determine the wine gallons (at 60 degrees Fahrenheit) of distilled spirits containing dissolved solids from the total weight of the liquid and its apparent proof (hydrometer indication, corrected to 60 degrees Fahrenheit). The proof gallons may then be found by multiplying the wine gallons by the true proof.

Example. 5,350 pounds of blended whisky containing added solids: Temperature °F 75.0° Hydrometer reading 92.0° Apparent proof 85.5° Obscuration 0.5° True proof 86.0°. 5,350.0 lbs.×0.12676 (W.G. per pound factor for apparent proof of 85.5°)=678.2 wine gallons. 678.2 W.G.×0.86=583.3 proof gallons.

§30.65 §186.65 Table 5, showing the weight per wine gallon (at 60 degrees Fahrenheit) and proof gallon at each percent of proof of spirituous liquor.

Table 5 This Table may be used to ascertain the weight of any given number of wine gallons (at 60 degrees Fahrenheit) or proof gallons of spirits by multiplying the pounds per gallon by the given number of gallons of the spirits. The Table should be especially useful where it is desired to weigh a precise quantity of spirits.

Example. It is desired to ascertain the weight of 100 wine gallons of 190-proof spirits:

6.79434×100 equals 679.43 pounds, net weight of 100 wine gallons of 190-proofs spirits.

Example. It is desired to ascertain the weight of 100 proof gallons of 190-proof spirits.

3.57597×100 equals 357.60 pounds, net weight of 100 proof gallons of 190-proof spirits.

The slight variation between this Table and Tables 2 and 3 on some calculations is due to dropping or adding of fractions beyond the first decimal on those Tables. This Table also shows the weight per wine gallon (at the prevailing temperature) corresponding to each uncorrected reading of a proof hydrometer.

§30.66 §186.66

Table 6, showing respective volumes of alcohol and water and the specific gravity in both air and vacuum of spirituous liquor.

This Table provides an alternate method for use in ascertaining the quantity of water needed to reduce the strength of distilled spirits by a definite amount. To do this, divide the alcohol in the given strength by the alcohol in the required strength, multiply the quotient by the water in the required strength, and subtract the water in the given strength from the product. The remainder is the number of gallons of water to be added to 100 gallons of spirits of the given strength to produce a spirit of a required strength.

Example. It is desired to reduce spirits of 191 proof to 188 proof. We find that 191-proof spirits contains 95.5 parts alcohol and 5.59 parts water, and 188-proof spirits contains 94.0 parts alcohol and 7.36 parts water.

95.5 (the strength of 100 wine gallons of spirits at 191 proof) divided by 94.0 (the strength of 100 wine gallons of spirits at 188 proof) equals 1.01.

7.36 (the water in 188 proof) multiplied by 1.01 equals 7.43.

7.43 less 5.59 (the water in 191-proof spirits) equal 1.84 gallons of water to be added to each 100 wine gallons of 191-proof spirits to be reduced.

This rule is applicable for reducing to any proof; but when it is desired to reduce to 100 proof, it is sufficient to point off two decimals in the given proof, multiply by 53.73, and deduct the water in the given strength. Thus, to reduce 112-proof spirits to 100 proof:

1.12×53.73-47.75 equals 12.42 gallons of water to be added to each 100 wine gallons of spirits to be reduced.

This Table may also be used to obtain the proof gallonage of spirituous liquor according to weight and percent of proof.

Example. It is desired to determine the number of gallons in 400 pounds of spirits of 141 percent of proof. Multiply the weight of one gallon of water in air by the specific gravity in air of the spirits-8.32823 by 0.88862-the product (7.40063) divided into 400 gives 54.049 wine gallons, which rounded to the nearest hundredth is 54.05 and multiplied by 1.41 gives 76.2 proof gallons. In rounding off where the decimal is less than five, it will be dropped; if it is five or over a unit will be added.

§30.67 §186.67 Table 7, for correction of volume of spirituous liquors to 60 degrees Fahrenheit.

Table 7 This Table is prescribed for use in correcting spirits to volume at 60 degrees Fahrenheit. To do this, multiply the wine gallons of spirits which it is desired to correct to volume at 60 degrees Fahrenheit by the factor shown in the Table at the percent of proof and temperature of the spirits. The product will be the corrected gallonage at 60 degrees Fahrenheit. This Table is also prescribed for use in ascertaining the true capacity of containers where the wine gallon contents at 60 degrees Fahrenheit have been determined by weight in accordance with Tables 2, 3, 4, or 5. This is accomplished by dividing the wine gallons at 60 degrees Fahrenheit by the factor shown in the Table at the percent of proof and temperature of the spirits. The quotient will be the true capacity of the container.

Example. It is desired to ascertain the volume at 60 degrees Fahrenheit of 1,000 wine gallons of 190-proof spirits at 76 degrees Fahrenheit: 1,000×0.991 equals 991 wine gallons, the corrected gallonage at 60 degrees Fahrenheit.

It will be noted the Table is prepared in multiples of 5 percent of proof and 2 degrees temperature. Where the spirits to be corrected are of an odd temperature, one-half of the difference, if any, between the factors for the next higher and lower temperature, should be added to the factor for the next higher temperature.

Example. It is desired to correct spirits of 180 proof at 51 degrees temperature:

1.006 (50°)−1.005 (52°)=0.001 divided by 2=0.0005

0.0005+1.005=1.0055 correction factor at 51°F.

Example. It is desired to correct spirits of 180 proof at 53 degrees temperature:1.005 (52°)−1.003 (54°)=0.002 divided by 2=0.001. 0.001+1.003=1.004 correction factor at 53°F.

Where the percent of proof is other than a multiple of five, the difference, if any, between the factors for the next higher and lower proofs should be divided by five and multiplied by the degrees of proof beyond the next lower proof, and the fractional product so obtained should be added to the factor for the next lower proof (if the temperature is above 60 degrees Fahrenheit, the fractional product so obtained must be subtracted from the factor for next lower proof), or if it is also necessary to correct the factor because of odd temperature, to the temperature corrected factor for the next lower proof.

Example. It is desired to ascertain the correction factor for spirits of 112 proof at 47 degrees temperature: 1.006 (46°)−1.005 (48°)=0.001 divided by 2=0.0005. 0.0005+1.005=1.0055 corrected factor at 47°F. 1.007 (115 proof)−1.006 (110 proof)=0.001. 0.001 divided by 5=0.0002 (for each percent of proof)×2 (for 112 proof)=0.0001. 0.0004=1.0055 (corrected factor at 47°F)=1.0059 correction factor to be used for 112 proof at 47°F.

Example. It is desired to ascertain the correction factor for spirits of 97 proof at 93 degrees temperature: 0.986 (92°)−0.985 (94°)=0.001 divided by 2=0.0005. 0.0005+0.985=0.9855 corrected factor at 93°F. 0.986 (95 proof)−0.985 (100 proof)=0.001. 0.001 divided by 5=0.0002 (for each percent of proof)×2 (for 97 proof)=0.0004. 0.9855 (corrected factor at 93°F)=0.0005=0.9851 correction factor to be used for 97 proof at 93°F.

Subpart F-Optional Gauging Procedures

§30.71 §Optional method for determination of proof for spirits containing solids of 400 milligrams or less per 100 milliliters. The proof of spirits shall be determined to the nearest tenth degree which shall be the proof used in determining the proof gallons and all fractional parts thereof to the nearest tenth proof gallon. The proof of spirits containing solids of 400 milligrams or less per 100 milliliters shall be determined by the use of a hydrometer and a thermometer in accordance with the provisions of §30.23. However, notwithstanding the provisions of §30.31, the proprietor may, at his option, add to the proof so determined the obscuration determined as prescribed in §30.32.

§30.72 Recording obscuration by proprietors using the optional method for determination of proof.

Any proprietor using the optional method for determination of proof for spirits containing solids of 400 milligrams or less per 100 milligrams as provided in §30.71 shall record the obscuration so determined on the record of gauge required by 27 CFR part 19.

Authority: 26 U.S.C. 7805. Source: T.D. ATF-198, 50 FR 8535, Mar. 1, 1985, unless otherwise noted. Editorial Note: Nomenclature changes to part 30 appear by T.D. ATF-438, 66 FR 5481, Jan. 19, 2001.

(c) Incorporation by reference. The "Gauging Manual Embracing Instructions and Tables for Determining Quantity of Distilled Spirits by Proof and Weight" (Publication 5110.6; November 1978) is incorporated by reference in this part. This incorporation by reference was approved by the Director of the Federal Register on March 23, 1981. This publication may be inspected at the National Archives and Records Administration

(NARA), and is available from the Superintendent of Documents, U.S. Government Printing Office, Washington, DC 20402. For information on the availability of this material at NARA, call 202-741-6030,

(Sec. 201, Pub. L. 85-859, 72 Stat. 1358, as amended (26 U.S.C. 5204); 80 Stat. 383, as amended (5 U.S.C. 552(a))) [T.D. ATF-198, 50 FR 8535, Mar. 1, 1985, as amended at 69 FR 18803, Apr. 9, 2004]

This chapter. Title 27, Code of Federal Regulations, Chapter I (27 CFR Chapter I).U.S.C. The United States Code. [T.D. ATF-198, 50 FR 8535, Mar. 1, 1985, as amended by T.D. ATF-438, 66 FR 5481, Jan. 19, 2001; T.D. TTB-44, 71 FR 16947, Apr. 4, 2006]

(Sec. 201, Pub. L. 85-859, 72 Stat. 1358, as amended (26 U.S.C. 5204)) [T.D. ATF-198, 50 FR 8535, Mar. 1, 1985, as amended by T.D. ATF-438, 66 FR 5481, Jan. 19, 2001], (Sec. 201, Pub. L. 85-859, 72 Stat. 1358, as amended (26 U.S.C. 5204)), (Sec. 201, Pub. L. 85-859, 72 Stat. 1358, as amended, 1362, as amended (26 U.S.C. 5211)), (Sec. 201, Pub. L. 85-859, 72 Stat. 1358, as amended 1362, as amended (26 U.S.C. 5204, 5211))

§5.37 Alcohol content.

(b) Tolerances. The following tolerances shall be allowed (without affecting the labeled statement of alcohol content) for losses of alcohol content occurring during bottling: (1) Not to exceed 0.25 percent alcohol by volume for spirits containing solids in excess of 600 mg per 100 ml; or (2) Not to exceed 0.25 percent alcohol by volume for any spirits product bottled in 50 or 100 ml size bottles; or (3) Not to exceed 0.15 percent alcohol by volume for all other spirits.

§5.47a Metric standards of fill (distilled spirits bottled after December 31, 1979).

(a) Authorized standards of fill. The standards of fill for distilled spirits are the following:

(1) For containers other than cans described in paragraph (a) (2), of this section-
1.75 liters,1.00 liter, 750 milliliters,500 milliliters,375 milliliters, 200 milliliters, 100 milliliters, 50 milliliters,
…….............355 milliliters, 200 milliliters, 100 milliliters, 50 milliliters
(b) Tolerances. The following tolerances shall be allowed:

(1) Discrepancies due to errors in measuring which occur in filling conducted in compliance with good commercial practice.

(2) Discrepancies due to differences in the capacity of bottles, resulting solely from unavoidable difficulties in manufacturing such bottles to a uniform capacity: Provided, That no greater tolerance shall be allowed in case of bottles which, because of their design, cannot be made of approximately uniform capacity than is allowed in case of bottles which can be manufactured so as to be of approximately uniform capacity.

(3) Discrepancies in measure due to differences in atmospheric conditions in various places and which unavoidably result from the ordinary and customary exposure of alcoholic beverages in bottles to evaporation. The reasonableness of discrepancies under this paragraph shall be determined on the facts in each case.

(c) Unreasonable shortages. Unreasonable shortages in certain of the bottles in any shipment shall not be compensated by overages in other bottles in the same shipment.

(Sec. 5, 49 Stat. 981, as amended (27 U.S.C. 203); 26 U.S.C. 5301) [T.D. ATF-25, 41 FR 10221, Mar. 10, 1976, as amended at 41 FR 11022, Mar. 16, 1976; 41 FR 11497, Mar. 19, 1976; T.D. ATF-35, 41 FR 46859, Oct. 26, 1976; T.D. ATF-62, 44 FR 71622, Dec. 11, 1979; T.D. ATF-146, 48 FR 43321, Sept. 23, 1983; T.D. ATF-228, 51 FR 16170, May 1, 1986; T.D. ATF-326, 57 FR 31128, July 14, 1992]

27 CFR Part 5 §5.23 Alteration of class and type.

(a) *Additions.* (1) The addition of any coloring, flavoring, or blending materials to any class and type of distilled spirits, except as otherwise provided in this section, alters the class and type thereof and the product shall be appropriately redesignated.

(2) There may be added to any class or type of distilled spirits, without changing the class or type thereof, (i) such harmless coloring, flavoring, or blending materials as are an essential component part of the particular class or type of distilled spirits to which added, and (ii) harmless coloring, flavoring, or blending materials such as caramel, straight malt or straight rye malt whiskies, fruit juices, sugar, infusion of oak chips when approved

by the Administrator, or wine, which are not an essential component part of the particular distilled spirits to which added, but which are customarily employed therein in accordance with established trade usage, if such coloring, flavoring, or blending materials do not total more than 21/2percent by volume of the finished product.

(3) "Harmless coloring, flavoring, and blending materials" shall not include (i) any material which would render the product to which it is added an imitation, or (ii) any material, other than caramel, infusion of oak chips, and sugar, in the case of Cognac brandy; or (iii) any material whatsoever in the case of neutral spirits or straight whiskey, except that vodka may be treated with sugar in an amount not to exceed 2 grams per liter and a trace amount of citric acid.

(b) *Extractions.* The removal from any distilled spirits of any constituents to such an extent that the product does not possess the taste, aroma, and characteristics generally attributed to that class or type of distilled spirits alters the class and type thereof, and the product shall be appropriately redesignated. In addition, in the case of straight whisky the removal of more than 15 percent of the fixed acids, or volatile acids, or esters, or soluble solids, or higher alcohols, or more than 25 percent of the soluble color, shall be deemed to alter the class or type thereof.

(c) *Exceptions.* (1) This section shall not be construed as in any manner modifying the standards of identity for cordials and liqueurs, flavored brandy, flavored gin, flavored rum, flavored vodka, and flavored whisky or as authorizing any product which is defined in §5.22(j), Class 10, as an imitation to be otherwise designated.

Extra Regulation.

Tax on adding Wine, Title 27 subpart C – Taxes Section 19.34

§ 19.34 Computation of effective tax rate. (a) The proprietor shall compute the effective tax rate for distilled spirits containing eligible wine or eligible flavors as the ratio of the numerator and denominator as follows: (1) The numerator will be the sum of: (i) The proof gallons of all distilled spirits used in the product (exclusive of distilled spirits derived from eligible flavors), multiplied by the tax rate prescribed by 26 U.S.C. 5001; (ii) The wine gallons of each eligible wine used in the product, multiplied by the tax rate prescribed by 26 U.S.C. 5041(b)(1), (2), or (3), which would be imposed on the wine but for its removal to bonded premises; and (iii) The proof gallons of all distilled spirits derived from eligible flavors used in the product, multiplied by the tax rate prescribed by 26 U.S.C. 5001, but only to the extent that such distilled spirits exceed 2 1/2% of the denominator prescribed in paragraph (a)(2) of this section. (2) The denominator will be the sum of: (i) The proof gallons of all distilled spirits used in the product, including distilled spirits derived from eligible flavors; and (ii) The wine gallons of each eligible wine used in the product, multiplied by twice the percentage of alcohol by volume of each, divided by 100. (b) In determining the effective tax rate, quantities of distilled spirits, eligible wine, and eligible flavors will be expressed to the nearest tenth of a proof gallon. The effective tax rate may be rounded to as many decimal places as the proprietor deems appropriate, provided that, such rate is expressed no less exactly than the rate rounded to the nearest whole cent, and the effective tax rates for all products will be consistently expressed to the same number of decimal places. In such case, if the number is less than five it will be dropped; if it is five or over, a unit will be added. (c) The following is an example of the use of the formula. BATCH RECORD Distilled spirits 2249.1 proof gallons. Eligible wine (14% alcohol by volume). 2265.0 wine gallons. Eligible wine (19% alcohol by volume). 1020.0 wine gallons. Eligible flavors 100.9 proof gallons. 2249.1($13.50) +2265.0 ($1.07)+1020 ($1.57)+16.6 1 ($13.50) = 2249.1+100.9+(2265.0×.28)+(1020×.38) $30,362.85+$2,423.55+$1,601.40+$224.10 = 2,350.0+634.2+387.6 $34,611.90 = $10.27, the effective tax rate. 3,371.8 1Proof gallons by which distilled spirits derived from eligible flavors exceed 2 1/2%) of the total proof gallons in the batch (100.9 ¥(21/2%)×3,371.8=16.6). (Sec. 6, Pub. L. 96–598, 94 Stat. 3488, as amended (26 U.S.C. 5010)) [T.D. ATF–297, 55 FR 18062, Apr. 30, 1990, as amended by T.D. ATF–307, 52736.

Notes

Bibliography Volumes 1 and 2

Because of the commonality of the resources used for both works; the bibliography is arranged topically and is the same for both volumes.

Osborne, NS, McKelvy, EC, and Bearce, HW. "Density and Thermal Expansion of Ethyl Alcohol and of its Mixtures with Water." *Bulletin of the [US] Bureau of Standards* 9 (1913): 327–474, http://nvlpubs.nist.gov/nistpubs/bulletin/09/nbsbulletinv9n3p327_A2b.pdf.

The original research on alcohol and water mixtures from which the U.S. Government developed its alcoholometric tables was conducted Osborne, McKelvy and Bearce on behalf of the United States National Bureau of Standards in 1910 and 1911. Their results were published in 1913, in five consecutive parts:

Part 1. Preparation of Anhydrous Ethyl Alcohol by E. C. McKelvy. Page 330

Part 2. Thermal Expansion of Mixtures of Ethyl Alcohol and Water by N.S. Osborne. Page 371.

Part 3. Density of Ethyl Alcohol and of its Mixtures with Water by N.S. Osborne. Page 405.

Part 4. A confirmatory series of experiments were carried out by H.W. Bearce. Page 429.

Part 5. An extensive bibliography is included beginning on page 436.

The quality of the original research is such that it is still cited in subsequent research.

This research and its results were subsequently developed by the U. S. Bureau of Alcohol Tobacco and Firearms (now the Tax and Trade Bureau) division of the U.S. Department of the Treasury into the alcoholometric Tables which have been in use ever since. Most recently, the Tables were published in 1978 as part of Title 27, Chapter I, Subchapter M, Part 186 of the Code of Federal Regulations. This gauging manual is known as Publication # 455 and alternately as ATF-P5110.6

There are seven Tables contained in the publication:

1. The true percents of proof spirit for any indication of the hydrometer at temperatures between 0° and 100°F.

2. Wine Gallons and Proof Gallons by Weight

3. Determining the number of Proof Gallons from the weight and proof of spiritous liquor

4. Showing the fractional part of a gallon per pound at each percent and each tenth percent of proof of spiritous liquor.

5. The weight per wine gallon (at 60°F) and proof gallon at each percent of proof of spiritous liquor

6. Respective volumes of alcohol and water and the specific gravity in both air and vacuum of spiritous liquor.

7. Correction of volume of spiritous liquors to 60°F.

The Gauging manual can be found here: https://www.ttb.gov/foia/gauging_manual_toc.shtml. The manual is now numbered as sections 30.1 through 30.71. I've cited the gauging manual using both the older numbering (§186) and the new (§30).

The TTB tables are available as PDF files for downloading, with each table linked into the section of the manual that describes its use. https://www.ttb.gov/foia/gauging_manual_toc.shtml#e

For unit conversion I have found that the computer program titled "Uconeer," published by Harvey Wilson and Katmar Software, contains the best and most consistent set of interrelated conversions. The conversions are available to eight decimal places if the particular converter requires them. They are also very closely reciprocal with one another. The software can be found at http://www.katmarsoftware.com.

Asimov, I. *The Collapsing Universe: The Story of Black Holes*. London: Hutchinson, 1977. ISBN 0-671-81738-8 1977.

Bureau International des Poids et Mesures (2006). "The International System of Units (SI)."

Carey, JS and Lewis, WK. "Studies in Distillation," *Ind. Eng. Chem.*, 1932, 24 (8), pp 882–883, doi: 10.1021/ie50272a011.

Committee on Revision of Methods (Doolittle, RE, chairman, Hartwell, BL, Hoover, GW, Patten, AJ, Seeker, AF, and Withers, WA. *Official and Tentative Methods of Analysis of the Association of Official Agricultural Chemists*, Second Edition, Revised to November 1, 1919. Association of Official Agricultural Chemists, 1921.

> This work is referred to as OMA. In this work, Table 10, which begins on page 389 reports the density of sugar and water mixtures on the basis of Specific Gravity at 20° C/4°C and 20°C/20°C as well in Degrees Baumé with a modulus of 145 in increments of .1 degrees of Brix or % by weight of sucrose.

Fine, RA, Millero, FJ. "Compressibility of water as a function of temperature and pressure," *The Journal of Chemical Physics*, 59(10), 1973: 5529-5536.
> Abstract: "The isothermal compressibility of water from 0 to 100° C and 0 to 1000 bar has been determined from Wilson's sound velocity measurements which have been normalized to Kell's 1 atm values. The isothermal compressibilies determined from the sound velocities have been fit, with a maximum deviation in compressibility of $\pm$ 0.016 × 10^{-6} bar^{-1}, to an extended bulk modulus equation $V_0 P/(V_0-V_P) = B + A_1 P + A_2 P_2$, where V_0 and V_P are the specific volume at an applied pressure of zero and P; and B, A_1, and A_2 are temperature dependent constants. Our specific volume results are in reasonable agreement with the work of Kell and Whalley at low pressures; however, our results at high pressures (1000 bar) disagree by as much as 169 ppm (the average deviation is approximately 115 ppm). A comparison of the compressibilities indicates a parabolic shift in Kell and Whalley's work with a maximum of approximately 0.205 × 10^{-6} bar^{-1} at 400 bar and 5°C. Since the velocity of sound data is extremely reliable ($\pm$ 0.2 m/sec) and the maximum error in the compressibilities derived from the sound data is within $\pm$ 0.016 × 10^{-6} bar^{-1}, our PVT results based upon the sound data are more accurate than any direct measurements made to date."

Fireman, P. *Distillery Operations*. Self-published, 2016. ISBN 978-0-9833376-4-5.

Flood, AE and Puagsa, S. "Refractive Index, Viscosity, and Solubility at 30°C, and Density at 25°C for the System Fructose + Glucose + Ethanol + Water," *J. Chem. Eng. Data*, 2000, 45 (5), pp 902–907, doi: 10.1021/je000080m. Publication Date (Web): August 16, 2000.

Glover, TJ. *Pocket Ref*, 2nd Ed. Anchorage, AK: Sequoia, 1997. ISBN 978-1-885071-00-2.

Glover, TJ. *Pocket Ref*, 4th Ed. Anchorage, AK: Sequoia, 2010. ISBN 978-1-885071-62-0.

International Alcoholmetric Tables. International Organization of Legal Metrology (French: Organization Internationale de Metrologie Legale or OIML). https://www.oiml.org/en/files/pdf_r/r022-e75.pdf/view.

> For reference the legal metrology tables known as the IAT or OIML tables can be consulted for the density and the corresponding alcohol values and are available on line. For some reason the file name is RO22e75. The General Formula itself is published in the Official Journal of the European Community as a Council Directive of July 27th, 1976 as the method of expressing alcoholic strength by volume or by mass on page No. L 262/149.

Kaye, GWC, and Laby, TH. *Tables of Physical and Chemical Constants*, Fourth Ed. London: Longmans, Green & Co., 1921.

Kyle, BG. *Chemical and Process Thermodynamics*, Third Edition. Prentice Hall, 1999. ISBN 978-0130874115.

Lawrence, Mark G. "The relationship between relative humidity and the dew point temperature in moist air: A simple conversion and applications," *Bull. Amer. Meteor. Soc.*, 86(2005): 225-233. doi: http://dx.doi.org/10.1175/BAMS-86-2-225.

Lide, DR (Ed.). *The Handbook of Chemistry and Physics*, 72nd edition. CRC Press, 1991. ISBN: 978-0849304729.

Lide, DR (Ed.). *The Handbook of Chemistry and Physics*, 89th edition. CRC Press, 2008. ISBN: 978-1420066791.

List, RJ (Ed.). "Acceleration of Gravity," *Smithsonian Meteorological Tables*, Sixth Ed. Washington, DC: Smithsonian Institution, 1968.

Mathlouthi, M., Reiser, P. (Eds.). *Sucrose: Properties and Applications*. Springer US, 1995. PDF file, doi: 10.1007/978-1-4615-2676-6, ISBN 978-1-4615-2676-6.

Taylor, BN. "The International System of Units (SI), 2008 Edition" National Institute of Standards and Technology, Publication 330, US Department of Commerce, 2008. PDF file, doi: 10.6028/NIST.SP.330e2008. Appendix 1 includes resolutions of the General Conference on Weights and Measures (French: CGPM) pertaining to the liter.

Travagli, V. "The Alcohol Dilution," paper, self-published, no date, https://www.scribd.com/doc/70865498/Alcohol-Dilution. His seven equations cited in Spedding, G, Weygandt, A, and Linske, M. "Alcohol Dilution Practices for Distillers: New and Older Approaches," *Artisan Spirit Magazine*, Winter 2016, pp. 65–70.

Tyco's Tables, Rochester, NY: Taylor Instrument Companies, 1918.

U.S. Standard Atmosphere, U.S. Government Printing Office, Washington, DC, 1976.

Alcohol Consumption:

West Virginia Code 17C-5-8 (2016 Edition) (blood alcohol level).

https://en.wikipedia.org/wiki/Blood_alcohol_content#Widmark_formula.

Wikipedia: https://en.wikipedia.org/wiki/Blood_alcohol_content#cite_note-PMC2724514-2 Retrieved, 3/14/17.

Wikipedia Solution to Widmark formula g/dl.

https://www.austintexas.gov/faq/what-does-blood-alcohol-concentration-bac-measure.

http://awareawakealive.org/educate/blood-alcohol-content.

Online Sources:

The engineering toolbox website also has some values that I have used or compared to: http://www/engineeringtoolbox.com.

Barometer Correction for Altitude: http://www.swaviator.com/html/issueJF03/Basics1203.html. General Aviation Rules of Thumb for pressure drop off with altitude = 1 inch Hg per 1,000 Feet.

Centrifugal Acceleration: Source for equation: http://www.funtrivia.com/askft/Question25977.html.

Compressibility of water: https://en.wikipedia.org/wiki/Properties_of_water#Compressibility.

Damp Air: http://www.conservationphysics.org/atmcalc/atmoclc2.pdf.

Damp Air: Meteorologist Jeff Haby. http://www.theweatherprediction.com/habyhints/260/.

Data source for Henry's law constants for methane, nitrogen, oxygen, and Carbon Dioxide http://thermo.mv.uni-kl.de/pdfs/2005_schnabel_fpe.pdf.

Definitions and History of the Litre: https://en.wikipedia.org/wiki/Litre, Retrieved 8/2/2017.

Density change with temperature formula: http://www.deltaenvironmental.com.au/ (Delta Environmental method).

Dew Point: http://iridl.ldeo.columbia.edu/dochelp/QA/Basic/dewpoint.html.

Dissolved Oxygen in water Article: http://www.waterontheweb.org/under/waterquality/oxygen.html.

Dissolved Oxygen in Water Table Generator. USGS https://water.usgs.gov/software/DOTABLES/.

Dissolved Oxygen Saturation Table: http://www.mainevlmp.org/wp-content/uploads/2014/01/Maximum-Dissolved-Oxygen-Concentration-Saturation-Table.pdf.

Gravity of the Earth: https://en.wikipedia.org/wiki/Gravity_of_Earth, Retrieved 4/12/201.

Gravity of the Earth: http://physics.stackexchange.com/questions/121775/are-we-slightly-lighter-during-the-day-and-slightly-heavier-at-night-owing-to-t. This discussion examines the effect of the sun's gravity and how our weight might vary when we are nearest to (day) and farthest away (night) from the sun.

History of the Inch: https://en.wikipedia.org/wiki/Inch#History, Retrieved 8/2/2017.

Henry's Law: http://en.wikipedia.org/wiki/Henry%27slaw. Retrieved 9/2014.

International Prototype Kilogram: https://en.wikipedia.org/wiki/Kilogram#Stability_of_the _international_prototype_kilogram.

Pycnometer and Relative Density: https://en.wikipedia.org/wiki/Relative_density, accessed April 24, 2017.

Rule of thumb for pumping material into wells: https://en.wikipedia.org/wiki/Pumping_(oil_well).

Tidal Force and Attraction: http://hyperphysics.phy-astr.gsu.edu/hbase/tide.html.

Tidal Force Calculator: http://keisan.casio.com/exec/system/1360312100.

Vapor Pressure Formula //www.srh.noaa.gov/epz/?n=wxcalc_vaporpressure.

Credit to Joseph Ervin for devising the polynomials used for TTB and IAT table analysis.

https://en.wikipedia.org/wiki/Alcohol_and_cancer
https://en.wikipedia.org/wiki/Color_of_water

General Resources

"SI Units" printed by the by the United States Department of Commerce (Publication 330).

Machinery's Hand Book 22nd Edition ISBN 0-8311-1155-0.

Jensen, WB. "The origin of alcohol proof," *J. Chem. Educ.* 2004; 81: 1258.

Owen, SC, "Alcohol." In: Rowe RC, Sheskey PJ and Weller PJ (eds.) *Handbook of Pharmaceutical Excipients*, 4th ed., Washington, DC, 2006.

Young, JA, "Ethyl alcohol," *J Chem Educ* 2004; 81: 1414.

European Pharmacopoeia, 5th Edition. Council of Europe. Strasbourg Cedex, F, 2004.

Rees, JA, Smith, I., Smith, B. (eds.) *Introduction to Pharmaceutical Calculation*. London, 2001.

Henley's Twentieth Century Book of Formulas, Processes and Trade Secrets. E-book available online (http://www.librum.us/) 1912, p. 703.

"Dilution and Concentration." Available online (http://pharmcal.tripod.com/ch8.htm#alcdil).

Ethyl Alcohol Handbook, Equistar, 6th Ed. Page 71.

United States Pharmacopoeia 29 - *National Formulary* 24. USP Convention, Inc. Rockville, MD, 2006 p. 3107.

Excel Solver Add-In: From Microsoft Office Online Help: "Solver is a Microsoft Excel add-in program you can use for what-if analysis. Use Solver to find an optimal (maximum or minimum) value for a formula in one cell, called the objective cell, subject to constraints, or limits, on the values of other formula cells on a worksheet. Solver works with a group of cells, called decision variables or simply variable cells that are used in computing the formulas in the objective and constraint cells. Solver adjusts the values in the decision variable cells to satisfy the limits on constraint cells and produce the result you want for the objective cell." Frontline Systems, Inc. P.O. Box 4288Incline Village, NV 89450-4288 (775) 831-0300 Web site: http://www.solver.com. E-mail: info@solver.com. Solver Help at www.solver.com.

Graphics Credits

Set of hydrometers: Payton Fireman, Own Work.

Reading a hydrometer: Joseph, Christopher (gen. ed.). *A Measure of Everything*. Firefly Books, 2005, 189.

Reading a hydrometer: http://www.brand.de/en/laboratory-instruments-liquid-handling-life-science/.

Water Molecule Ball & Stick: Benjah-bmm27, Own work, Public Domain, https://commons.wikimedia.org/w/index.php?curid=1997535.

Ethanol Diagram: Benjah-bmm27, Own work, Public Domain, https://commons.wikimedia.org/w/index.php?curid=2067606.

Ethanol Molecule Ball & Stick: Public Domain, https://commons.wikimedia.org/w/index.php?curid=1587150.

Broad spectrum density diagram: penmai.com.

Pycnometer: https://commons.wikimedia.org/w/index.php?curid=7139587.

Notes

About the Author

Payton D. Fireman founded The West Virginia Distilling Company, LLC, in 1998. It was West Virginia's first legal distillery licensed since Prohibition. Payton was born in Boston in 1958 and raised in New York City as well as Lake Placid, NY. He attended high school in Clearwater, Florida. Moving to West Virginia in 1976 he enrolled in West Virginia University and completed his law degree in 1983. Payton is a member of both the West Virginia and Florida Bar Associations. In 1998 he began building his distillery. His company websites are www.mountainmoonshine.com and www.alcoholblending.com. Payton recently published a book on how to operate a small distillery under the title of *Distillery Operations*, ISBN: 978-0-9833376-4-5. The book's focus is on the actual mechanical, biological and chemical aspects of operating a distillery and outlines the procedures and processes necessary to produce quality beverage grade alcohol on a commercial but still artisanal scale.

https://www.amazon.com/Distillery-Operations-How-Run-Small/dp/0983337640/ref=sr_1_2?s=books&ie=UTF8&qid=1466436047&sr=1-2&keywords=Distillery+Operations

In conjunction with the distillery operations book, Payton recently uploaded to his YouTube channel, MoonshineDistilling, a two-hour video chronicling an entire year's production cycle at his distillery. This video complements the operations book by demonstrating the processes and procedures discussed in the book.

Distillery Operations Complete Video: https://youtu.be/4qAgrDDVHfo

Payton has also written, published and maintained two alcohol industry related computer programs.

1. Alcohol Blending Software: To improve upon the government regimen for alcohol blending and accounting, Alcohol Blending Software (ABS), is a program which can be used to manage virtually every part of a distiller's business. http://mountainmoonshine.com/alcoholblendingsoftware/orderinginformation.html

2. Distillery Operations Workbook 1.0: This workbook contains all of the spreadsheets which were created in order to write the Distilling Operations textbook. The software is fully functional and users can modify the input values for each worksheet to meet their own particular needs.

http://mountainmoonshine.com/distilloperationsbook.html

Payton Despard Fireman
Outside Warhol Museum, Pittsburgh, PA,
2010